Groundwater Contamination in the United States

Second Edition

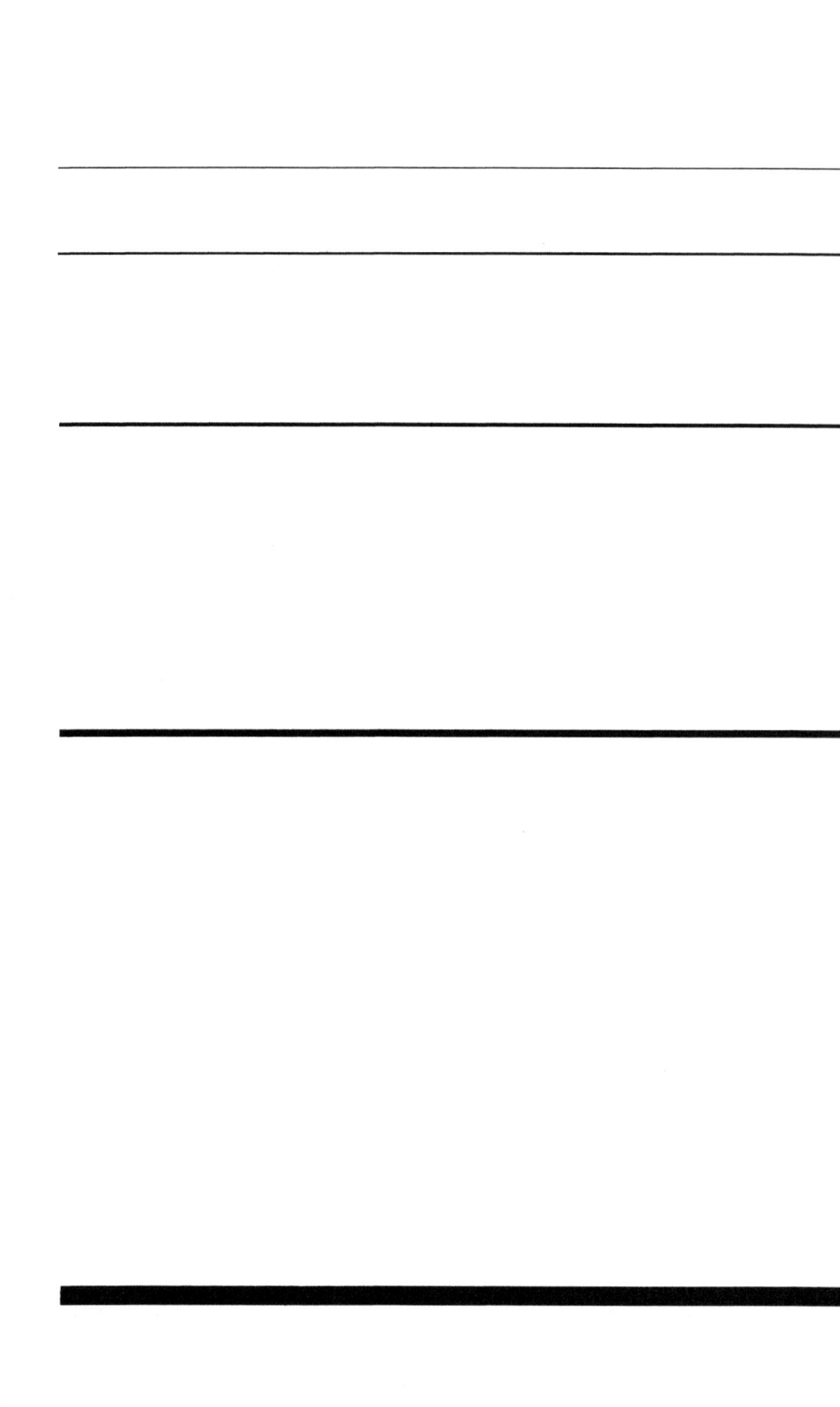

Groundwater Contamination
in the United States
Second Edition

Ruth Patrick

Emily Ford   John Quarles

Original Edition by Veronica I. Pye,
Ruth Patrick, and John Quarles

University of Pennsylvania Press   Philadelphia

Design by Tracy Baldwin

Library of Congress Cataloging-in-Publication Data

Patrick, Ruth.
  Groundwater contamination in the United States.

  Original ed. by Veronica I. Pye, Ruth Patrick, and
John Quarles.
  Bibliography: p.
  Includes index.
  1. Water, Underground—Pollution—United States.
I. Ford, Emily. II. Quarles, John. III. Pye,
Veronica I. Groundwater contamination in the United
States. IV. Title.
TD223.P368 1987   363.7′394   87-19143
ISBN 0-8122-8079-2
ISBN 0-8122-1256-8 (pbk.)

Printed in the United States of America

Contents

Contents

Contents

Four

The Severity of Groundwater Contamination   *104*

Five

**The Effects of Groundwater Contamination on
Public Health   *114***

Contents

Contents

Contents

Contents

Contents

Thirteen

State and Local Groundwater Quality Protection Programs   *396*

Contents

Tables

Tables

Figures

Figures

Figures

Acknowledgments

The authors of both the first and second editions of this book wish to thank the members of the Environmental Assessment Council of the Academy of Natural Sciences, Philadelphia and its Advisory Committee and their colleagues, for their guidance and reviews of the two editions. In addition, for the second edition as well as the first, we are indebted to the state and federal personnel, too numerous to mention individually, who willingly made available information on groundwater quality that formed the basis for this assessment. Drafts of the report were reviewed by persons familiar with one or more of the topics covered. For review and comment on the second edition we are grateful to Stanford Bach, John Cherry, Edwin H. Clark, III, Harold F. Elkin, David W. Miller, Marian Mlay, and Heather Wick. For the chapter "Federal Statutes Relevant to the Protection of Groundwater," we are grateful to Leslie Ritts, assistant to John Quarles, and to John Quarles.

We wish particularly to thank Robert G. Donlop for his advice and guidance during the preparation of this book. We wish to thank Linda Babcock and Linda Waters for help in typing the manuscript, Jennifer M. Malloy and Elaine Schmerling for their technical assistance, and Su-Ing Yong for her art work.

Financial support for this project was provided by funds remaining from those provided from the first edition, from the Pew Charitable Trusts and Mr. Laurance S. Rockefeller, and from other accrued monies.

Preface

This is the second edition of the book, *Groundwater Contamination in the United States*, by V. I. Pye, R. Patrick, and J. Quarles. The purpose of this edition is to include the facts necessary to understand various types of groundwater contamination and how to mitigate their effects. Since the first edition, our knowledge about groundwater contamination and how to mitigate and prevent contamination has greatly improved.

This book discusses the characteristics of groundwater and its contamination. It includes the types of aquifers, the factors affecting the flow of groundwater, and its natural characteristics. It describes how groundwater is contaminated and the types of contamination most commonly found. What is known about the effects of various types of contaminants on human health and the environment are briefly discussed.

Various proposed strategies for preventing and managing groundwater contamination are set forth. Federal laws that pertain to groundwater contamination are given with particular emphasis on laws and regulations enacted since 1982. When compared with the first edition, much more attention is given to strategies and management of aquifers adopted by the states and the laws and regulations that are in place to prevent or at least mitigate contamination.

Preface

These studies have been sponsored by the Environmental Assessment Council of the Academy of Natural Sciences of Philadelphia. The purpose of the Council is to identify environmental problems before they become environmental issues, and to gather scientific and other data which relate to the problems. The Council also seeks to bring together the laws and regulations related to the problems. The various chapters in this second edition that contain new material have been reviewed by various colleagues (and by the Advisory Committee of the Environmental Assessment Council).

The members of the Council are Robert G. Dunlop, Caryl Haskins, Richard E. Heckert, George Lamb, Charles F. Luce, Ruth Patrick (Chair), Glenn Paulson, William Reilly, Harlan Snider, Abel Wolman, and M. Gordon Wolman. The members of the Advisory Committee of the Environmental Assessment Council are Edwin H. Clark, II, Joseph V. D'Ambrisi, James Marum, Harold F. Elkin, Richard D. Johnson, Thomas E. Lovejoy, III, Glenn Paulson, John Quarles, James C. Hildrew, Cy Rhode, John R. Cooper, George Wills, and M. Gordon Wolman.

Overview

The interest in groundwater contamination has increased greatly since the first edition was issued in 1983. The public sector as well as the federal and state governments have wrestled with the complex issues related to groundwater contamination and its control. Much research has been carried out to try to find physical, chemical, and biological methods by which the adverse effects of contamination can be mitigated.

Many types of remedial action have been developed. Some of these have been put into practice and others are just being introduced. They range from the construction of barriers to prevent the contaminated water from reaching other parts of the aquifers to various types of physical, chemical, or biological *in situ* treatment to mitigate the effects. Other means to mitigate the effects of contamination use various methods of extraction and treatment of the contaminated water. Such means may involve new methods for biological treatment; the use of reverse osmosis to separate the contaminant from the water; and the venting of volatiles from the water and subsequently capturing the toxic substances.

Groundwater Contamination in the United States: An Executive Summary of the Technical Report

This study of groundwater contamination in the United States is undertaken to identify what is known about the severity, extent, and sources of the problem, to determine what legal and regulatory framework is currently available to protect and manage groundwater and, finally, to provide some options, together with their consequences, for dealing effectively with both present and future problems rather than providing recommendations for action. It also points out gaps in that knowledge.

Contamination has occurred in most regions of the country and will probably continue to occur. Once contamination is detected there is a choice as to whether to take action and, if so, what form that action should take. The type of actions vary with the type of contamination and its hydrogeologic and geographic setting. There are regulatory agencies at the federal, state, and local levels that have requirements that may prevent or correct many types of groundwater contamination. In consideration of the future quality of groundwater the question is whether or not to take action that would completely or partially prevent such contamination.

Importance of the Groundwater Resource

More than 50% of the population depends upon groundwater as its primary source of drinking water. Approximately 75% of American cities derive their supplies, in total or in part, from groundwater. The states vary in their dependence upon groundwater. Those in the arid West, where there are few permanent rivers, lakes, or streams, are much more reliant on groundwater than the states of the more humid East. In 1980, 88 billion gallons of groundwater were used in the United States per day, and 68% of this total was used for irrigation. The total use in 1950 was 34 billion gallons per day (bgd), and in 1970, 68 bgd. (Solley et al., 1983).

What is Groundwater?

Groundwater, as considered here, is water that occurs in saturated non-consolidated geologic material (sand or gravel) and in fractured and porous rock. These saturated strata are called aquifers. Groundwater comprises approximately 4% of the water contained in the hydrologic cycle, which consists of the movement of water between oceans and other bodies of surface water, the atmosphere, and the land. Groundwater is recharged by precipitation.

Types of Aquifers

The two main types of aquifers are unconfined and confined. An unconfined aquifer is not overlain by impermeable material, and precipitation may percolate down to the water table. Water in the aquifer is usually at atmospheric pressure. The amount of water in storage is dependent on seasonal cycles of natural recharge. A perched aquifer may occur where there is a limited layer, or lens, of impermeable material above the water table, forming a zone of saturation above it.

The second major category of aquifers is the confined aquifers, some of which are artesian. Confined aquifers are bounded top and bottom by layers of relatively impermeable geologic formations termed aquitards. Artesian aquifers contain water that is under pressure greater than atmospheric. Some have discrete and variable recharge areas where the geologic material of the aquifer forms an outcrop at the surface. A few aquifers have no recharge areas at all.

The areal extent of aquifers may be regional or local, and some, such as the Ogallala in the Midwest, may extend across several states. This fact is important when one comes to consider the institutional organization for the management and protection of groundwater.

Most aquifers occur at variable depths within 2,500 feet of the land surface. Estimates of water in storage within 2,500 feet of the land surface within the conterminous United States range from 33 quadrillion gallons to 100 quadrillion gallons. It would be possible to drill a well with a yield sufficient for domestic use in almost every region of the country.

3

Factors Affecting the Natural Quality of Water

The natural quality of groundwater is affected by the geological characteristics and climatic conditions—in particular, the temperature and rainfall—of the region. When evaporation and transpiration exceed precipitation, as in the arid Southwest, the recharge of groundwater takes place only in "wet," or multicycle years. Some aquifers, such as the Ogallala, have little, if any, recharge. In the East, where precipitation usually exceeds evaporation, unconfined aquifers are easily recharged. The natural chemical content of groundwater is influenced by the type and depth of soils and by the subsurface geologic formations through which the groundwater passes. The leaching of the soil and rocks is influenced by the pH of the precipitation and by the metabolic products of subsurface organisms. In the aquifer the water reacts with the geologic material, usually increasing its content of total dissolved solids (TDS). U.S. Public Health Service drinking-water standards consider 500 parts per million (ppm) TDS the maximum permissible, although the Safe Drinking Water Act regulations consider water with up to 10,000 ppm TDS as potentially potable. The hardness of water reflects its calcium and magnesium content and may render the water unfit for domestic or industrial use. Water from some aquifers may be unusable due to high salinity or to the presence of naturally occurring toxic substances such as arsenic and radionuclides. Natural groundwater may be unfit for certain uses or be of the highest quality that can be used for drinking water with no pretreatment.

Relationships Among Groundwater Quality and Quantity, Surface Water, and Land Use

In the arid areas, where evaporation and transpiration can greatly exceed precipitation, the mineralization of groundwater due to leaching is a significant cause of contamination; evapotranspiration may further concentrate the salts. Pumping at greater than recharge rates can increase the degradation of groundwater. If pumping lowers the water level in the vicinity of coal mines, the exposed sulfur-bearing rocks may be subject to oxidation. Should the water levels increase, the oxidized sulfur-compounds may dissolve and increase the acidity of the water. In coastal areas, where there is a high demand for water, groundwater pumping may cause saltwater intrusion into potable aquifers.

4

Most of the low flow of streams is supplied by groundwater. If the aquifers are overpumped and recharge does not occur, the stream flow may cease during periods of drought. Moreover contamination of groundwater by contaminated streams may occur if a negative pressure develops in the aquifer from the use of groundwater.

The recharge of aquifers may be greatly reduced by paving the land surface in recharge zones or by altering the vegetation cover so that precipitation runs off instead of infiltrating. Siting of contaminating activities in recharge areas without careful consideration of practices that could eliminate or reduce contamination may cause a threat to the quality of groundwater and hence its usability.

Thus, the relation between groundwater quantity and quality, between groundwater and surface water, and between groundwater and land use patterns must be taken into consideration in any protection and management strategy.

Sources of Groundwater Contamination

There are three main mechanisms by which the chemical composition of groundwater may be changed: by natural processes, by man's waste-disposal practices such as those for sanitary wastes, liquid industrial wastes, solid wastes, and radioactive wastes, and by spills, leaks, and agricultural activities and other sources unrelated to disposal.

Types of Groundwater Contamination

The types of contaminants which emanate from anthropogenic sources are extremely varied. They range from simple inorganic ions such as the nitrate from septic tanks, feedlot wastes, and the use of fertilizer, chloride from highway deicing salts, saltwater intrusion, and certain industrial processes, heavy metal ions from many different industrial processes such as plating works, to complex synthetic organic compounds resulting from industrial and manufacturing processes and from the use of pesticides and household cleaning fluids, e.g., trichloroethylene. Compounds of low solubility may yield solutions that are toxic or offensive. The chemical composition of wastes deposited in landfills or surface impoundments is often known; nevertheless, when the constituents of such wastes interact, new compounds may be formed. Many industrial waste-disposal practices now include the stabilization of wastes, thereby rendering them less active chemically. It is still possible, however, that leachate production may transform some of the constituents.

5

Attenuation and Movement of Contaminants in Groundwater

A contaminant usually moves from the surface of the land, percolating downward through the non-saturated zone into a semi-saturated or saturated zone and into the aquifer. Contaminant movement through the subsurface is determined by its solubility in water, its density, and the various attenuation processes that may occur below the surface. The attenuation of contaminants may take place by biological activity, dilution, volatilization, mechanical filtration, precipitation, buffering, neutralization, ion exchange, radioactive decay, hydrolysis, complexation, and adsorption. These processes may reduce or increase the severity or toxicity of the contaminant, and may also alter its rate of movement. Different soils, and soil conditions have different capacities for attenuation, and affect various chemicals to different degrees. At present these attenuation processes are being researched and our knowledge is limited. Once in the aquifer the contaminant movement is mainly governed by advection, dispersion, and diffusion; the flow pattern is often laminar. Because contaminants slowly mix with water and turbulence is lacking, a well defined plume may be formed until contaminants completely diffuse into the aquifer. The movement of contaminants is usually but not always with the direction of flow. The miscibility of contaminants and their specific gravity are also important characteristics in determining movement. These factors determine whether contaminants float on top of the water table or sink.

6

Risks Posed by Different Types of Contaminants

The degree of risk posed by contaminants varies according to many factors. These include the volume and toxicity of the contaminant, its concentration in the aquifer, its persistence in the environment, and the degree of human and environmental exposure to the contaminant. In addition, the number of persons affected, or likely to be affected, over time and the percentage of available groundwater both locally and regionally should be taken into consideration. If the contaminants in the groundwater exceed the standards set for drinking water by the federal government, for example, then the water is hazardous for the use for which it was designated under the standards. These standards include, however, only a limited number of chemicals, and thus they do not necessarily protect humans or the environment against either the short-term or the long-term effects of every contaminant that might be found in groundwater. Drinking water standards are designed to protect human health; they may not protect other organisms.

One of the main criteria usually used in judging risk is the contaminant's effect upon public health. The epidemiological evidence for an association between chronic disease and the use of contaminated groundwater is inconclusive but suggestive of a link with certain diseases. More is known about the acute effects, with regard to disease caused both by pathogenic microorganisms and by chemicals, of ingesting contaminated groundwater. Some of the synthetic organic chemicals that have been detected in groundwater have not been tested for their potential to affect human health, and thus we do not know whether they pose a hazard.

The effects upon the environment of contaminated groundwater are even less well documented than its effects on human health, although it is well established that certain contaminants may cause stress to vegetation or even its destruction, and that they probably adversely affect animals. For example, water for irrigation should be relatively free of boron and chloride, both of which have a deleterious effect upon plant growth. Many plants selectively accumulate heavy-metal ions from water. Water for human use should ideally be of the highest quality, but livestock may also be particularly sensitive to contaminants at concentrations that are not injurious to humans.

Another way to assess the potential danger of contaminated groundwater is to gauge the persistence of its deleterious effects on the environment. For example, cadmium, which is toxic and believed to be carcinogenic, has an infinite half-life and is not degraded. Some inorganic and organic compounds may be degraded by biological action in the soil zone, so that only harmless end products reach the aquifer. All of the heavy-metal ions and some synthetic organic chemicals associated with manufacturing and service industries would probably not be degraded.

7

Geographical Extent of the Problem

No comprehensive national survey of groundwater contamination has been undertaken, for the most part because it would be extremely expensive and time-consuming. There have been many federal and state-run surveys, conducted on a regional and local level, on ambient-water quality, drinking-water supplies, and potential sources of contamination. Incidents of contamination have been reported from every state, and the incidents are beginning to be detected with increasing frequency (Office of Technology Assessment [OTA], 1984). From the assessments listed above, it may be concluded that problems indigenous to one area may not occur in another, but that several sources of groundwater contamination occur at a high or moderate degree of severity in each area studied. The four types of pollutants most commonly reported (heavy metals, organics, microorganisms, and nitrates) are probably more of a reflection of the prevailing monitoring practices at the time the surveys were made than they are of the actual incidence of the contaminants that most threaten the quality of groundwater.

For this report we selected nineteen states for which we had case histories of incidents of contamination (AZ, CA, CT, FL, ID, IL, MA, MT, ND, NE, NJ, NM, OR, PA, RI, SC, TX, VT, and WA). These states represent areas of differing levels of industrialization, agricultural activity, climatic conditions, and population density.

Analysis of these states indicates that in many of them industrial and municipal wastes were among the most frequently reported sources of contamination. The industrial Northeast predictably had more problems associated with industrial wastes, petroleum products, and landfill leachate. California, Florida, Idaho, Nebraska, and Texas reported problems arising from agricultural practices, whereas California, New Mexico, and Texas reported problems with the disposal of oil-field brines. These summaries include, of course, only the known incidents of contamination; a comprehensive national survey might well uncover other important sources of contamination, or different frequencies from the same sources. The problems encountered to date vary regionally and depend upon the type and degree of industrial and agricultural activities carried out and the population density. The known incidents vary from a low to a high frequency of occurrence of contamination, and they fall into geographical patterns.

8

Severity of the Problem

In order to determine the overall magnitude of the problem, it is necessary to assess the severity of the contamination of groundwater and this can be undertaken in several ways. If the concentration of contaminants exceeds the numerical standards set for a particular use, as, for example, for drinking water, the problem could be called severe. The severity depends on the population affected by well closings, the nationwide percentage of aquifers affected, and the percentage of the known reserves of groundwater that are contaminated. The degree of hazard posed by the contaminants—for example, their toxicity—enters into the evaluation, as does the cost and feasibility of finding an alternative water supply. Many of the data required to assess the severity of the problem by these methods are unavailable.

Since no systematic national survey of the contamination of groundwater has been carried out, the only estimates of the order of magnitude of the amount of groundwater currently contaminated rely upon surveys of the numbers of certain types of potential sources of contamination. EPA did such an estimate for certain selected point sources, such as surface impoundments and landfills, which it considered the most important sources, and also for secondary sources, such as subsurface disposal systems and petroleum exploration and mining. The agency concluded that up to 1% by area of usable surface aquifers near the land surface may be contaminated by these sources. Lehr (1982) estimated that up to 2% could be contaminated. It should be pointed out that these are nationwide estimates for the entire United States. These estimates did not include contamination from non-point sources, such as agricultural activities. In certain areas, especially those with a high population density, a much higher percentage of the groundwater may be contaminated.

9

Monitoring for Groundwater Contamination

Monitoring groundwater quality and aquifer conditions can bring to light contamination before it becomes a significant problem. The appropriate design of monitoring system depends upon the hydrological, geological, and soil characteristics of the site, pollution sources, population density, and climate of the region. The four main types of monitoring systems are those which measure quality in relation to numerical standards, source monitoring for effluent quality and quantity, case-preparation monitoring for enforcement-action data, and research monitoring.

In order that the samples of water withdrawn for analysis are not altered chemically, well-casing materials and sampling procedures must be carefully chosen. Detailed knowledge of groundwater flow-patterns, natural water quality, and the attenuation capacities of the soil and geologic strata would enhance the siting of monitoring facilities so that they produce the most useful data with the minimum of damage and expense. The chemicals analyzed should be the best indicators for the suspected type of groundwater contamination under investigation. They should include organic chemicals from natural and man-made origins as well as inorganic ions, such as chloride, nitrate, and heavy metals.

The sophistication of present-day analytical methods may have outstripped our ability to interpret what they reveal and our ability to determine the significance of low-concentrations of contaminants on the environment and on public health.

Measures for Protecting Groundwater

The protection of groundwater from contamination may be cost-effective for valuable aquifers. Many federal and state regulations can be used for this purpose, as summarized below, although the majority of regulations were not specifically designed with groundwater protection as their primary goal. The most efficient way of preventing important aquifers from becoming contaminated is by protecting their recharge areas. For most confined aquifers it would be necessary to prohibit the siting of contaminating activities in specified recharge areas or to ensure that potentially contaminating activities use adequate methods to prevent groundwater degradation. For unconfined aquifers this might be difficult because of their extensive recharge areas. Protection might entail careful regional and local planning since recharge areas may cross county or state boundaries. Other protective measures can be taken, including the careful restriction of groundwater pumping in areas where salt water can intrude into potable aquifers, the capping and sealing of dry boreholes that traverse one or more aquifers and offer a conduit for contaminants, and the continued stringent supervision of the drilling and maintenance of injection wells and water wells.

10

Aquifer Rehabilitation

Remedial action for contaminated aquifers is complex, time-consuming, and extremely expensive. It is feasible under certain circumstances, assuming no additional contamination occurs, and the choice of remedial action determined upon is dictated by the time period during which contamination has taken place, the type and behavior of the contaminants, and the hydrogeology of the site. Often it is more cost-effective to locate a new source of water than to attempt treatment. Contamination that occurs in shallow plumes in unconsolidated rock may be controlled by the excavation and removal of the plume, once the problem of the final disposal of the contaminated material has been solved. There are numerous remedial alternatives: in situ treatment methods, withdrawal treatment techniques, and final disposal. In-situ techniques include the detoxification, stabilization, and immobilization of contaminants and require the use of biological cultures, chemical reactants, or sealants. The aim in each case is to detoxify or stabilize the contaminants or to form an impervious barrier around the plume in order to prevent its migration. Methods for withdrawal, treatment, and final disposal may use collection or withdrawal wells, subsurface gravity collection-drains, impervious grout-curtains, and cut-off trenches. Treatment of the water removed could be by reverse osmosis, ultrafiltration, ion-exchange resins, wet-air oxidation, ozonation and ultraviolet radiation, coagulation and precipitation, aerobic biological treatment, or the use of activated charcoal filters.

Costs, ranging from several thousands to several billions of dollars, have to be estimated on a case-by-case basis and are determined by many factors specific to the site. For many problems pertaining to groundwater pollution, the prevention of contamination is far better than curative action.

11

EPA 1984 Ground-Water Protection Strategy

In 1984, the EPA set forth a Ground-Water Protection Strategy which addressed four major objectives needed for effective groundwater protection: the need to strengthen state groundwater programs; the need to address groundwater contamination problems of national concern; the need to create a policy framework for guiding EPA programs; and the need to strengthen internal groundwater organization. In order to strengthen state groundwater programs EPA has provided states with funds to: establish or begin state strategies; support necessary program development and planning; create needed data systems; assess legal and institutional impediments to comprehensive state management; and develop state regulatory programs such as permitting and classification. To address groundwater contamination problems of national concern EPA has undertaken several reviews of a number of groundwater contamination sources that have appeared to be a widespread problem, for example leaking underground storage tanks, land disposal facilities, and non-point source pollution. In the strategy, EPA also recognized that in order to effectively protect groundwater, guidelines were greatly needed for the assurance of consistency within EPA's groundwater protection programs. The strategy's framework is based on the rationale that not all groundwater is of equal value, and the extent of protection should be based on its vulnerability to contamination and its respective value. Lastly, in order to strengthen internal groundwater organization, EPA has established the Office of Groundwater Protection to oversee the implementation of this strategy.

Classification of Aquifers

The classification of aquifers is a management tool that is in use in some states and under consideration in others. It establishes water-quality goals for each aquifer and identifies the standards or controls necessary to assure that those goals are met. A policy that chooses selectively to protect different aquifers at different levels of quality is dependent upon a classification system. Because hydrogeologic conditions, water quality, and present and future groundwater use vary from state to state, there is no aquifer classification system in use that could serve as a model for other states wishing to develop such a system. The policies of preserving and restoring aquifers for drinking water and of prohibiting the degradation of existing water quality would not require classification systems.

The benefits provided by using a system of aquifer classification include legal protection for valuable aquifers, a basis for siting potentially contaminating activities in low-risk areas, and guidance for planning programs which would protect the quality of water at all levels of government. The deterrents to establishing a classification system include the difficulty of delineating aquifer boundaries, the difficulty of gaining public acceptance for degradation zones or waste-receiving zones, and the difficulty of projecting groundwater use, needs, and availability for the future.

Federal Regulations Applicable to Groundwater

Although there is no single statute specifically aimed at protecting groundwater, there are many programs that provide some measure of protection. The most significant federal statutes affecting groundwater are the Resource Conservation and Recovery Act (RCRA), the Safe Drinking Water Act (SDWA), the Clean Water Act (CWA), and the Comprehensive Environmental Response, Compensation and Liability Act (CERCLA or "Superfund Law"). Statutes less directly related to groundwater include the Toxic Substances Control Act (TSCA), the Surface Mining Control and Reclamation Act (SMCRA), the Federal Insecticide, Fungicide and Rodenticide Act (FIFRA), and the National Environmental Policy Act (NEPA).

The Resource Conservation and Recovery Act (RCRA) of 1976, along with the 1984 Hazardous and Solid Waste Amendments (HSWA) to RCRA, provides a comprehensive framework for the management of municipal and industrial solid and hazardous wastes. RCRA requires owners and operators of facilities that treat, store, or dispose of hazardous wastes to obtain RCRA permits and meet groundwater standards at facility boundaries. It aims to protect groundwater affected by facility operations and correct past contamination that occurred at the sites.

The Safe Drinking Water Act (SDWA) of 1974 contains programs designed to provide for sanitary drinking water supplies, protect sole-source aquifers, and establish a program to control the underground injection of wastes. The 1986 amendments to the Safe **13** Drinking Water Act contain provisions that require states to establish "Wellhead Protection Areas" around public drinking water well fields.

The Clean Water Act of 1977 (CWA) has limited statutory authority to implement groundwater protection measures. Its primary goal is the reduction and control of discharge of pollutants into the nation's navigable water. Section 104, however, requires that EPA establish a program to equip and maintain a water-quality surveillance system for both surface and groundwater.

The Comprehensive Environmental Response, Compensation and Liability Act of 1980 (CERCLA, or "Superfund Law") is essentially a groundwater remediation law. It enables the federal government to respond immediately to releases or threatened releases of hazardous substances into the environment which pose a threat of imminent and substantial danger to public health. Contained within the act is the National Contingency Plan (NCP), which outlines response action for the release of hazardous substances and oil spills. The act also includes criteria for determining priorities for remedial actions among these releases in the Hazardous Ranking System (HRS). The HRS "score" is then used by EPA to determine whether sites should be included in the National Priority List. As of January 1987, there were 703 sites on the list, and a proposed 248 to be added.

Many of the statutes outlined above contain broad imminent-hazard provisions that enable EPA to take immediate action to restrain activities posing a threat to any feature of the environment, including groundwater.

State and Local Groundwater Policies and Management Strategies

Within the last decade states have made large-scale efforts to prevent, abate, and monitor groundwater pollution. In the past these efforts have unfortunately received less attention and fewer program resources than corresponding surface-water programs. More recently, however, the groundwater problem has received national attention, and a thrust toward groundwater protection is taking place within many state environmental institutions. Because of the variation in the natural quality of groundwater within each state, the regional characteristics of man-made contamination, and the industrial goals for development, each state has been given **14** primary responsibility to establish, implement, and enforce groundwater protection programs. There are essentially three basic types of policies that states have chosen to implement (either alone or in combination): non-degradation, limited degradation, or differential protection. A non-degradation policy calls for the protection of groundwater at its existing quality. A limited degradation policy strives for minimum degradation of existing quality but allows for degradation up to a given standard. A differential protection policy, which is a use oriented policy, calls for different levels of protection based on groundwater characteristics, current needs, and anticipated uses.

State regulations that affect the quality of groundwater fall into five broad categories:

- those dealing with particular sources of pollution, such as septic systems and waste disposal sites
- those establishing and implementing water quality standards for aquifer water, in which aquifers may be classified according to natural characteristics or to current or projected uses
- those regulating chemicals in effluent that may enter groundwater.
- those regulating well construction and use
- those regulating the use of land in areas overlying critical aquifer-recharge zones.

As with federal regulations, the state and local controls that effectively protect groundwater often are not designed for that purpose and have several other regulatory objectives.

Conclusions

- Groundwater is essential to the continuance of our present quality of life and for industry and agriculture.
- The quality of groundwater varies considerably because of natural and anthropogenic contamination.

- Estimates of the percentage of usable groundwater near the surface that is contaminated range from less than 1% to 2%, which only takes into account contamination from certain point sources. In addition it should be emphasized that any estimate that is a nationwide average is not very meaningful because most areas would exceed or fall below the estimate. In addition, contamination is most likely to exceed these estimates in areas of high population density, where the number of persons affected may be large.

  15

- Contamination incidents have been reported from every state in the nation.
- Sources of contamination are varied and include both point and nonpoint sources and may be planned or inadvertent.
- The contaminants range in type from simple inorganic ions to complex synthetic organic chemicals.
- The types of contamination vary from one region of the country to another and are influenced by climate, population density, intensity of industrial and agricultural activities, the hydrogeology of the region, and the status and enforcement of federal and state regulations that can be used to protect groundwater.
- The methods for assessing the severity of the problem of groundwater contamination are varied and include its effects on the environment and on human health, the margin by which water standards are exceeded for a particular use, the percentage of usable groundwater contaminated, and the ease and cost of finding an alternative water supply.
- The risks posed to humans and the environment vary according to the volume and toxicity of the contaminants, their concentration and persistence in the aquifer, and the degree of exposure to them.
- The impact of contamination on humans may be measured by well closings and illness caused by chemical or pathogenic contamination.
- The lack of comprehensive national surveys of the extent of groundwater contamination and the fact that few of the contaminants have been tested for their effects upon human health, make it difficult to assess, in quantitative terms, the national risk of using contaminated groundwater.
- Aquifer rehabilitation is feasible in certain cases but is expensive and time-consuming, with no guarantee of complete success.
- Aquifer classification based on water quality and present and

projected use might be a tool for groundwater management and protection.

- Groundwater management must recognize the relation among land use, surface water, and groundwater.
- The prevention of the contamination of useful, potable aquifers is far superior to subsequent curative efforts. This can, in part, be achieved by regional planning and the enforcement of federal and state regulations designed to protect groundwater.

Two

The Groundwater Resource

- Groundwater occurs in two types of aquifers, confined and unconfined, and is usually within 2,500 feet of the ground surface.
- The susceptibility of an aquifer to recharge varies greatly.
- Aquifer water mixes very slowly compared with surface water.
- Groundwater varies greatly in quantity and quality.
- Groundwater is a source of drinking water for more than 50% of the population of the conterminous United States. It is used in large quantities for irrigation.

What is Groundwater?

Groundwater may be defined as subsurface water that occurs beneath the water table in soils and geologic forms that are fully saturated (Freeze and Cherry, 1979). It is an integral part of the hydrologic cycle, and any approach to groundwater problems should recognize this. Groundwater is not only an important natural resource but also an essential part of the natural environment. Much of the folklore and many of the widely held misconceptions about the nature of groundwater—for example, that it occurs in underground lakes, rivers, and veins and can be detected by listening for the noise—are dispelled by Lehr and Pettyjohn (1975) in their description of groundwater and its flow patterns.

The global hydrologic cycle consists of the movement of water between the oceans and other surface water, the atmosphere, and the land. The process is illustrated by Figures 2-1 and 2-2. A part of the precipitation, either as rainfall or melting snow, infiltrates the ground and percolates down through the unsaturated soil, known as the aerated zone, to the zone of saturation or the water-table level. Table 2-1, an estimate of the water balance of the world, shows the relative volumes of water contained in each part of the hydrologic cycle. Only 2.7% of all the water on the planet is fresh, and of that only 0.36% is easily available to users (Leopold, 1974). It has been estimated that about 40,000 billion gallons per day (bgd) pass over the conterminous United States as water vapor. Approximately 10% of this, 4,200 bgd, is precipitated as rain, snow, sleet, or hail, equivalent to an average uniform annual rainfall of 30 inches nationwide. About two-thirds of the precipitation evaporates or is transpired by vegetation. The remaining 1,450 bgd accumulates in ground and surface waters, flows to the sea or across national boundaries, is consumed, or evaporates from reservoirs (U.S. Water Resources Council, 1978a).

18

Table 2-1. Estimate of the Water Balance of the World

Parameter	Surface Area (km²) × 10⁶	Volume (km³) × 10⁶	Volume (%)	Equivalent Depth (m)*	Residence Time
Oceans and Seas	361	1370	94	2500	~4000 yrs
Lakes and Reservoirs	1.55	0.13	<0.01	0.25	~10 yrs
Swamps	<0.1	<0.01	<0.01	0.007	1–10 yrs
River Channels	<0.1	<0.01	<0.01	0.003	~2 wks
Soil Moisture	130	0.07	<0.01	0.13	2 weeks–1 yr
Ground water	130	60	4	120	2 wks–10,000 yrs
Icecaps and Glaciers	17.8	30	2	60	10–1000 yrs
Atmospheric Water	504	0.01	<0.01	0.025	~10 days
Biospheric Water	<0.1	<0.01	<0.01	0.001	~1 wk

* Computed as though storage were uniformly distributed over the entire surface of the earth

Source: R. Allan Freeze, John A. Cherry. *Groundwater* (Englewood Cliffs, N.J.: Prentice Hall, Inc., 1979), p. 84. Reprinted by permission of the publisher.

Of the 1,450 bgd only 675 bgd are usually available for intensive beneficial uses.

The ability of an aquifer to store and transmit water is a function of its permeability and porosity (Freeze and Cherry, 1979; U.S. EPA, 1977). When the saturated substratum is sufficiently permeable to store and transmit significant quantities of water, the geological formation is called an aquifer. There are two main types of aquifers, confined and unconfined. An unconfined or water-table aquifer contains water under atmospheric pressure; the upper surface of the water is called the water table and may rise and fall according to the volume of water stored, which is dependent upon seasonal cycles of natural recharge. A perched aquifer is one in which a limited layer, or lens, of impermeable material occurs above the water table, forming a thin zone of saturation above it. It is a type of water-table aquifer. The second major category of aquifer is the confined, or artesian, aquifer. These are bounded top and bottom by layers of relatively impermeable geologic formations termed aquitards, or confining layers (Figure 2-1). The aquifer is completely saturated with water that is under greater than atmospheric pressure. An artesian aquifer is not recharged everywhere uniformly, but in one or more general recharge areas. Water levels in non-pumping wells of unconfined aquifers correspond to the level of the water table and therefore vary according to the volume of water in storage. Water levels in non-pumping wells tapping confined aquifers are dependent upon the artesian pressure of the water in that aquifer, and in some cases the water may exceed the top of the well casing, thus causing a flowing well. The hypothetical projection of such water levels is known as the potentiometric surface (Figure 2-1). The aquitards, or confining layers, are not totally impermeable and permit some recharge or discharge to lower confined aquifers. Aquifers can occur in unconsolidated materials such as sand and gravel or in consolidated material or bedrock. The latter may consist of carbonate rock, volcanic rock, or fractured igneous, metamorphic, and sedimentary rocks. Sand and gravel aquifers usually contain the most groundwater, but high-yield wells can also occur in carbonate and volcanic rocks.

19

Groundwater Movement

Groundwater moves in response to gravity, pressure, and friction, the first two driving the water, the latter resisting motion. Due to the complexity of the channels through which groundwater flows, it is difficult to construct a model of the movement of groundwater at a microscopic level. The French engineer Darcy, however, formulated an empirical law in the mid-nineteenth century that effectively averages the microscopic complexities, providing a macroscopic model for groundwater movement. Darcy's Law relates the rate at which groundwater flows across a surface to the rate of change of energy of the groundwater along the flow path. Under ideal

20 homogeneous isotropic geologic groundwater conditions, the average linear groundwater velocity ($\bar{v}$) can be expressed by the Darcy relation (Cherry, 1981):

$$\bar{v} = - [K \, dh/dl]/n$$

where: $\bar{v}$ = the average linear groundwater velocity
K = hydraulic conductivity
dh/dl = hydraulic gradient
n = transport porosity

For both granular and fractured media, the same equation is used to compute values for average linear groundwater velocity. The great difference is in the value of n. For granular media n nearly represents the total porosity; in fractured rock or clay n represents the total void space in connected fractures within a unit volume of media (Cherry, 1984). The value of n in granular material is usually between 0.3 and 0.45; in fractured material n is usually very small, of the order of 10^{-2} or 10^{-4}. When the values of K and dh/dl are similar in granular and fractured media, the average linear groundwater velocity may be orders of magnitude larger in fractured media than in granular media. The velocity of groundwater in a horizontal sand and gravel aquifer is usually in the range of 0.05 to 1 meter per day (Cherry, 1984).

Darcy's Law is central to the derivation of equations used to model the flow of groundwater. Mathematical models, based on the physics of groundwater flow and boundary conditions imposed by the groundwater basin in question, usually take the form of a boundary-value problem. Bear (1972) discusses the application of the three methods used for solving such boundary-value problems, namely analytical, analog, and numerical. Analytical methods result in explicit mathematical expressions of the solution and are useful for problems where the governing equations may be simplified and the boundaries of the basins idealized. Analog models are scaled physical models usually involving the construction of electronic equipment to simulate groundwater flow (Freeze and Cherry, 1979). Numerical mathematical techniques are the basis for digital

computer simulation of transient groundwater flow in aquifers.
Digital simulation of groundwater flow requires an expertise in
computer programming and its use is bound to digital computers.
However, it is more flexible in its ability to handle irregular
groundwater boundaries and variations of groundwater movement
through time and space than are the other two modeling techniques
(Freeze and Cherry, 1979). The two most widely used numerical
methods for solving groundwater equations are the finite-difference
and finite-element techniques.

21

The main difference between groundwater and surface water is
that the movement of groundwater takes place very slowly. Mixing
of groundwater is slow, in contrast to the good mixing potential in
the turbulent flow of most surface water. This factor is extremely
important when considering the fate of contaminants.

Occurrence and Natural Quality of Groundwater

Most aquifers occur within 2,500 feet of the surface of the land.
They may be thick or thin, extensive or local, very near the land
surface or at considerable depths. It has been estimated that 30%
of the stream flow of the United States is supplied by groundwater
that emerges as natural springs or other seepage. In certain areas
in times of drought, the stream flow in low-flow months may be
largely provided by groundwater (U.S. Water Resources Council,
1978a). The interrelatedness of surface water and groundwater is
further underlined by the fact that under certain conditions seepage
from lakes, rivers, streams, reservoirs, and canals may recharge
aquifers.

The quantity of groundwater in storage is much greater than the
volume of surface water available in streams and lakes. It
comprises more than 96% of all the fresh water in the United
States, the remaining 4% occurring in lakes, rivers, and streams
(Weimar, 1980). Estimates of groundwater within 2,500 feet of the
land surface in the conterminous United States range from 33
quadrillion to 59 quadrillion gallons (U.S. Water Resources
Council, 1978a) to as much as 100 quadrillion gallons (Lehr, 1982).
As a point of comparison, Lake Michigan contains 1.3 quadrillion
gallons of water. Not all of the groundwater in storage is available
for use, however, because some is bound to soil particles. The cost
of extraction may also restrict its use in certain areas. The
geographical extent of the main groundwater resources in the
United States is shown in Figure 2-3. It would be possible to drill a

well with a yield sufficient for domestic use in almost every region of the country (U.S. EPA, 1977). The country has been divided up into 10 groundwater regions, as shown in Figure 2-4 (Lehr et al., 1976). This is considered the best broad classification of the groundwater situation in the conterminous United States. The division is based on the types of aquifers.

1. The western mountain ranges consist mainly of igneous, metamorphic, and consolidated sedimentary rocks. Most of the groundwater occurs in rock fractures. The large amount of precipitation that falls in this area recharges aquifers in adjacent regions.
2. Alluvial basins consist of valleys surrounded by mountains in the arid southwest. Water levels in many aquifers have declined over recent years due to the use of large volumes of groundwater for agricultural irrigation.
3. The Columbia Lava Plateau consists of lava flows and unconsolidated sediments. In some areas there are large supplies of groundwater.
4. The Colorado Plateau has high, dry plateaus of sedimentary shale and sandstone and has a scarcity of productive aquifers.
5. The High Plains are extensive and semi-arid to sub-humid. The unconsolidated sedimentary rocks form the Ogallala Aquifer. Due to extensive agricultural irrigation, withdrawals have greatly exceeded groundwater recharge in some areas.
6. The glaciated central lowlands consist of glacial drift, sand, and gravel forming major groundwater reservoirs. In some heavily populated or highly industrialized areas the groundwater has been overdeveloped or polluted.
7. The unglaciated Central Region is composed of horizontal consolidated sedimentary rocks, with limestone and sandstone formations providing the major aquifers. The yields are low to moderate and adequate for domestic supply but would not be sufficient for irrigated agriculture.
8. The unglaciated Appalachians have high-yielding limestone and sandstone aquifers, and lower yielding shale aquifers. The water table may occur at a considerable depth below the surface. This region has an abundance of surface water year round.
9. The glaciated Appalachians region is similar to the unglaciated Appalachians.
10. The Atlantic and Gulf Coast Plain has an abundance of groundwater and surface water. The geology consists of

unconsolidated gravel, sand, silt and clay, and limestone.
Supplies of groundwater may not be abundant in some
areas of Texas, Louisiana, Mississippi, and Alabama.
Florida has a prolific aquifer.

Unfortunately there is more than one classification system for
dividing the country up into water regions. The system used by the
U.S. Water Resources Council (1978b) is shown in Figure 2-5. For
the purpose of compiling and analyzing water-resources data for
both surface water and groundwater, the U.S. Water Resources
Council divided the nation into 21 major water-resources regions,
18 within the conterminous United States, and the other 3 being
Alaska, Hawaii, and the Caribbean area. These 21 regions are
further subdivided into 106 subregions. The regional divisions are
areas that contain either the drainage area of a major river or the
combined drainage of a series of rivers.

The EPA used another classification system in compiling data on
groundwater pollution for various regions of the United States
(Fuhriman and Barton, 1971; Miller et al., 1974; Miller et al., 1977;
Scalf et al., 1973; and Van der Leeden et al., 1975). The EPA
system most closely resembles that described by Lehr et al.
(1976), which is summarized above.

Along with the geological characteristics of the groundwater
regions, climatic conditions—in particular, precipitation and
temperature—have effects upon the quantity and quality of the
groundwater available. Variations from average precipitation may
cause droughts or floods. Figure 2-6 shows the average annual
precipitation in the United States. Water is lost from the land to the
atmosphere by evaporation and transpiration, or evapotranspiration.
When evapotranspiration exceeds precipitation, groundwater
recharge by percolation does not occur (U.S. EPA, 1977). In the
arid areas of the Southwest, for example, annual precipitation is
less than 10 inches, and potential evaporation could exceed this 4
to 20 times. In such areas groundwater recharge occurs mainly in
wet, or multi-year, cycles. In the eastern states, the annual
precipitation exceeds the evaporation, thus providing surpluses
that contribute to stream flow. Regional variations in temperature
may also affect groundwater in ways unrelated to surface
evaporation. For example, frozen ground does not permit infiltration
of rain, thus causing floods and preventing groundwater recharge.

The United States Geological Survey (USGS) published summary
appraisals of the nation's groundwater resources between 1974 and
1982 for each of the regions that were briefly outlined above in the
U.S. Water Resources Council scheme. A listing of the titles is

given in Table 2-2. These geological survey professional papers list groundwater usage for each region by categories of use (e.g., municipal water supply and agricultural irrigation). They list the natural groundwater quality and any problems that may be encountered due to the presence of natural iron in the water or to its salt content and hardness. In addition, they list present and potential anthropogenic threats to the quality of groundwater.

The National Water Summary 1984, also put together by the USGS, contains descriptions of the occurrence, use, and general quality of groundwater resources of each state. Each summary contains: (1) the physiographic, hydrologic, and geologic framework of the state's groundwater system; (2) a description of the principal aquifers; (3) groundwater withdrawals and water-level trends; (4) a description of the state's groundwater management program(s); (5) maps showing geographic distribution of the principal aquifers; (6) figures illustrating the areal distribution of

Table 2-2. Summary Appraisals of the Nation's Groundwater Resources, by Region

(A) Ohio—R. M. Boyd, Jr., 1974.

(B) Upper Mississippi—R. M. Boyd, Jr., 1975.

(C) Upper Colorado—Don Price and Ted Arnow, 1974.

(D) Rio Grande—S. W. West and W. L. Broadhurst, 1975.

(E) California—H. E. Thomas and D. A. Phoenix, 1976.

(F) Texas–Gulf—E. T. Baker, Jr. and J. R. Wall, 1976.

(G) Great Basin—Thomas E. Eakin, Don Price, and J. R. Harrill, 1976.

(H) Arkansas–White–Red—M. S. Bedinger and R. T. Sniegrocki, 1976.

(I) Mid-Atlantic—Allen Sinnott and Elliott M. Cushing, 1978.

(J) Great Lakes—William G. Weist, Jr., 1978.

(K) Souris–Red–Rainy—Harold O. Reeder, 1978.

(L) Tennessee—Ann Zurawski, 1978.

(M) Hawaii—K. J. Takasaki, 1978.

(N) Lower Mississippi—J. E. Terry and C. T. Bryant, 1979.

(O) South Atlantic–Gulf—D. J. Cederstrom, E. H. Boswell, and G. R. Tarver, 1979.

(P) Alaska—Chester Zenone and Gary S. Anderson, 1978.

(Q) Missouri Basin—O. James Taylor, 1978.

(R) Lower Colorado—E. S. Davidson, 1979.

(S) Pacific Northwest—Bruce L. Foxworthy, 1979.

(T) New England—Allen Sinnott, 1982.

(U) Caribbean—Fernando Gómez-Gómez and James E. Heisel, 1980.

Source: U.S. Geological Survey Professional Paper 813. U.S. Government Printing Office.

major groundwater withdrawals and trends in water levels; and (7) selected references (Heath, 1985).

The quality of groundwater is often described in terms of hardness and salinity. Hardness reflects its calcium and magnesium content and is usually expressed as the equivalent amount of calcium carbonate. It can be viewed as a measure of usefulness for domestic and industrial purposes. Naturally occurring hardness values are shown in Figure 2-7.

Water quality, however, is usually defined in terms of the **25** concentration of its chemical constituents (Tables 2-3 and 2-4). As water moves through the hydrologic cycle (Figures 2-1 and 2-2), it interacts with the atmosphere, soils, and subsurface geologic formations, all of which affect its chemical composition. Thus, there is a natural background level for the chemical content of the water, and this level varies regionally and may be subsequently augmented from industrial and domestic sources. Freeze and Cherry (1979) discuss the natural chemical content of groundwater in detail. Rainwater is saturated with oxygen, nitrogen, and carbon dioxide gases and is usually slightly acidic, having a pH of about 5.6. The acidity may be increased by industrial pollutants, namely, the oxides of sulfur and nitrogen. The more acid the rainwater, the more likely it is to react with the geologic materials with which it comes into contact.

Rainwater percolating through the soil may increase in acidity due to the biological processes that occur in that zone. Plant and microbial respiration produce carbon dioxide, which increases acidity. It would be possible for the percolating water to become supersaturated with carbon dioxide. Acidity may also be increased by products of decomposition such as humic and fulvic acids, nutrient uptake by roots, and nitrifying bacteria, but these are minor

Table 2-3. Simple Groundwater Classification Based on Total Dissolved Solids

Category	Total Dissolved Solids (mg/l or g/m^3)
Fresh water	0–1000
Brackish water	100–10,000
Saline water	10,000–100,000
Brine water	More than 100,000

Source: Freeze and Cherry, 1979.

Table 2-4. Classification Based on Dissolved Inorganic Constituents in Groundwater

Major Constituents (greater than 5 mg/l)

Bicarbonate	Silicon
Calcium	Sodium
Chloride	Sulfate
Magnesium	Carbonic acid

Minor Constituents (0.01–10.0 mg/l)

Boron	Nitrate
Carbonate	Potassium
Fluoride	Strontium
Iron	

Trace Constituents (less than 0.1 mg/l)

Aluminum	Molybdenum
Antimony	Nickel
Arsenic	Niobium
Barium	Phosphate
Beryllium	Platinum
Bismuth	Radium
Bromide	Rubidium
Cadmium	Ruthenium
Cerium	Scandium
Cesium	Selenium
Chromium	Silver
Cobalt	Thallium
Copper	Thorium
Gallium	Tin
Germanium	Titanium
Gold	Tungsten
Indium	Uranium
Iodide	Vanadium
Lanthanum	Ytterbium
Lead	Yttrium
Lithium	Zinc
Manganese	Zirconium

Source: S. N. Davis and R. J. M. DeWiest, *Hydrogeology* (New York: John Wiley and Sons, Inc., 1966). Reprinted by permission of the publisher.

factors compared with the amount of carbon dioxide produced by respiration. Thus, the water undergoes a chemical change during its passage through the soil to the underlying water-bearing formations.

In the saturated zone, the water reacts with the geologic formations, increasing the content of total dissolved solids. Thus, the chemical quality of groundwater depends both upon its age and the geological formations encountered in its flow history. Carbonate formations would increase the magnesium and calcium content and also the concentration of bicarbonate due to the dissolution of calcite ($CaCO_3$) and dolomite ($Mg \cdot CaCO_3$). Aluminosilicate minerals would increase the concentrations of sodium, potassium, magnesium, calcium, and silicon hydroxide ($Si(OH)_4$) in the water.

27

In consolidated deposits consisting of minerals from a range of sedimentary, igneous, and metamorphic sources, the order of encounter of the groundwater with the different assemblages determines the chemical constituency at a particular point in time. Sulfate-bearing minerals such as gypsum and anhydrite, although they occur less frequently than the carbonate and crystalline formations, are characterized by high solubilities. Thus, in older groundwater that has encountered sulfur-bearing minerals, the sulfate anion may dominate the bicarbonate anion. In very deep, old groundwater, the chloride anion may dominate both the bicarbonate and sulfate ones due to the presence of readily soluble minerals such as halite ($NaCl$) and sylvite (KCl).

The U.S. Public Health Service (USPHS) standard for total dissolved solids (TDS) in drinking water is 500 ppm, although the 1974 Safe Drinking Water Act considers waters containing up to 10,000 ppm TDS as potential sources of drinking water (U.S. EPA, 1977). The four types of naturally occurring groundwater that often exceed 10,000 ppm TDS are connate water, intruded seawater, magmatic and geothermal water, and water affected by salt leaching and the products of evapotranspiration. In many areas of the United States, groundwater can be used for drinking with no pretreatment, in which case it is said to be a raw resource. In other areas pretreatment to correct hardness or color may be necessary. Where contamination by pathogenic organisms is a potential threat, chlorination may be indicated. Water from some aquifers is unusable due to its salinity or to the presence of naturally occurring toxic substances such as arsenic or radionuclides. Most of the freshwater aquifers are underlain by brackish or saline aquifers, and in general, salinity may be said to increase with depth.

Use of Groundwater

Groundwater is a major natural resource in the United States and is often more easily available than surface water. It is estimated that more than 50% of the population uses groundwater as its primary source of drinking water (Heath, 1985; U.S. EPA, 1985c). Figure 2-8 shows early estimates of population served by groundwater and surface water for drinking purposes. These estimates are from unpublished data for 1970 from the USGS (U.S. EPA, 1977). They indicate that groundwater delivered by community systems supplied 29% of the population, and an additional 19% had its own domestic wells. Approximately 36% of the municipal public drinking-water supplies came from groundwater. Of the rural population, 95% was dependent upon groundwater for drinking purposes in 1970. Of the total groundwater usage, 60% was used for public supplies and 40% for rural supplies (U.S. EPA, 1977). EPA's National Statistical Assessment of Rural Water Conditions (U.S. EPA, 1984b) estimated that 72.3% of all major rural household supplies was extracted from groundwater. This dependence varied regionally, with 87.5% of north-central rural households and 56.7% of western households dependent upon groundwater for water supplies. Table 2-5 illustrates groundwater withdrawals for rural domestic supply as a percentage of total freshwater use. This table indicates that 97% of water withdrawn for the nation's domestic rural use is supplied by groundwater. Figure 2-9 shows groundwater as a percentage of total water use for all purposes in 1980. It is clear that the states west of the Mississippi River, in the area where irrigated agriculture is prevalent, depend heavily on groundwater. Arkansas, Nebraska, Colorado, and Kansas use more than 90% of their groundwater for agricultural activities (U.S. EPA, 1985c). Irrigation was the principal use of all groundwater withdrawn across the country in 1980 (Heath, 1985). The more humid eastern portion of the country is far less dependent on groundwater. Figure 2-10 shows the actual usage figures for the various states in millions of gallons per day, for public water systems (1980) and total groundwater usage for all purposes (1980). The western states, again, are the main users of groundwater.

The most recent figures for groundwater use are from 1980 USGS data. Table 2-6 shows the historical trends in groundwater use as a percentage of total withdrawals. The total use of fresh groundwater showed an increase of 22% from 1970 to 1975 and an increase of only 7% from 1975 to 1980. This change in the rate of

Table 2-5. Groundwater Withdrawals as a Percentage of Total Freshwater Withdrawals for Rural Domestic Supply

State	Percentage	State	Percentage
Alabama	100	Montana	94
Alaska	99	Nebraska	100
Arizona	100	Nevada	94
Arkansas	100	New Hampshire	98
California	93	New Jersey	100
Colorado	36	New Mexico	97
Connecticut	100	New York	89
Delaware	100	North Carolina	100
District of Columbia	0	North Dakota	100
Florida	100	Ohio	90
Georgia	100	Oklahoma	83
Hawaii	90	Oregon	87
Idaho	96	Pennsylvania	100
Illinois	97	Rhode Island	100
Indiana	90	South Carolina	100
Iowa	100	South Dakota	94
Kansas	86	Tennessee	100
Kentucky	91	Texas	84
Louisiana	100	Utah	90
Maine	98	Vermont	85
Maryland	100	Virginia	100
Massachusetts	100	Washington	78
Michigan	100	West Virginia	95
Minnesota	100	Wisconsin	100
Mississippi	100	Wyoming	92
Missouri	74		
		Total national percentage	97

Source: After Heath, 1985.

increased use may be attributable to the fact that high demand for groundwater may decrease the ease and efficiency of its withdrawal, and in some cases its quality. Escalated costs resulting from these factors and the cost of the water itself have perhaps influenced users, especially irrigators, to be more efficient with groundwater use (Solley et al., 1983). Actual water-use figures for 1975 and 1980 (including saline groundwater withdrawals and water used for the generation of thermoelectric power) are given in Table 2-7. Figure 2-11 shows, for comparison, the historical trends in the use of groundwater and surface water, and total water use.

**Table 2-6.   Historical Trends in U.S. Groundwater Use as a
Percentage of Withdrawals: 1950–1980**

	1950	1960	1970	1975	1980
Total Fresh Groundwater Withdrawals (bgd)	34	50	68	82	88
Public Supplies (%)	12	13	14	13	14
Rural Supplies (%)	8	6	5	5	5
Irrigation (%)	62	68	66	69	68
Industry (%)	18	13	15	14	13

Note: May not total 100% due to rounding.

Sources: Murray and Reeves, 1972, 1977; Makichan and Kammerer, 1961; Soiley et al., 1983.

Tables 2-8 and 2-9 give an idea of the way water is used in a typical community. In an average American community, the average per-capita water consumption is approximately 159 gallons per day (Last, 1980). In an American home, depending on the nature of the residence and the climate, exterior residential use may range from 5% to more than 150% of the interior use, averaging about 75%. This is primarily for watering lawns. Personal use of water, excluding laundry, toilet flushing, and so forth, accounts for 40% of the internal residential use and averages out to 6.3 gallons per capita per day.

The safe yield of a groundwater basin is defined as the amount of water that can be withdrawn annually without producing an undesired effect (Todd, 1959). Undesired results include depletion of the groundwater reserves, intrusion of water of an undesirable quality, contravention of water rights, the deterioration of the economic advantages of pumping, excessive depletion of stream flow, and land subsidence (Domenico, 1972; Freeze and Cherry, 1979). Any withdrawal in excess of a safe yield is termed an overdraft. It has been suggested that the optimal yield would be a more useful concept and should be determined by an optimal groundwater-management scheme that includes and best meets economic and social objectives associated with the use of water. Under certain conditions, optimal yields may involve the mining of groundwater to depletion or complete conservation (Freeze and Cherry, 1979).

Increased demands on groundwater have strained the supply in certain regions, resulting in reduced artesian pressure, land

Table 2-7. U.S. Groundwater Use: 1975, 1980

	1975	1980
Population served by groundwater for public supplies	64,700,000	73,700,000
Total groundwater withdrawn (mgd) for public supplies	11,000	12,000
Rural use of groundwater (mgd)		
Domestic	2,700	3,300
Livestock	1,200	1,200
Total withdrawn for irrigation (mgd) to irrigate a total of 54,000,000 acres (1975) and 63,125,000 acres (1980)	57,000	60,000
Groundwater withdrawals for industrial use (mgd)		
Fresh	11,000	10,000
Saline	980	930
Water used for generation of thermoelectric power	1,390	1,600
Total withdrawals of groundwater (mgd)	85,270	89,030

Sources: Murray and Reeves, 1977; Soiley et al., 1983.

31

subsidence, reduced spring and stream flow, and the intrusion of salt water. This kind of groundwater overdraft is occurring in the High Plains from Texas to Nebraska, parts of California, Arizona, Louisiana, Florida, New Mexico, Arkansas, Wisconsin, Illinois, and North Carolina. It has been reported that groundwater levels of the Carrizo Aquifer in south-central Texas have declined as much as

Table 2-8. Allocation of Water Use in U.S. Communities

Use	Percentage
Residential	40
Commercial	15
Industrial	25
Public	5
Unaccounted for	15
Total	100

Source: J. M. Last, in Maxcy–Rosenau, eds. *Public Health and Preventive Medicine*, 11th ed. (Norwalk, Conn.: Appleton-Century-Crofts, 1980), p. 976. Reprinted by permission of the publisher.

Table 2-9.   Allocation of Interior Residential Water Use

Use	Percentage
Drinking and cooking	5
Bathing	30
Toilet flushing	40
Laundry	15
Dishwashing	5
Miscellaneous	5
Total	100

Source: J. M. Last, in Maxcy–Rosenau, eds. *Public Health and Preventive Medicine*, 11th ed. (Norwalk, Conn.: Appleton-Century-Crofts, 1980), p. 976. Reprinted by permission of the publisher.

400 feet since 1930 because of overpumping (Texas Department of Water Resources, 1984b). Figure 2-12 shows the extent of the areas affected by groundwater overdraft and related problems. Figure 2-13 shows areas across the country where water table decline or artesian water level decline has occurred in excess of 40 feet in at least one aquifer since development began (Mann, 1985). In the Second National Water Assessment, the U.S. Water Resources Council (1978a) lists actions that could be taken before pumping becomes uneconomical in the areas with declining water levels. These include locating alternate water sources, developing artificial recharge of the aquifers, relocation of water-requiring activities, and reduction of water use through better water management. Groundwater overdraft is considered to be serious when it continues indefinitely (U.S. GAO, 1980), causing the exhaustion of the resource, associated economic and social dislocations, and perhaps general weakening of a region's economy. Such potential problems are developing in California, the High Plains regions, Florida, New Mexico, Colorado, and Arizona, but many states are successfully dealing with the problem (U.S. GAO, 1980). Although this book deals with changes in groundwater quality, changes in quantity through overdraft are briefly mentioned here, as they can affect the quality of the remaining groundwater in the region.

Profile of an Aquifer

In order to provide a picture of the importance of groundwater, it is useful to describe a specific aquifer, its characteristics and uses, and the problems that affect it as a natural resource. A study by the High Plains Study Council provides such data for the Ogallala Aquifer. The Ogallala is an exceptional aquifer in both extent and volume, running as it does through parts of eight states. It is an integral part of the burgeoning economy of those states, as well as of the environment. Because of its unusual size and certain other characteristics, however, it should not be viewed as a representative aquifer.

The Ogallala Aquifer is a major source of groundwater in the High Plains region, underlying a land area of approximately 134,000 square miles (Weeks, 1986), three times the size of New York State, or about the size of California (Bittinger, 1981). The formation extends from southern South Dakota to northwestern Texas and transects portions of six other states: Nebraska, Wyoming, Colorado, Kansas, Oklahoma, and New Mexico (Figure 2-14). It contains an estimated quadrillion gallons, or 3.25 billion acre-feet, of water, the equivalent of Lake Huron (Stengel, 1982). The aquifer fuels a $30 billion-a-year agricultural economy (B. Turner; Camp, Dresser & McKee; personal communication); irrigates over 12 million acres of farmland (Bittinger, 1981), or 20% of the irrigated acreage in the United States (Banks, 1981); and helps support a population of nearly half a million people (Press and Siever, 1978). Furthermore, 40% of the beef consumed in this country is fattened in the aquifer area (Banks, 1981). Such heavy exploitation of the vast groundwater resources in the Ogallala has resulted in serious depletion in some areas. These problems are now the subject of intensive study.

The Ogallala was formed during the early Pleistocene time, some two million years ago, from glacial outwash of the ancestral Rocky Mountains that consisted of gravel, sand, and finer debris that was caught up by streams of meltwater running away from the glacier (Press and Siever, 1978). The outwash settled unevenly throughout the High Plains, giving the aquifer water-storage capacities that varied according to the depth and content of the outwash. In Nebraska, where two-thirds of the Ogallala's waters lie and where the aquifer is 1,000–1,500 feet thick, the aquifer has a considerably greater storage capacity per unit of area than it does in the South Plains of Texas, where it is less than 100 feet thick (H. O. Banks; Camp, Dresser & McKee; personal communication). At present, there are an estimated 2 billion acre-feet of water in storage in Nebraska (H. O. Banks; Camp, Dresser & McKee; personal communication) compared with 350 million acre-feet in Texas and 100 million acre-feet in Oklahoma (Scalf et al., 1973).

33

Annual withdrawals from the entire aquifer average 5 or 6 million acre-feet (U.S. Water Resources Council, 1980).

The Ogallala's extensive groundwater resources were virtually unknown until the early part of this century. The first wells were drilled into the formation over 90 years ago (Press and Siever, 1978), but the land was used primarily for cattle grazing and dryland wheat production well into the 1900s (Warren et al., 1982). The start of irrigated farming, with the use of high-capacity pumps, after World War II brought rapid economic growth but also caused the water table of the aquifer to decline steadily, in some areas by as much as 100 feet (Press and Siever, 1978), and even to the point of drying up. Gaines County, Texas, has dried up its groundwater resources entirely (H.O. Banks; Camp, Dresser & McKee; personal communication). The aquifer is very slow to recharge. Because of its high rate of evaporation and its high percentage of impermeable soils, the water is not replenished at the rate at which it is used. The depth of the water table nearly everywhere in the Ogallala formation is 50 feet or more below the roots of plants. Plants get the first opportunity at capturing any infiltrating rainfall, making that portion unavailable for pumping (U.S. Water Resources Council, 1980). Overall, the amount of water being overdrawn annually from the aquifer exceeds 3 million acre-feet (Mapp and Eidman, 1976). Annual recharge from precipitation averaging 12 to 22 inches per year (Warren et al., 1982), irrigation return-flows, and some stream-bed percolation is estimated to be 0.27 million acre-feet (Bekure, 1971). The state of Texas has estimated that, at the current rate of use and without an imported water supply, approximately 40% of the now-irrigated acreage in the High Plains of West Texas will have to revert to dryland farming or be abandoned by the year 2000, and 60% will have to be abandoned by the year 2020 (Banks, 1981).

The depletion of the Ogallala began to attract public attention in the early 1970s (Banks, 1981). Members of Congress representing the Great Plains states of Colorado, Kansas, Nebraska, New Mexico, Oklahoma, and Texas encouraged the development of legislation that would mandate intensive study of the problem. A $6-million study was authorized by public law in 1976, and the High Plains Study Council was formed under contract with Camp, Dresser & McKee. The Council approved a study design with the following objectives in 1977 (Warren et al., 1982):

1. to determine the potential development alternatives for the High Plains,
2. to identify and describe policies and actions required to carry out promising development strategies, and

3. to evaluate the local, state, and national implications of these alternative development strategies.

It is a comprehensive resource and economic-development study of the area served by the Ogallala (Banks, 1981). Major emphasis is being given to improving the water supply by local conservation and by improved practices of irrigation and agricultural management and to interbasin transfers of surplus water from adjacent basins, the Missouri River, and streams in Arkansas. Effective, comprehensive, interstate management of the **35** Ogallala groundwater resource is a further major element of the study.

Currently there is little or no public support for comprehensive, interstate management of the aquifer, although interdependencies are widely recognized (Banks, 1981). Laws in the states vary widely. Oklahoma and Texas are the only states of the High Plains with private ownership of groundwater. The groundwater in the six other states is dedicated to the people of the state, and allocation is administered by state and local officials. In all eight states, well registration or permitting systems exist (Gutentag, et al., 1984).

Contamination of the Ogallala aquifer does not, to date, appear to be a major problem or concern (B. Turner; Camp, Dresser & McKee; personal communication). There has been some contamination around major population centers like Lubbock, Texas, but the depletion problem is of more immediate concern (Figure 2-15). A study by Camp, Dresser & McKee predicts that 5.1 million acres of irrigated land in six Great Plains states will dry up by the year 2020 (Stengel, 1982). A continuation of present trends will mean the loss of 1.6 million irrigated acres in Kansas, 1.2 million in Texas, 260,000 in Colorado, 224,000 in New Mexico, and 330,000 in Oklahoma. Twenty recommendations have been made for remedial action, but only one is an actual cure. The Army Corps of Engineers has proposed a system of huge canals that would import water from South Dakota, Missouri, and Arkansas. The cost of the project (from $3.6 billion to $22.6 billion) is prohibitive, however, and stopgap efforts like conversion to dryland farming will probably be the most immediate solutions to the depletion problem.

Uses of the Ogallala Groundwater Resources in a Few Selected States

In New Mexico the aquifer is heavily developed from southern Quay County to southern Lea County (Scalf et al., 1973). The thickest and most productive part of the aquifer is in northern Lea County. Individual well yields in this area can be as much as 1,600 gallons per minute. In southern Lea County, where the aquifer is thinner, wells yield from 300 to 1,000 gpm, but water levels have been declining at a rate of almost 3 feet per year since 1950.

In Oklahoma the Ogallala is the most important aquifer, underlying most of the panhandle (Scalf et al., 1973). Deposits of sand, gravel, and minor amounts of clay store high-quality

36   groundwater. The aquifer is as much as several hundred feet thick, contains more than 100 million acre-feet of available water, and supplies most of the water requirements of the panhandle. It is used to irrigate 135,000 acres, to supply water for the industrial needs of the Keyes helium plant and the natural gas industry in the panhandle, and to supply all the public and domestic needs in the area.

In Texas the aquifer formation consists of interfingered and intergraded lenses and layers of sand, gravel, silt, clay, and caliche, a crust of calcium carbonate (Scalf et al., 1973). The High Plains is divided by the Canadian River into the northern High Plains and the southern High Plains. The northern area comprises approximately 9,300 square miles. The zone of saturation in most places is 100 to 500 feet thick. Irrigation has developed more slowly here than in the southern High Plains, which is the area of greatest groundwater development in Texas. This area includes about 25,000 square miles and has the most serious depletion problems of the entire aquifer. According to a 1985 USGS report, groundwater withdrawals in the southern High Plains have been averaging 6,500 million gallons a day (mgd) (USGS, 1985). The rate of recharge has been 125 to 134 mgd.

The Texas Department of Water Resources has made estimates and projections of groundwater availability in the Ogallala through the year 2030 (Texas Department of Water Resources, 1984c). It has determined the annual effective recharge to be 143 billion gallons. The amount of water in recoverable storage as of 1980 was determined to be 125,609.27 billion gallons. (Recoverable storage is defined as that portion of the underground reservoir capacity that, it is estimated, is physically capable of yielding water economically.) The projected annual average of groundwater availability was then determined through 2030. (Annual average of groundwater availability is defined as the estimated sustainable annual yield, or effective recharge, plus that amount of water that can be recovered from storage over a specified period of time without causing irreversible harm, such as land-surface subsidence or water-quality deterioration.)The projections are as follows:

1990	2,132.17 billion gallons
2000	2,678.33 billion gallons
2010	2,495.95 billion gallons
2020	1,959.99 billion gallons
2030	1,491.09 billion gallons

The remaining recoverable storage in 2031 is projected to be 49,696.48 billion gallons (Texas Department of Water Resources, 1984c).

Figure 2-1. Illustration of Relationships Within the Hydrologic System
Source: U.S. EPA, 1977

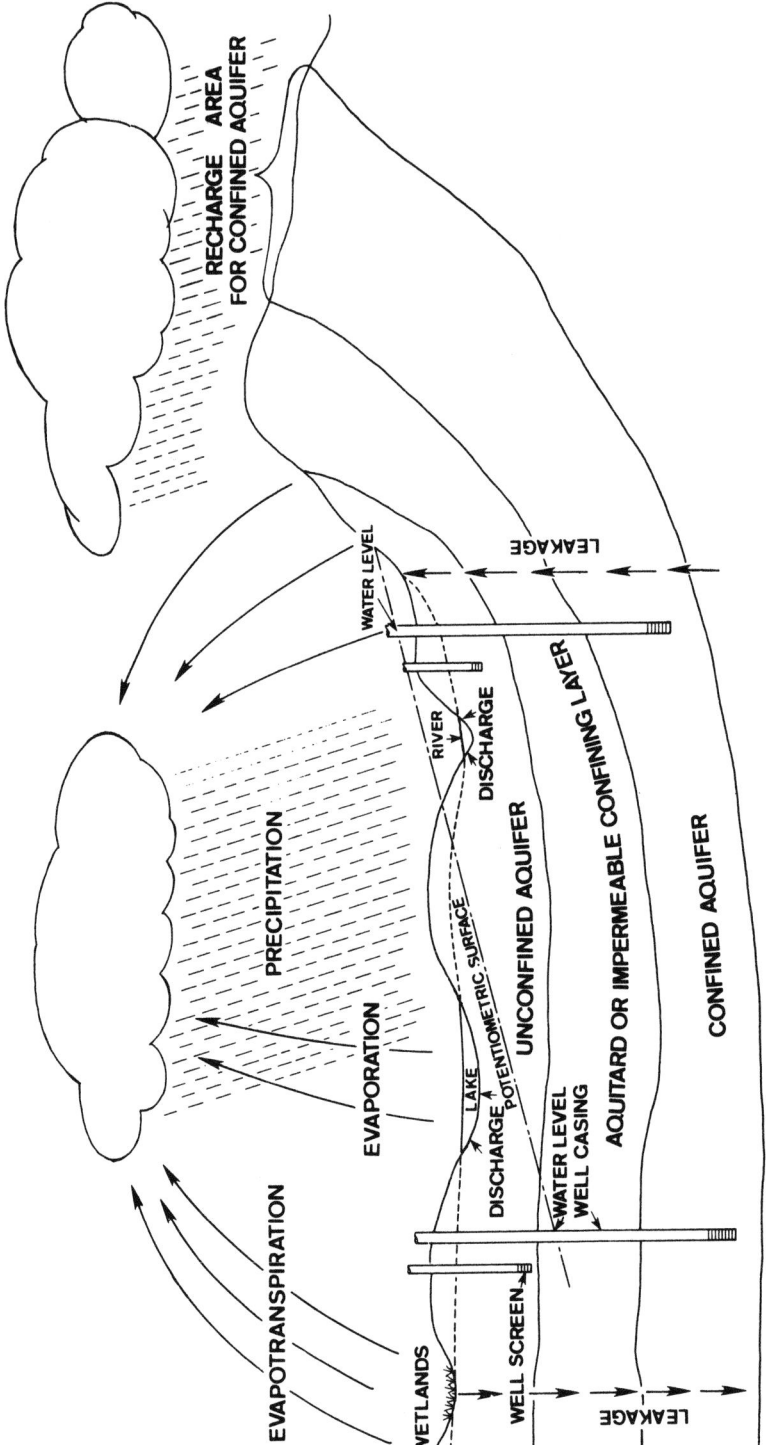

Figure 2-2. Geochemical Cycle of Surface Water and Groundwater
Source: Lehr et al., 1976

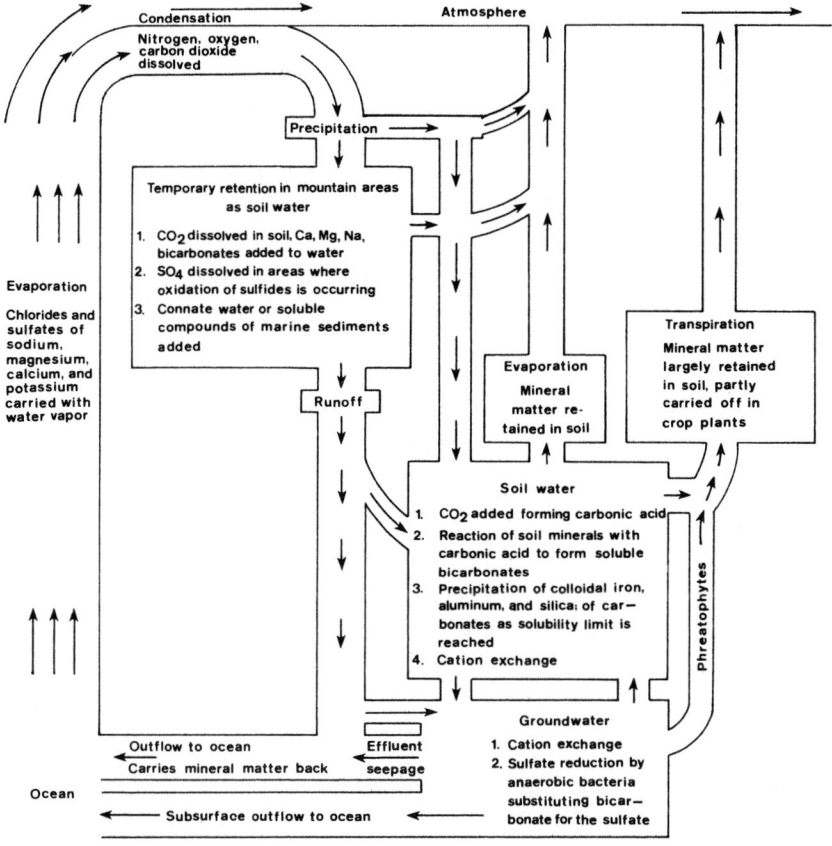

Figure 2-3. Groundwater Resources of the United States
Source: U.S. Water Resources Council, 1978b

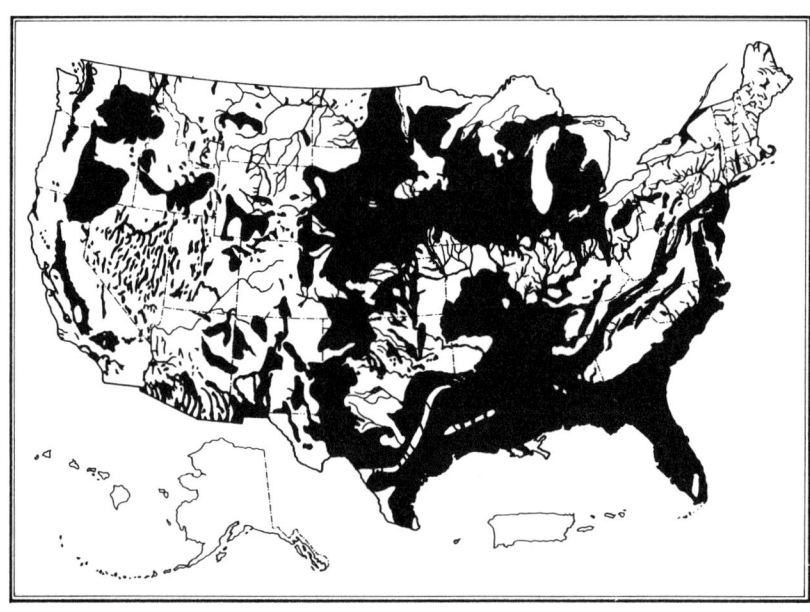

	Watercourse related aquifers
■	Areas of extensive aquifers that yield more than 50 gallons per minute of fresh water
□	Areas of less extensive aquifers having smaller yields

Figure 2-4. Ten Major Groundwater Regions of the United States, except Alaska and Hawaii
Source: Lehr et al., 1976

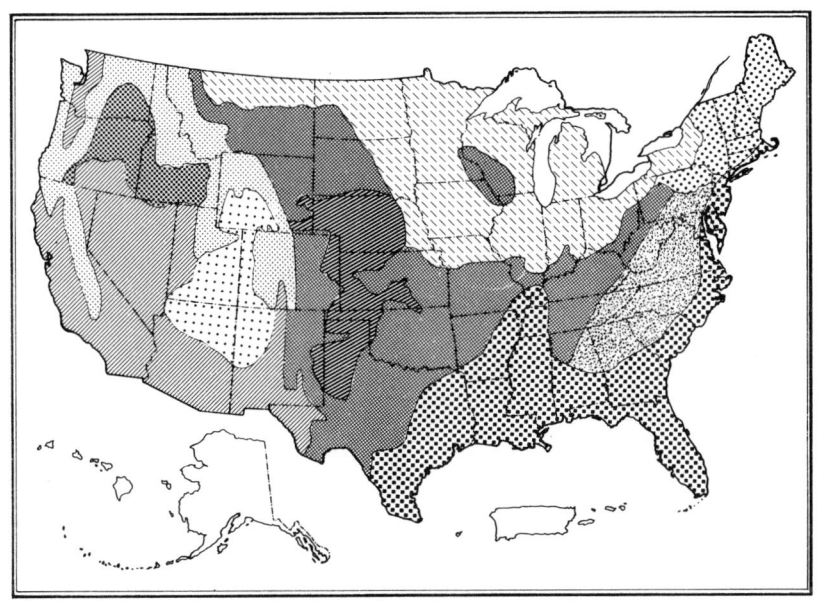

Western Mountain Ranges
Alluvial Basins
Columbia Lava Plateau
Colorado Plateau
High Plains

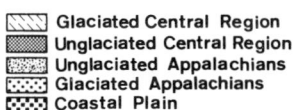
Glaciated Central Region
Unglaciated Central Region
Unglaciated Appalachians
Glaciated Appalachians
Coastal Plain

Figure 2-5. Water Resources Regions
Source: U.S. Water Resources Council, 1978

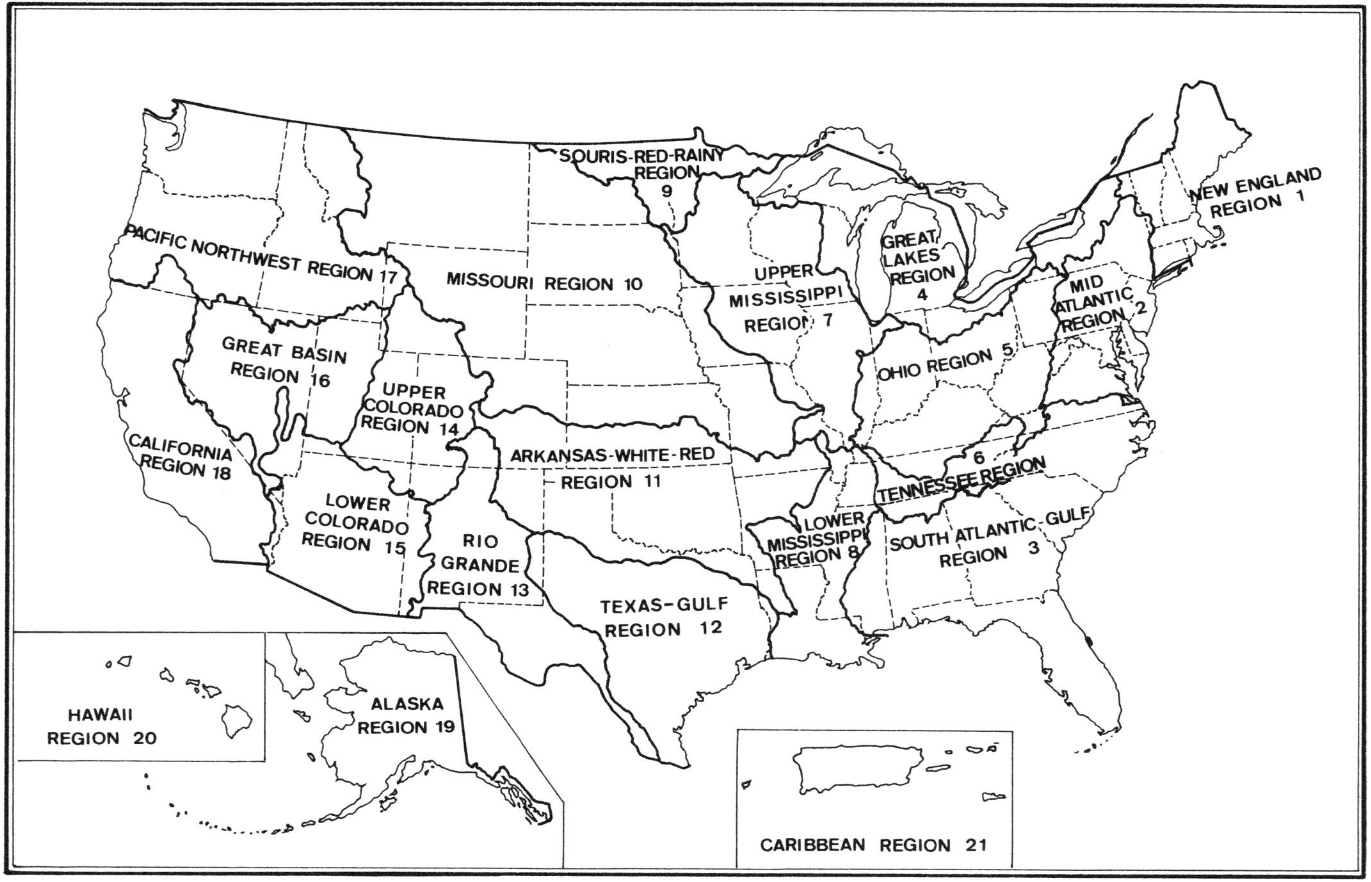

Figure 2-6. Average Annual Precipitation
Source: U.S. EPA, 1977

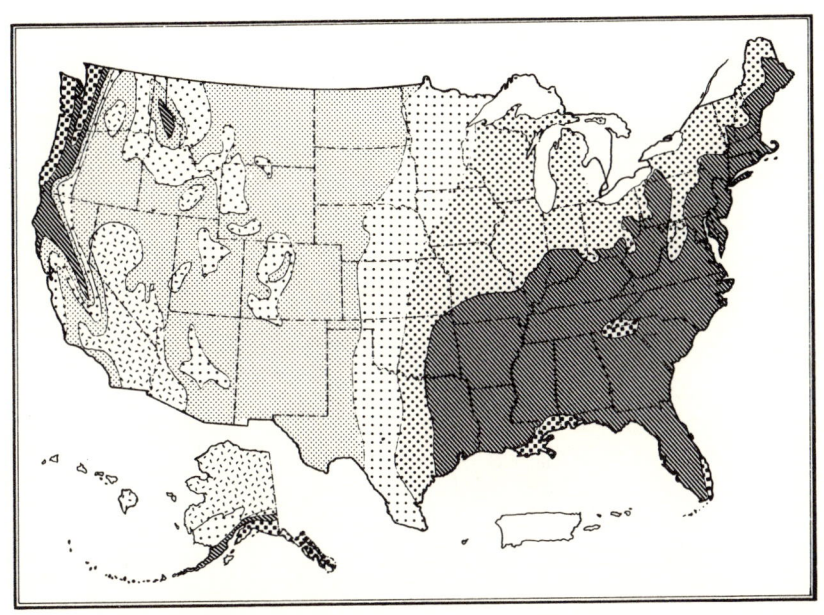

**Average Annual
Precipitation in inches**

0–10   30–40
10–20   40–60
20–30   60 and over

Figure 2-7. Hardness of Groundwater
Source: Geraghty et al., *Water Atlas of the United States,* 1973.
Used by permission of the Water Information Center, Inc., Syosset,
N. Y.

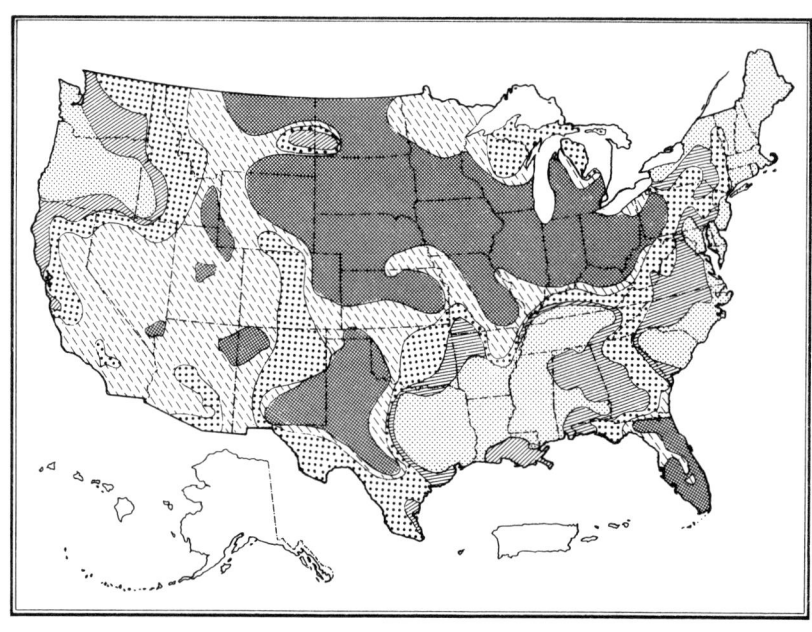

Hardness as $CaCO_3$ in parts per million

	< 80		180–240
	80–120		> 240
	120–180		

Figure 2-8. Population Served by Source and Supply, 1970
Source: U.S. EPA, 1977

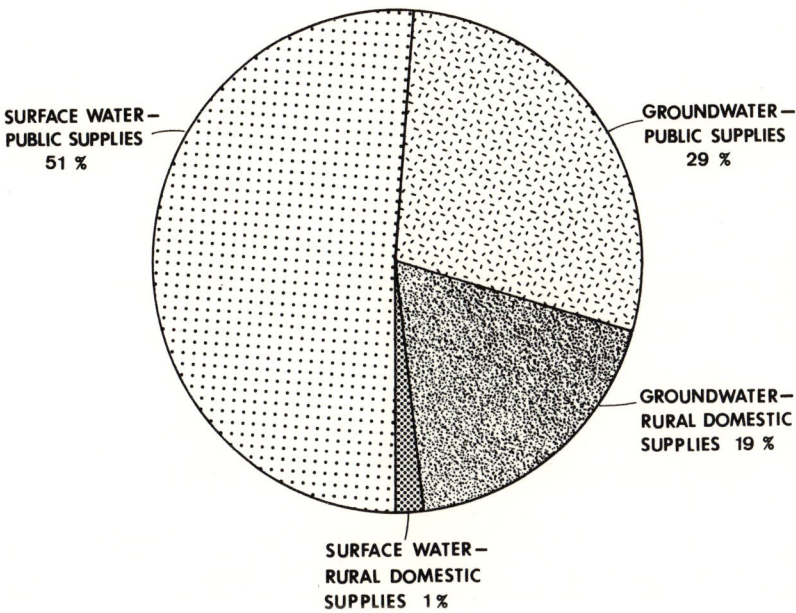

SURFACE WATER –
PUBLIC SUPPLIES
51 %

GROUNDWATER –
PUBLIC SUPPLIES
29 %

GROUNDWATER –
RURAL DOMESTIC
SUPPLIES 19 %

SURFACE WATER –
RURAL DOMESTIC
SUPPLIES 1 %

Figure 2-9. Groundwater Use as a Percentage of Total Water Use, 1980
Source: National Water Summary, USGS, 1985

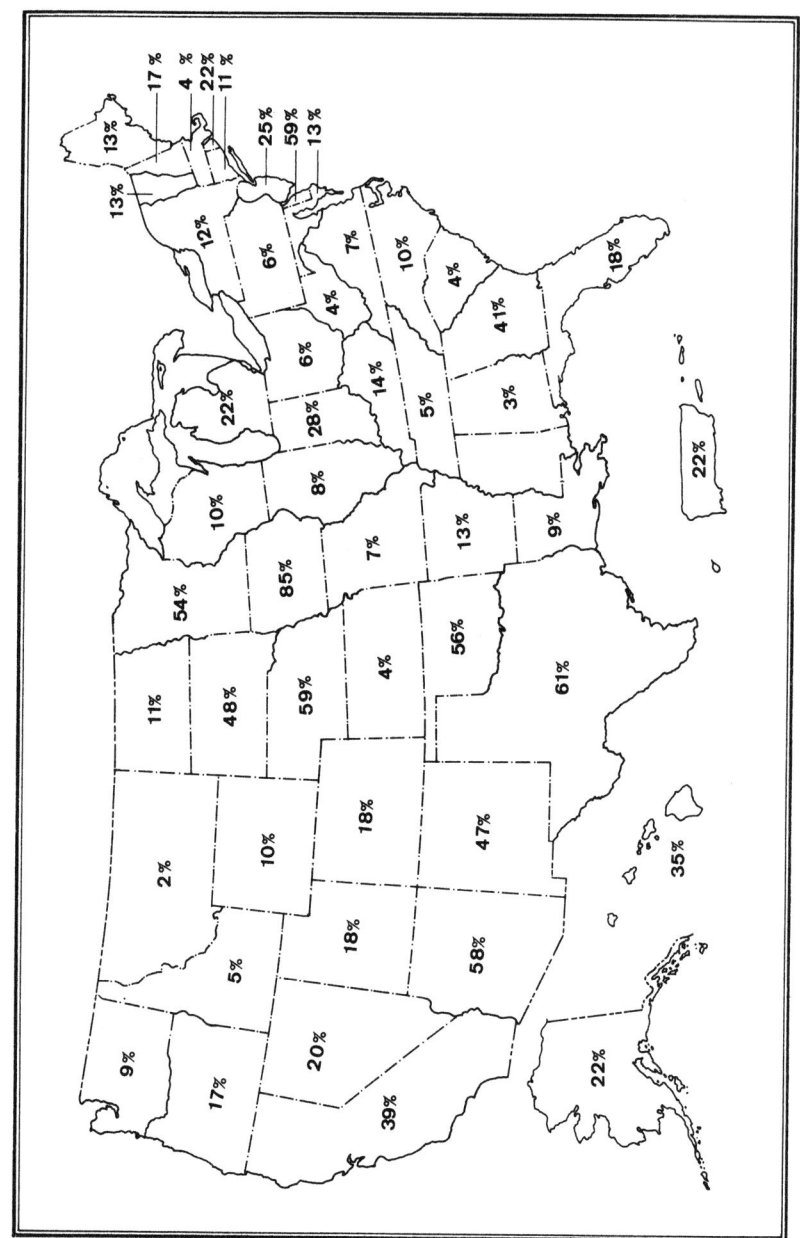

Figure 2-10. Groundwater Usage in the United States, mgd
Source: National Water Summary, USGS, 1985

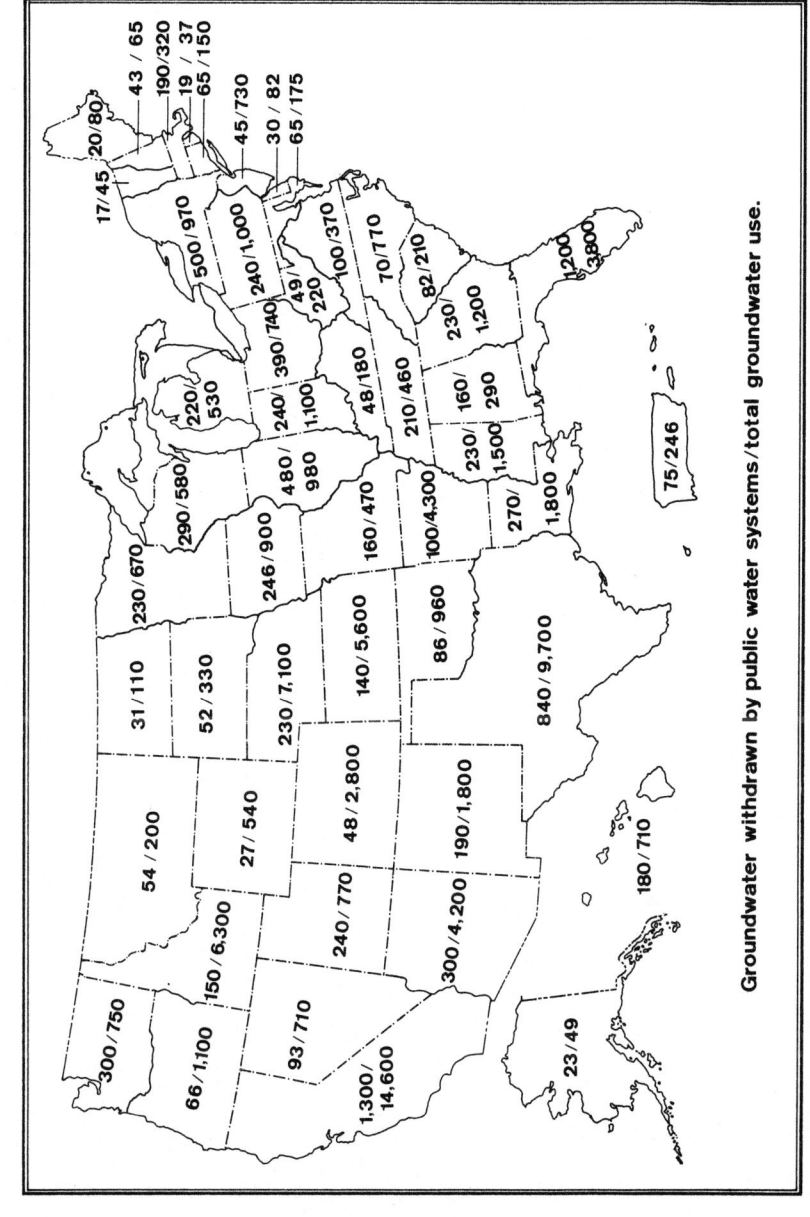

Groundwater withdrawn by public water systems/total groundwater use.

Figure 2-11. Trends in Water Withdrawals for Public Supplies, Rural Supplies, Irrigation, and Self-Supplied Industry, 1950–1980
Source: Water Use Data, USGS, 1983

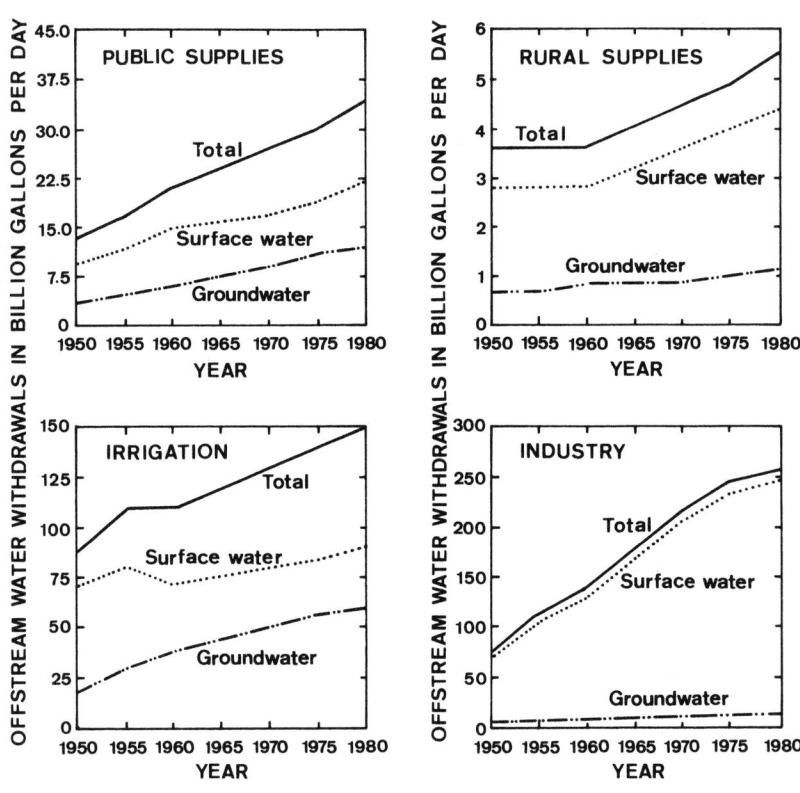

Figure 2-12. Groundwater Overdraft and Related Problems
Source: U.S. Water Resources Council, 1978a

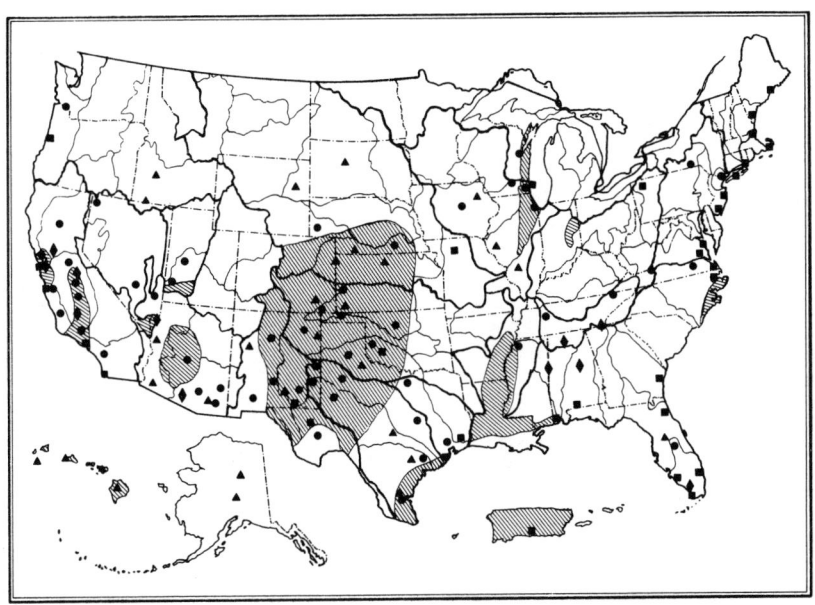

AREA PROBLEM

▨ Area in which significant groundwater
   overdraft is occurring
□ Unshaded area may not be problem-free, but
   the problem was not considered major

BOUNDARIES

— Water resources region
— Subregion

SPECIFIC PROBLEMS (as identified by Federal
and State/Regional study teams)

● Declining groundwater levels
▲ Diminished springflow and streamflow
♦ Formation of fissures and subsidence
■ Saline water intrusion into fresh-water
   aquifers

Figure 2-13. Areas of the Water Table Decline or Artesian Water Level Decline in Excess of 40 Feet Since Development Began

Source: National Water Summary, USGS, 1985

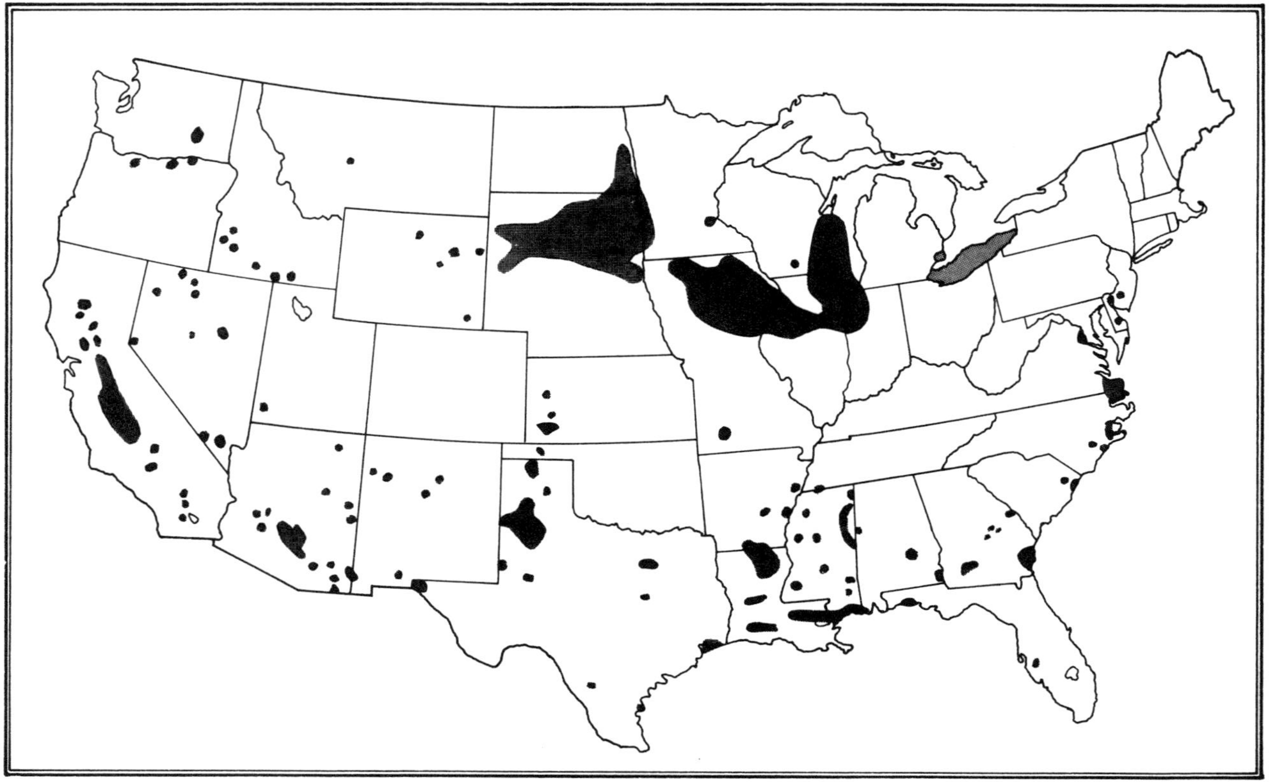

Figure 2-14. Ogallala Aquifer
Source: Banks, 1981

Figure 2-15. Projected Groundwater Depletion from the Ogallala Aquifer, 1977–2020
Source: High Plains Associates: Camp Dresser & McKee Inc., Black & Veatch, and Arthur D. Little, Inc., "Six State High Plains-Ogallala Aquifer Regional Resources Study, March, 1982," Austin, Tex. Used by permission.

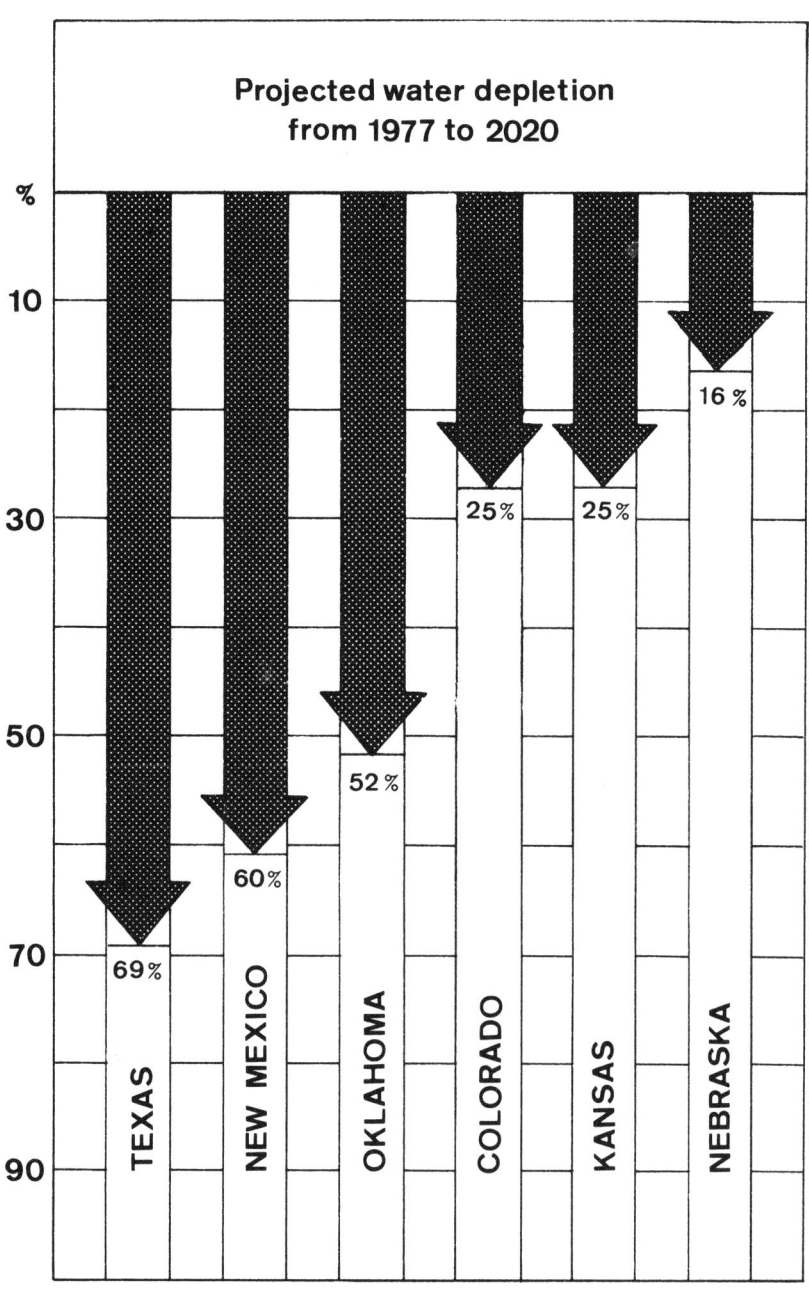

Three

Groundwater Contamination

- Groundwater contamination incidents have been reported from all parts of the United States.
- More contamination is likely to be discovered in the future, especially if a comprehensive national survey is undertaken.
- There are many and varied sources of contamination: natural and anthropogenic, point and non-point, planned and inadvertent.
- Types of contaminants found in groundwater range from simple inorganic ions, such as chloride, nitrate, and heavy metals, to complex synthetic organic chemicals.
- The types of reported contamination problems, and their sources, vary from one region of the country to another and are influenced by climate, population density, intensity of industrial and agricultural activities, and the hydrogeology of the region.

The mobility of our society, as well as the distribution of industry and agriculture, depends upon an available supply of clean water. Nevertheless, instances of groundwater contamination have been found in most sections of the country (Kerns, 1977; U.S. EPA, 1980a, 1980c, 1985b; OTA, 1984). For the purpose of this book groundwater contamination will be defined as the addition to water of elements, compounds, or pathogens that alter its composition.

The pollution of groundwater occurs when the concentrations of the contaminants render the water unfit for present and future uses. Figure 3-1 shows a geographic distribution of groundwater pollution, from the Second National Water Assessment.

One of the major difficulties with groundwater contamination is that it occurs underground, out of sight. The sources of pollution are not easily observed, nor is the effect of pollution often seen until irreversible damage has occurred. There are no obvious warning signals such as fish kills, discoloration, or stench that are often early indicators of surface-water pollution. Where contamination affects pumping wells, such indicators may occur, although many commonly found contaminants are both colorless and odorless, which makes detection difficult. They occur in concentrations that, if ingested, may cause long-term chronic illness, rather than acute poisoning. Many chemicals now found in groundwater have not been tested for their effects on human health. The tangible effects of groundwater contamination usually come to light long after the incident causing the contamination has occurred. The long time lag between occurrence and detection is a major problem.

Groundwater can be contaminated by a variety of compounds, both natural and man-made. Contamination due to man has occurred for centuries, but industrialization, urbanization, and increased population have greatly aggravated the problem in some areas. Miller (1981a), Braids (1981a) and Saar (1985) summarized the mechanisms of contamination and attenuation. Freeze and Cherry (1979) give a detailed account of these mechanisms, as well as of the movement of a contaminant in an aquifer.

Movement of Contaminants

A contaminant usually enters the groundwater system from the surface of the land, percolating down through the aerated soil and non-saturated zone. The root zone may extend two or three feet into the soil as shown in Figure 3-2. Many reductive and oxidative biological processes take place in this zone that may degrade or biologically change the contaminants. Plant uptake can remove certain heavy metals; microbial fixation and other biological processes can also remove a fraction of the contaminants, the size of the fraction being dependent on the nature of the contaminant. For example, depending upon conditions, iron and manganese solubility can be affected by microbial oxidation and reduction in the soils. In the deeper geologic material that consists mainly of humus and weathered rocks, there is a lessening of such biological processes. Attenuation of contaminants in this zone may occur by surface adsorption as ions in the contaminants are attracted to the charges on the clay particles. Other contaminants may be removed by insoluble organic matter complexing, giving rise to complexed humic acids. Microbial action may influence redox potentials and cause the release of inorganic ions during decomposition (Braids, 1981a). The susceptibility of different contaminants to differential attenuation varies. Present knowledge indicates that more different attenuation processes are active in the non-saturated zone than in the aquifer; and that the amounts of degradation are larger (Braids, 1981a). However, many significant bio-conversions are being discovered in anaerobic subsurface environments. In fact many compounds that persist under aerobic conditions are metabolized in anaerobic conditions (Suflita, 1987).

55

In aquifers composed of unconsolidated granular media, contaminant movement is determined by groundwater flow rates and flow paths which are governed by advection, diffusion, and dispersion, by the density of the contaminant and biological, chemical and physical interactions with the material through which it flows. Cherry (1984) states that "Contaminant migration in fractured media is determined by advection and dispersion in the fractures and by diffusive mass transfer between the flowing water in the fractures and the stagnant water in the porous but low permeable matrix."

Specific contaminants move in different ways. Liquid petroleum hydrocarbons, for example, would be expected to move downward through the aquifer; however, they may spread out on the top of the aquifer if they are present in sufficient quantities. They usually do not move very far from the point of entry because of the limiting effect of surface tension between the liquid and the water-coated grain of the unconsolidated material. However, dissolved chemicals from these petroleum hydrocarbons move with the groundwater.

Industrial liquids that are halogenated hydrocarbons have a viscosity less and a density much greater than those of water. The fractions that dissolve in water move with the groundwater. However, they may move downward, settle out in the aquitard, and (depending on the slope) may move downslope and not in the direction of the bulk flow of the groundwater (Cherry, 1984).

The migration of contaminants through unconsolidated granular media is better understood than is their migration through fractured material. Unfortunately, contaminant migration through subsurface materials is a highly complex and poorly understood science. Much of the quantitative data obtained for the prediction of contaminant migration from theoretical models and laboratory data is of questionable use when applied to the field. Problems arise mainly from variable geologic conditions (heterogeneous, anisotropic, and fractured subsurface materials), and from the many types and characteristics of the contaminants themselves (Cherry, 1984).

56

Prediction of plume movement and behavior is made difficult because geologic formations are not often uniform. See Figure 3-3. Layered beds and lenses may cause fingering or separation of the plume. Different geologic materials may retard or enhance the movement of the plume. The nature of the plume is further affected by the reactivity of the constituents of the contaminant. Concentrations can be altered by a variety of chemical and biochemical reactions, including adsorption-desorption reactions, acid-base reactions, solution-precipitation reactions, oxidation-reduction reactions, ion pairing, and microbial cell synthesis (Freeze and Cherry, 1979). The plume continues to move with groundwater flow unless it is blocked and eventually reaches points of groundwater discharge such as streams, wetlands, lakes, and tidal waters. Because very little dilution takes place, concentrations of contaminants are often much higher in groundwater than in surface water (Miller, 1981a).

Sources of Contamination

Sources of groundwater contamination may be divided into three main categories:
1. natural pollution
2. waste-disposal practices
3. non-disposal sources due to human activities.

The degree of threat posed by these sources of contamination depends upon the concentration of the contaminant, its toxicity at that concentration, the volume of groundwater affected, the hydrogeological conditions, the uses made of the water from that particular aquifer, the population affected by such uses, and the availability of an alternate water supply.

Groundwater-pollution problems and their sources have been the object of numerous summaries. An incomplete but illustrative listing includes Keeley, 1976; Fuhriman and Barton, 1971; van der

Leeden et al., 1975; Miller et al., 1977; Scalf et al., 1973; Miller et al., 1974; U.S. EPA, 1980a, 1978a, 1978b; U.S. GAO, 1978, 1980; U.S. Water Resources Council, 1978a, 1978b, 1978c. A comprehensive survey of sources of contamination due to waste-disposal practices was completed for the 1977 Report to Congress (U.S. EPA, 1977). However, the most comprehensive listing of sources thus far is that put together by the Office of Technology Assessment (1984). OTA identified 33 sources of known groundwater contaminants and categorized them according to the nature of their release (Table 3-1).

Table 3-1. Sources of Groundwater Contamination

Category I—Sources designed to discharge substances
Subsurface percolation (e.g., septic tanks and cesspools)
Injection wells
  Hazardous waste
  Non-hazardous waste (e.g., brine disposal and drainage)
  Non-waste (e.g., enhanced recovery, artificial recharge, solution mining, and in-situ mining)
Land application
  Wastewater (e.g., spray irrigation)
  Wastewater byproducts (e.g., sludge)
  Hazardous waste
  Non-hazardous waste

Category II—Sources designed to store, treat, and/or dispose of substances; discharge through unplanned release
Landfills
  Industrial hazardous waste
  Industrial non-hazardous waste
  Municipal sanitary
Open dumps, including illegal dumping (waste)
Residential (or local) disposal (waste)

Category III—Sources designed to retain substances during transport or transmission
Pipelines
  Hazardous waste
  Non-hazardous waste
  Non-waste
Materials transport and transfer operations
  Hazardous waste
  Non-hazardous waste
  Non-waste

Category IV—Sources discharging substances as consequence of other planned activities
Irrigation practices (e.g., return flow)
Pesticide applications
Fertilizer applications
Animal feeding operations
De-icing salts applications
Urban runoff
Percolation of atmospheric pollutants
Mining and mine drainage
  Surface mine-related
  Underground mine-related

Table 3-1.   Sources of Groundwater Contamination (*continued*)

Surface impoundments	Category V—Sources providing
Hazardous waste	conduit or inducing discharge
Non-hazardous waste	through altered flow patterns
Waste tailings	Production wells
Waste piles	Oil (and gas) wells
Hazardous waste	Geothermal and heat recovery
Non-hazardous waste	wells
Materials stockpiles (non-waste)	Water supply wells
Graveyards	Other wells (non-waste)
Animal burial	Monitoring wells
Aboveground storage tanks	Exploration wells
Hazardous waste	Construction excavation
Non-hazardous waste	
Non-waste	
Underground storage tanks	Category VI—Naturally occurring
Hazardous waste	sources whose discharge is
Non-hazardous waste	created and/or exacerbated by
Non-waste	human activity
Containers	Groundwater–surface water
Hazardous waste	interactions
Non-hazardous waste	Natural leaching
Non-waste	Salt-water intrusion/brackish water
Open burning and detonation sites	upconing (or intrusion of other
Radioactive disposal sites	poor-quality natural water)

Source: Office of Technology Assessment, 1984

58

Changes in the Composition of Groundwater Due to Natural Processes

All groundwater contains some dissolved salts. Mineralization of groundwater due to leaching is a significant source of saline groundwater in the arid areas of the country and is greatest in the areas of lowest precipitation (Figure 2-6). In areas where the water table is near the surface, evaporation and transpiration further concentrate the salts in the remaining water. Natural leaching has been a significant source of contamination in the arid southwest and south-central area comprising Arizona, California, Nevada, Utah, Arkansas, Louisiana, New Mexico, Oklahoma, and Texas. In this area, natural leaching has exceeded in importance anthropogenic sources of groundwater contamination (Fuhriman and Barton, 1971; Scalf et al., 1973; U.S. EPA, 1985c,e). Many of the inorganic substances found in groundwater are largely the result of natural contamination, although human sources contribute significant amounts as well. Table 3-2 lists 37 inorganic substances that have been found in groundwater. Chlorides are a widespread contaminant. Sulfates, nitrates, fluorides, and iron are also common natural contaminants that occur in localized natural deposits, often causing the groundwater in those areas to exceed EPA standards. Radioactivity from uranium deposits has caused problems in Texas, Illinois, Iowa, Oklahoma, Nevada, Arkansas, and New Mexico (Scalf et al., 1973; EPA, 1985c,e). Arsenic may be a local problem in the thermal springs of the northwest region (van der Leeden et al., 1975). Other examples of such localized problems occur throughout the country. Water from fault zones or of volcanic origin can contain high levels of salts or toxic chemicals. As mentioned in the section characterizing groundwater, the aquitards that confine the aquifers are not totally impermeable. A change in the composition, thickness, or continuity of the aquitard may permit leakage from one aquifer to another. The leaking of a saline aquifer into a potable aquifer may cause deterioration of the quality of water in the potable aquifer. Although natural alteration of the chemical composition of groundwater cannot be prevented, it may be fit for some uses with or without pretreatment.

59

Contamination Due to Waste-Disposal Practices

The 1984 OTA report to Congress *Protecting the Nation's Groundwater from Contamination* had the best estimate of the number of sources of groundwater contamination from waste disposal at the time that this report was compiled. The major sources described in this report and elsewhere will be summarized below. It should be noted that many of the numbers given are estimates and that definitive data on waste-disposal practices often were not, and are not, readily available.

Waste from natural and manufactured products is often stored on or beneath the land surface. In fact, burying wastes seems an ingrained cultural phenomenon (Braids, 1981a). Such burial of wastes is now regulated by the statutes described in Chapters 12 and 13 of this book. Sources of contamination are derived from all aspects of our lives, including industry, agriculture, and government. One estimate is that between 50,000 and 75,000 chemicals are now being used and distributed through the

Table 3-2.   Inorganic Substances Found in Groundwater

Substance	Concentration (mg/l)
Aluminum	0.1–1,200
Ammonia	1.0–900
Antimony	—
Arsenic	0.01–2,100
Barium	2.8–3.8
Beryllium	less than 0.01
Boron	—
Cadmium	0.01–180
Calcium	0.5–225
Chlorides	1.0–49,500
Chromium	0.06–2,740
Cobalt	0.01–0.18
Copper	0.01–2.8
Cyanides	1.05–14
Fluorides	0.1–250
Iron	0.04–6,200
Lead	0.01–5.6
Lithium	—
Magnesium	0.2–70
Manganese	0.1–110
Mercury	0.003–0.01
Molybdenum	0.4–40
Nickel	0.05–0.5
Nitrates	1.4–433
Nitrites	—
Palladium	—
Potassium	0.5–2.4
Phosphates	0.4–33
Selenium	0.6–20
Silver	9.0–330
Sodium	3.1–211
Sulfates	0.2–32,318
Sulfites	—
Thallium	—
Titanium	—
Vanadium	243.0
Zinc	0.1–240

— indicates that the substance has been detected in groundwater but no concentrations have been reported

Source: OTA, 1984.

environment and that an additional 700 to 800 are being added each year (Henderson et al., 1984). In 1984, the OTA listed more than 200 substances that have been detected in groundwater; these include 175 organic chemicals and 50 inorganic chemicals, organisms, and radionuclides. For disposal, however, most are reacted with other chemicals, immobilized, or otherwise treated and burned. The ways waste-disposal practices may impinge upon the purity of groundwater are shown in Figure 3-4.

61

Individual Sewage Disposal Systems

The 1980 census estimated that there were approximately 22 million septic systems serving nearly one-third of the nation's population (U.S. EPA, 1986d). Annual flow to an individual septic tank ranges from approximately 49,000 to 75,000 gallons per household. The total annual flow to all domestic systems can be estimated to be from 1.08 trillion to 1.65 trillion gallons (OTA, 1984), making these systems the highest-ranking source of wastewater discharged directly into groundwater. The three methods of on-site domestic waste disposal are septic tanks, with their subsurface disposal system; the less satisfactory cesspool, commonly found in older installations where there is a deep layer of permeable soil; and finally the pit privy. Properly constructed septic-tank systems and privies permit effective treatment of human waste. Cesspools work only in coarse or highly fissured materials and have a high potential for contamination. The septic tank has several important functions. It separates solid and liquid wastes, stores solids and floatable materials, and treats aerobically both stored solids and nonsettleable material (Canter et al., 1982). Soils with slow percolation rates, shallow soils, soils over permeable bedrock, and soils with permanent or periodic high water tables are not suitable for the use of conventional septic tank systems.

The leach beds for septic tanks typically consist of 500 to 600 square feet of drain bed, located 18 to 24 inches below the surface of the ground. The effluent is thus distributed over a wide area, and the shallow depth of the bed permits some evaporation and some uptake of contaminants by plants. Because it is near the surface, the drain field operates under aerobic conditions, with the result that good quality water reaches the groundwater. It is important that a zone of unsaturated soil occur between the leach bed and the water table so that the effluent from the septic tank is not discharged directly into the groundwater and hence into the aquifer.

The degree of potential risk posed by these systems depends in large measure upon their design and installation and upon the hydrogeology of the area. When sited correctly they operate efficiently, renovating wastewater and returning good quality water to the watershed. When operating incorrectly, they can allow pollutants to enter the groundwater. This would be particularly troublesome in areas where drinking water wells tap the same aquifer in the vicinity of the leach bed.

62

The advantages of septic tanks include the following (Canter et al., 1982):

- Minimal maintenance is required.
- The cost of individual or community septic tanks is less than the cost of central wastewater collection and treatment plants.
- It is a low-technology system.
- The energy requirements are low when compared with centralized wastewater-treatment facilities.

The disadvantages include:

- The potential for groundwater contamination exists when the tanks are incorrectly sited, with regard to density of the individual units or soil characteristics or both.
- The systems require proper maintenance. Failure to provide this results in system overflows or pollution of nearby wells.
- Septic-tank cleaners may severely contaminate groundwater.

The typical composition of domestic sewage is shown in Table 3-3. The constituents that pose the greatest threat to the quality of groundwater are nitrates, phosphates, heavy metals, inorganic ions (Na^+, Cl^-, $SO_4^=$, K^+, Ca^{++}, and Mg^{++}), and pathogenic organisms. In addition, toxic synthetic organic chemicals are becoming a more significant hazard due to the increased use over the last ten years of cleaning products containing such chemicals. The geographic density of the use of on-site disposal systems is shown in Figure 3-5. It is difficult to show by mapping how regions are affected by sources of contamination. All contamination incidents are indicated by points on a map, but not all the groundwater in the area indicated is contaminated. Since the most important factors affecting the contamination of groundwater by sewage are the density of the individual units and the soil characteristics, the greatest potential for contamination is in the eastern region of the country. A septic-tank density of greater than 40 per square mile constitutes a region of potential contamination to groundwater (U.S. EPA, 1977).

Table 3-3. Characteristics of Domestic Sewage

Constituent	Concentration* Typical Domestic Sewage
Total Suspended Solids	200
Conductivity	700
Chemical Oxygen Demand (COD)	500
Biochemical Oxygen Demand (5-day BOD)	200
Total Organic Carbon (TOC)	200
pH	8.0
Alkalinity (as CaCO)	100
Acidity (as CaCO)	20
Total Phosphorus	10
Total Nitrogen	40
Chloride	50
Calcium	50
Magnesium	30
Iron	0.1
Manganese	0.1

* mg/l except for conductivity (micro-mhos/cm), and pH (pH units).

Source: U.S. EPA, 1973.

Land Disposal of Solid Wastes

According to the Resource Conservation and Recovery Act of 1976, solid wastes are defined as any garbage, refuse, or sludge from waste-treatment plants, water-supply treatment plants, or air-pollution control facilities, and any other discarded materials, including solid, liquid, semi-solid, or contained gaseous materials, resulting from industrial, commercial, mining, and agricultural operations and from community activities. They do not include solid or dissolved material from domestic sewage, irrigation return-flows or industrial discharges that are point sources, or products and by-products of the nuclear industry. Solid wastes are considered hazardous when their quantity, concentration, or physical, chemical, or infectious characteristics cause, or significantly contribute to, an increase in mortality or an increase in serious illness, or pose present or potential hazards to human health or the environment when improperly treated, stored, transported, disposed of, or otherwise managed.

The solid portion of household wastes contains a high proportion of putrescible matter that is broken down by biodegradation, releasing carbon dioxide and methane gas (UNESCO, 1980). The leachate contains high concentrations of sulfate, chloride, and ammonia. Abundant quantities of cellulose from paper products may retard the movement of halogenated hydrocarbons by absorption. Solid commercial wastes have a similar composition to household wastes. They may, however, contain greater quantities of oils, phenols, and hydrocarbon solvents. Phenols are the most resistant to biological breakdown and may be leached. Domestic and commercial wastes are serious groundwater contaminants because of the numerous dissolved constituents and the high biological oxygen demand (BOD) of may of the constituents. Table 3-4 is a summary of leachate characteristics based on 20 samples from municipal solid wastes. The table does not include synthetic organic chemicals.

The solid components of industrial wastes vary with the source of production (UNESCO, 1980). Cyanide wastes are produced in metallurgical operations; sulfite wastes come from paper and pulp manufacturing; mercury is a waste product in the electrical industry; and the petrochemical industry produces several solid residues ranging from polychlorinated biphenyls (PCBs) to pesticides and herbicide residues to phenol-rich tar wastes. The amount of solid waste disposed of by various manufacturing industries is listed in Table 3-5 (U.S. Department of Commerce, 1986).

Other serious solid-waste contaminants can result from thermal-power generation from the burning of coal, forming fly ash which, due to its high surface area volume ratio, is fairly reactive. Leaching of the deposited fly ash may initially give rise to high concentrations of arsenic, chromium, selenium, and chloride. Another contaminating product resulting from the burning of coal is the sludge formed by the aqueous scrubbing of flue gases. Sludges typically contain cyanide and heavy metals and are of low pH unless neutralized by the addition of lime. It has been found that mixtures of sludge, fly ash, and lime rapidly set into a low-permeability solid which leaches less readily.

Contamination occurs when precipitation infiltrates and percolates through solid wastes at poorly designed land-disposal sites, forming a leachate of dissolved minerals, heavy metals, and organic chemicals.

Leachate formed at disposal sites which are located in wetlands, flood plains, or where there is a shallow water table and that are

Table 3-4. Summary of Leachate Characteristics Based on 20 Samples from Municipal Solid Wastes

Components	Median Value (ppm)[a]	Ranges of All Values (ppm)[a]	
Alkalinity (CaCO₃)	3,050	0	− 20,850
Biochemical Oxygen Demand (5-day BOD)	5,700	81	− 33,360
Calcium (Ca)	438	60	− 7,200
Chemical Oxygen Demand (COD)	8,100	40	− 89,520
Copper (Cu)	0.5	0	− 9.9
Chloride (Cl)	700	4.7	− 2,500
Hardness (CaCO₃)	2,750	0	− 22,800
Iron, Total (Fe)	84	0	− 2,820
Lead (Pb)	0.75	<0.1	− 2.0
Magnesium (Mg)	230	17	− 15,600
Manganese (Mn)	0.22	0	− 125
Nitrogen (NH₄)	218	0.06	− 1,106
Potassium (K)	371	28	− 3,770
Sodium (Na)	767	0	− 7,700
Sulfate (SO₄)	47	1	− 1,558
Total Dissolved Solids (TDS)	8,955	584	− 44,900
Total Suspended Solids (TSS)	220	10	− 26,500
Total Phosphate (PO₄)	10.1	0	− 130
Zinc (Zn)	3.5	0	− 370
pH	5.8	3.7	− 8.5

[a] Where applicable

Source: U.S. EPA, 1977.

65

constructed without natural or artificial barriers is likely to contaminate groundwater.

The types of land disposal sites for solid wastes include dumps, landfills, sanitary landfills, and secured landfills—listed in increasing order of ability to protect the environment from the adverse effects of their use. Landfill regulations promulgated under the Resource Conservation and Recovery Act (RCRA) were issued in July 1982. The Comprehensive Environmental Response, Compensation and Liability Act (CERCLA) requires remedial actions at inactive sites that pose a threat to the environment and to human health. The U.S. EPA estimates that there are 93,000 landfills in the United States (U.S. EPA, 1985c). Approximately

Table 3-5. Solid Waste Disposal by Selected Manufacturing Industry Group: 1975 to 1980*

Industry Group	1975	1980	1983 Total	Hazardous	Non-hazardous
All Industries	139.1	149.9	89.0	8.0	81.0
Food	12.6	14.4	9.6	0.2	9.4
Lumber and Wood	8.1	5.9	4.0	(Z)	4.0
Paper	9.1	12.3	13.7	0.1	13.5
Chemicals	38.7	43.4	18.8	3.6	15.2
Petroleum	2.0	4.9	3.6	1.5	2.1
Stone, Clay, Glass	11.3	13.3	6.2	0.2	5.9
Primary Metal	42.7	37.5	17.7	1.0	16.7
Fabricated Metals	1.9	1.9	2.0	0.2	1.7
Machinery (excluding electrical)	2.7	3.0	1.8	0.2	1.7
Electrical Equipment	1.5	2.1	1.7	0.3	1.4
Transportation Equipment	3.8	4.2	3.0	0.3	2.7

* In millions of short tons. Excludes recovered materials. Data include both wet- and dry-weight figures. Excludes apparel and other textiles and, beginning 1980, establishments with fewer than 20 employees.
(Z) less than 50,000 short tons

Source: U.S. Department of Commerce, 1986.

75,000 of these are on-site industrial landfills and 18,500 receive municipal wastes (U.S. EPA, 1985c). In 1983, another survey estimated that there are between 15,000 and 20,000 active landfills in the nation, including municipal landfills, open dumps, and non-hazardous waste facilities, operating, abandoned, or closed. (OTA, 1984).

Hazardous wastes may be disposed of in a landfill, dump, or impoundment. EPA requires that site locations and amounts and types of wastes stored be reported, under CERCLA by owners and operators of inactive sites, and under RCRA by present owners and operators of active privately owned industrial sites. About 199 active hazardous-waste landfill facilities (OTA, 1984) and approximately 20,000 abandoned hazardous-waste sites (EPA, 1985c) are known to exist across the country.

The amount of solid waste generated in this country has been estimated to be enough to cover 400 acres of land to a depth of 10 feet each day (Schneider, 1972). It is estimated that in 1980, 3.74 pounds of solid waste were generated per person per day, amounting to 155.6 million tons of municipal wastes nationwide for the year (U.S. Department of Commerce, 1986). OTA (1984) also estimates that 30 million tons of solid wastes generated by utilities and 1 million tons of municipal sludge are disposed of in landfills per year. Between 40 million and 140 million wet tons of non-hazardous industrial solid wastes are disposed of annually (OTA, 1984). In 1981 hazardous-waste facilities received 14.7 billion gallons of both liquid and solid hazardous wastes (CEQ, 1984).

67

Collection, Treatment, and Disposal of Municipal Wastewater

The collection, treatment, and disposal of municipal wastewater may be a problem in urban areas. Municipal wastewaters include domestic wastewater, storm water and its associated debris, and industrial wastes. They are collected by sanitary sewer systems and transported to treatment sites. Approximately 170 million Americans were served by sewer systems in 1984 (U.S. Department of Commerce, 1986).

Leaks may occur in the system due to design, age, disruption by tree roots, seismic activity, or poor construction. The lagoons and ponds used in wastewater treatment operate under anaerobic, aerobic, or facultative conditions. The total volume of sewage handled by one of these plants is estimated to be approximately 0.667 mgd (OTA, 1984). Lagoons are often constructed without adequate seals, thus promoting leakage under certain geologic conditions and posing a threat to groundwater. Another possible route of groundwater contamination is by the land disposal of treated wastewater, which could alter the chemical constituents of the natural groundwater. Land disposal methods include agricultural irrigation, rapid infiltration ponds, overland runoff, and discharge into dry stream beds and ditches. Methods involving spraying onto land are not suitable for use in freezing winter months. Figure 3-6 shows the geographic distribution of various densities of population that are served by municipal sewage-treatment facilities that utilize the land disposal of effluent.

Industrial and Other Wastewater Impoundments

Under the Safe Drinking Water Act of 1974, EPA undertook a national survey of surface impoundments, which was carried out from 1976 to 1978, and the interim report was published in 1978 (U.S. EPA, 1978b). The final report, released in 1983, contains data from 1978 through 1980. These reports define surface impoundments as depressions in the land (pits, ponds, lagoons, and pools) containing liquid, semi-solid, and solid wastes. In the final report EPA stated that the number of sites located in the United States totaled approximately 80,263 and contained more than 180,000 impoundments. Thirty-one percent of the sites were oil and gas brine pits, 25% municipal sites, 18% agricultural sites, 15% industrial sites, and 9% mining sites. Table 3-6 shows EPA's assessment of these sites. At the time of the survey it was estimated that approximately 70% of the industrial sites, 84% of the agricultural sites, and 78% of the municipal sites remained unlined. More than 98% of the sites were located in areas with proximity to potential water supplies. Approximately 15% of the sites (excluding oil- and gas-related facilities) contained hazardous wastes (U.S. EPA, 1983). New Mexico was estimated to have the largest number of sites, with 12,901, and Rhode Island the fewest, with 44. Table 3-7 shows the total number of impoundments and impoundment sites by state. The regulations under RCRA may have affected the validity of these estimates, but no recent figures are available.

Table 3-6. Summary Statistics for Located Active Surface Impoundment Sites

Category	Located Sites*	Assessed Sites*	Located Impoundments*
Industrial	11,760 (15%)	8,662 (28%)	27,912 (15%)
Municipal	19,746 (25%)	10,822 (34%)	37,185 (21%)
Agricultural	14,850 (18%)	6,646 (21%)	19,437 (11%)
Mining	7,364 (9%)	1,552 (5%)	25,038 (14%)
Oil & Gas Brine Pits	24,990* (31%)	3,354 (11%)	65,488 (36%)
Other	1,553 (2%)	350 (1%)	5,913 (3%)
Total	80,263	31,386	180,973

* SIA site numbers for the mining and oil & gas brine pit sites are usually related to lease or field data, not actual ownership, and should not be referred to as the actual number of legal sites. The number of located impoundments would be a closer approximation for these two categories. *Located sites*: Total number of facilities identified in the inventory. *Assessed sites*: Total number of facilities evaluated in the inventory. *Located impoundments*: Total number of impoundments identified in the inventory. The number is larger than located sites since many facilities had more than one impoundment.

Source: U.S. EPA, 1983.

Table 3-7. Estimated Number of Located Impoundments and Impoundment Sites for All Categories, by State

State	Sites	Active Impoundments	Sites	Abandoned Impoundments
Alabama	1,063	1,696		
Alaska	129	179	27	31
Arizona	641	1,615	20	48
Arkansas	861	6,806	12	24
California	1,750	7,577	31	101
Colorado	1,267	3,631		
Connecticut	305	758	83	188
Delaware	52	131		
Florida	3,644	5,610		
Georgia	1,363	1,640	65	70
Hawaii	94	296		
Idaho	253	538		
Illinois	4,471	6,677	591	745
Indiana	1,897	2,678		
Iowa	1,784	2,539		
Kansas	1,970	3,328		
Kentucky	640	922		
Louisiana	984	2,804	49	77
Maine	132	319	41	133
Maryland	443	836	18	25
Massachusetts	341	2,211	38	208
Michigan	2,187	3,505	76	135
Minnesota	1,952	2,733	135	160
Mississippi	1,306	1,667	52	58
Missouri	3,657	4,683	507	507
Montana	668	1,266		
Nebraska	1,632	3,254		
Nevada	216	541	10	382
New Hampshire	152	274	2	4
New Jersey	311	896	46	129
New Mexico	12,901	17,746	39	116
New York	1,070	1,837	40	52
North Carolina	709	1,051	26	30
North Dakota	667	1,168		
Ohio	3,287	16,537		
Oklahoma	2,935	4,538		
Oregon	370	714		
Pennsylvania	7,346	34,224	234	627
Rhode Island	44	133	3	13
South Carolina	1,868	2,259		
South Dakota	606	926		

Table 3-7. Estimated Number of Located Impoundments and Impoundment Sites for All Categories, by State (*continued*)

State	Sites	Active Impoundments	Sites	Abandoned Impoundments
Tennessee	746	1,325	59	151
Texas	3,672	10,740	170	365
Utah	211	2,345		
Vermont	214	274	10	17
Virginia	1,212	1,975	30	62
Washington	403	1,049		
West Virginia	688	1,861	66	186
Wisconsin	1,022	1,720	49	82
Wyoming	1,146	1,738		
Total	79,818	180,328	2,519	4,726

Source: U.S. EPA, 1983.

Impoundments—which are used for the treatment, storage, or disposal of wastes—are natural or man-made, and may or may not be lined. Discharging impoundments are designed to discharge regularly into bays, oceans, lakes, or streams. The liquid in non-discharging impoundments is lost by evaporation or by seepage into the soil. The types of wastes received by these sites range from sewage wastes, industrial wastes—including those from air scrubbers, cooling-tower blow-down, and ash residues—and oil- and gas-extraction wastes to animal-feedlot and other agricultural wastes. Of these, the largest users of surface impoundments are the oil and gas industries (U.S. EPA, 1978a and 1978b; Inside EPA, 1980). Evaporation ponds were used extensively in the South for oil-brine disposal. Evaporation ponds usually lose more fluid by infiltration than by evaporation, especially in the more humid regions of the country (Miller, 1981a; Pettyjohn, 1972). EPA promulgated regulations for hazardous-waste surface impoundments in July 1982. In addition, state regulations exist in most oil-producing states.

The chemical character of the waste varies enormously and may include suspended and dissolved solids, pathogenic organisms, oil and grease, detergents, heavy metals, and toxic organic chemicals. It may have a high biological oxygen demand (BOD), chemical oxygen demand (COD), and total organic carbon (TOC). The pH also varies. Not all surface impoundments contain hazardous wastes. Some are used as holding ponds prior to the treatment of the fluids. Because the chemical character of the wastes varies, only a minimal number of impoundments would hold all of the contaminants listed. The type of groundwater contamination known to occur as a result of leaking impoundments reflects the nature of the fluids in the impoundments. Dissolved materials, both inorganic and organic, move into groundwater by direct seepage, while solids may be leached by precipitation or waste fluids. As with all sources of contamination, the sorptive capacity and low permeability of some soils may slow or impede the movement of some of the contaminants.

71

Land Spreading of Sludge

When the wastewater from municipal and industrial wastes is treated (see above), a residue of sludge remains. Beginning in November 1981, all facilities disposing of sludge considered hazardous on land were required to have a groundwater-monitoring system in place under Resource Conservation and Recovery Act regulations. In 1982 there were at least 2,463 publicly owned treatment facilities applying liquid and thickened sludge on land surfaces, and an additional 485 facilities in operation or under construction using sludge spray irrigation practices (OTA, 1984). Of the 6.8 million dry tons of sludge produced by municipalities in 1982, 24%–29% was spread directly onto crops (OTA, 1984). The purpose of land treatments is to biodegrade the organics and to immobilize the inorganics. The potential for contamination exists, however, since the chemical constituents and viruses in sludge may be leached by precipitation after the sludge is spread on the land. Sludge from manufacturing industries may contain either useful agricultural chemicals or toxic compounds, and it varies from degradable to very refractory. Sludge is stabilized and may be dewatered before land application. The higher the proportion of clay and organic colloids in the soil, the greater the capacity for the heavy-metal ions present in sludge to be immobilized. The main hazardous constituents of the sludges include heavy metals and organic chemicals, such as dyes, inks, oils, pesticides, detergents, organic solvents, and polynuclear aromatic hydrocarbons (PAHs); organic chemicals are typically trace contaminants. Sludges from the municipal treatment of wastewater or the biotreatment of industrial wastewater include biological components such as bacteria, viruses, fungi, algae, protozoans, rotifers, and other parasites such as worms and flukes. The constituents that may be found in various types of industrial wastes are enumerated in the 1977 Report to Congress (U.S. EPA, 1977). As of 1985, EPA reported seven states that had experienced groundwater contamination due to land spreading of sludges (U.S. EPA, 1985c).

72

Land farming of sludge has been used for nearly 30 years by the petroleum industry. From its research on this method of disposing of oil sludge, the Solid Waste Management Committee of the American Petroleum Institute concludes that if the area farmed is carefully matched to the amount of organic material that must be degraded, the type of land farming practiced, and the degradation rate of the sludge, then the underlying soil and groundwater will not be significantly contaminated (Knowlton and Rucker, 1979).

Brine Disposal Associated With the Petroleum Industry

Groundwater contamination associated with oil production has been documented in at least 21 states (OTA, 1984). During 1983 the total production of crude petroleum was more than 3 billion barrels, valued at more than $83 billion (U.S. Dept. of Commerce, 1986). The principal oil-producing states in 1983 were Texas, Arkansas, Louisiana, and California. Brief mention of the disposal of brines by the use of surface impoundments was made above. Oil production is usually accompanied by the production of saline wastewater in amounts that vary with production procedures. One estimate is that an old well may produce 100 barrels of brine for each barrel of oil (U.S. EPA, 1977). Others estimate that the production of 8 million barrels of crude oil produces 30 million barrels of saltwater and that the ratio of brine to crude oil recovered is 10:1 (Keeley, 1977). Still others estimate a ratio of 4:1 brine to oil (OTA, 1984).

In the early days of oil and gas production, the brine was often disposed of in unlined pits. The extensive use of such "evaporation" pits that leaked their contents has caused groundwater contamination in the oil-producing regions (Scalf et al., 1973; Fuhriman and Barton, 1971; U.S. EPA, 1977). Use of unlined pits has been completely eliminated in some states and partially eliminated in others. For example, unlined pits were banned in Texas in 1969. Other states permit the use of small pits in remote areas if they pose no threat to groundwater.

Present practices for brine disposal include injection of the brine into deep underground formations that are deemed unsuitable for other purposes and reinjection of brines into oil-producing formations to enhance oil recovery. These injection practices, if not properly controlled, pose a potential threat to groundwater.

Past oil production practices, including well drilling, completion, and abandonment practices, also were not as strictly controlled as they are now. Some wells were improperly equipped, and abandoned wells were improperly plugged, thereby providing a potential for groundwater contamination. In 1983 abandoned production wells numbered around 1.2 million (OTA, 1984). The need to protect groundwater was recognized, however, and state oil and gas regulatory agencies now have regulations concerning brine injection, well completion, and plugging practices. The Interstate Oil Commission sponsored a study in 1978 that concluded that in Texas, Arkansas, Louisiana, New Mexico, and Oklahoma, reinjection of brines had resulted in fewer instances of groundwater contamination than had the use of evaporation pits. It therefore believes that the practice does not pose a major threat to the groundwater supplies.

73

Most oil-field brines are corrosive, but the corrosion rate of old and new well casings is unpredictable, and frequent monitoring and surveillance is therefore carried out. Although brine is the main threat to the quality of groundwater related to oil production, minor threats are also posed by the oil and gas themselves, by drilling fluids, by chemicals used in treating wells, by corrosion inhibitors, and by other chemicals (U.S. EPA, 1977). Oil drilling may entail drilling through freshwater aquifers, thus raising the potential for contamination.

Disposal of Mine Wastes

Mining creates conditions and products conducive to groundwater contamination. The recharge and movement of groundwater may be affected by the mining process. Domestic mining for minerals other than organic fuels was more than a $20-billion industry in 1983 (U.S. Department of Commerce, 1986). Domestic mining for minerals other than organic fuels was a $10-billion industry in 1974 (U.S. EPA, 1977). Coal has enjoyed a comeback as an energy source due to the goal of increased self-sufficiency in energy, and it is estimated that between one-third and one-half of the world's total reserves are found in the United States. Figure 3-7 shows the distribution of coal fields in the conterminous United States. Techniques of surface and underground mining pose different threats to groundwater quality, although the threat from waste disposal from the two is similar. Although such mining changes the environment, good reclamation may stabilize some of these changes without causing serious damage to the groundwater (Thompson, 1977). Sand, clay, gravel, and stone quarrying are not considered a serious threat to groundwater quality (U.S. Congressional Research Service, 1980). Threats from the remaining categories of mines include those posed by leachate from tailings piles and drainage from strip, surface, and underground mines (van der Leeden et al, 1975; OTA, 1984). Dewatering of mines, involving a regional lowering of the water table, may have three results: a decrease in the amount of groundwater supplied, a lowering of the water level below the intakes of productive wells, and the oxidation of exposed minerals (Miller et al., 1974). Coal is often found in conjunction with iron pyrites (FeS_2) and other sulfides, and the oxidation of these sulfides and the presence of percolating waters produce sulfuric acid and an increase in Fe^{++} and $SO_4^{=}$, leading to acidic groundwater, a serious problem in the coal-mining areas of Pennsylvania and Appalachia. It is very costly either to neutralize acid mine drainage or to backfill abandoned mines in order to prevent the problem. Sealing the mines with air- or watertight seals has not proved very successful (Miller et al., 1974). A report by the National Academy of Sciences summarizes these and other effects that coal mining may have on groundwater quality (NAS, 1981b).

Metal mining produces drainage waters that have lower concentrations of iron and sulfate than do colliery waters, but that have higher concentrations of dissolved heavy metals due to the lower pH caused by the oxidation of metal sulfides. Wastewater from metal mining may also contain organic flocculants from the on-site processes of screening and dressing the ores. The wastes from such operations, which can be both toxic and radioactive, may contaminate groundwater. The tailings ponds used in the waste-disposal operations of mines may contribute contaminated leachate.

The waste products of quarrying for stone, lime, gypsum, and the like are generally inert, and their only contributions to the water that percolates through them is the possibility of an occasional increase in suspended-solids if crushed material is present. Quarrying does, however, increase the potential for local groundwater contamination

74

in other ways. The removal of soil reduces the possibility of an attenuation of any spills or leakages that may occur in the area and increases the potential for contamination, especially if the bedrock is fractured.

Because of the serious nature of surface-water contamination by mining activities, more attention has been given to it than to groundwater contamination, although aquifers in regions with a long history of mining have been written off as sources of water supply (U.S. EPA, 1977). Figure 3-8 shows the states in which significant volumes of wastewater are discharged from mining and ore-processing operations (excluding coal and petroleum), and Figure 3-9 shows similar data for coal mining and processing, both for 1972. Since then coal mining on a large scale has been initiated in the West, particularly in North Dakota, Wyoming, and Montana. The EPA (1977) estimated that 3.6 million tons of acid may be generated annually from the approximately 200,000 acres used across the nation for disposal of coal-mining wastes. The EPA (1977) also estimated that 10% of the acid enters groundwater each year. Mining operations generate approximately 2.3 billion tons of total waste material annually, including radioactive and non-radioactive waste piles and tailings (OTA, 1984).

The most serious threat to groundwater is posed by the metal mines, particularly uranium and copper mines, which produce waste containing dissolved toxic materials (e.g., arsenic, sulfuric acid, copper, selenium, and molybdenum), as well as radioactive materials.

Deep-Well Disposal of Liquid Wastes

The injection of a variety of liquid wastes into deep wells to avoid having them contaminate the biosphere was becoming a widely adopted waste-disposal practice in the late 1970s (Freeze and Cherry, 1979), mainly because there were stringent regulations governing disposal into surface waters. Such wells are less widely accepted now in certain states (J. W. Keeley, Kerr Environmental Research Laboratory, personal communication). It has been estimated that there are more than 221,000 wells in the United States injecting liquid wastes underground (U.S. Library of Congress, 1983): approximately 280 are used for hazardous wastes; 140,000 are used either for brine disposal or for the injection of fluids used in enhanced oil-recovery process; 40,000 are used for agricultural drainage, urban runoff and sewage disposal; and at least 12,000 are in operation for solution mining (OTA, 1984).

Because of the hazardous nature of the injected fluid—often radioactive, toxic chemical, petrochemical, and pharmaceutical wastes that are difficult to treat—deep-well disposal may pose a great threat if improperly practiced. Eighteen states have already experienced contamination through injection wells (U.S. EPA, 1985c). The Underground Injection Control Program of the Safe Drinking Water Act, which prohibits direct injection into drinking-water aquifers, and regulations issued under the Resource Conservation and Recovery Act provide such controls. All injection wells are in the depth range of 660 to 13,200 feet, most being between 990 to 6,600 feet. Injection zones are usually located in sandstone, basalt, and carbonate rocks. Injection occurs under pressure, and injection rates vary from 500 to 370 gallons per minute. Injection causes hydrodynamic changes in the aquifer, including the formation of a mound in the potentiometric surface extending in the direction of the regional flow in the aquifer (Freeze and Cherry, 1979). This is essentially the reverse of what happens when a well is pumped in a confined aquifer. Until 1976, documented cases of disposal-system failure were rare, but they may increase if underground injection becomes more common and occurs over longer periods of time. Underground injection in the vicinity of old unplugged wells, the locations of which are often not known, may lead to the upward leakage of the wastes (Van Everdingen and Freeze, 1971; OTA, 1984). These sources estimate that there may be more than one million unplugged, unlocated wells in North America that were used for purposes other than disposal. Another hazard associated with the practice of waste injection but only indirectly linked to groundwater contamination is the triggering of earthquakes due to increased pore-water pressures along fault lines (Freeze and Cherry, 1979). The most publicized occurrence, at the Rocky Mountain Arsenal in Colorado, occurred after injection of chemical-warfare waste, using a 12,000-foot well, into fractured pre-Cambrian rocks. The earthquakes thus caused stopped when injection ceased (Ballentine et al., 1972; van der Leeden et al., 1975). Other problems may arise if an injection well becomes clogged with injected suspended solids, corrosion products, or precipitation forming between injection fluid and connate water in the aquifer. Insufficient treatment of injected water may lead to bacterial clogging, and swelling of the mineral constituents of the geologic formation (Lehr, 1975). The Report to Congress (U.S. EPA, 1977) stated that injection wells can cause

76

groundwater contamination through the following mechanisms:
- direct injection into drinking-water aquifers
- leakage into potable aquifers due to well-construction faults or failure
- leakage through confining beds by hydraulic fracturing or insufficient thickness
- displacement of saline water into potable aquifers
- migration to potable water zone of the same aquifer.

The report concluded, however, that properly constructed, sited, and monitored waste injection wells can be operated with little danger of groundwater contamination.

77

Disposal of Wastes from Animal Feedlots

Groundwater contamination by the manure from high-density animal-feeding operations is a relatively new phenomenon. In addition to the threat to groundwater from the vast quantities of manure thus generated, there is also a threat from the food additives, such as hormones and antibiotics, which the manure may contain (Keeley, 1977). As recently as 1977 it was claimed that feedlots pose no danger of groundwater contamination in the southeastern states due to the soil's ability to denitrify the wastes (Miller et al., 1977). Feedlots for poultry and hog farms in Arizona, California, Nevada, and Utah were thought to lead to water-quality problems from bacteria, viruses, nitrate concentrations and with general color, taste, and odor (Fuhriman and Barton, 1971). The main beef-raising areas, however, are the Corn Belt and the High Plains. Until the 1950s and 1960s, beef was raised on pasture, and the wastes were easily assimilated naturally and posed no problem. The need for more meat at reasonable cost led to the establishment of feedlots, with capacities of between 1,000 and 50,000 head of cattle. During the four to five months each animal spends in the lot, it produces 0.5 tons of manure (dry weight). This can lead to high nitrate concentrations in the groundwater under some conditions. Where the feedlots are located in areas with a deep water table, the risk of contamination is minimized (Scalf et al., 1973). The distribution of cattle-feeding operations in the United States is shown in Figure 3-10. The main poultry-rearing region is the Mid-Atlantic and South, in Delaware, Maryland, Virginia, Arkansas, Georgia, and Mississippi, for hogs it is the Midwest, and for sheep the Far West. Of the potential contaminants in manure, nitrate is the most important, as it is soluble in water and its concentration remains unchanged while passing through the soil. The other contaminants from manure—bacteria and phosphate—are highly attenuated by the soil and pose less threat. Feedlot management, including stocking rate, density, and manure removal, plays an important role in the protection of groundwater quality. Heavy manure accumulations may produce an impermeable mat, which in turn produces anaerobic conditions that favor denitrification. Thus nitrate is volatilizied, and little infiltration takes place, minimizing the nitrate-contamination potential (U.S. EPA, 1977). Concentrated animal feeding operations are regulated by the Federal Water Pollution Control Act Amendments of 1972 and may require a permit as a point source under the National Pollutant Discharge Elimination System (NPDES).

78

Contamination From Radioactive Sources

Radionuclides from natural sources, radioactive fallout, nuclear testing and accidents, nuclear fuel cycling, commercial and industrial products and wastes pose a threat to groundwater quality. Almost all groundwater contains some amount of naturally occuring radioactive substances. Types and levels vary geographically and depend on local geology. The highest concentrations of natural radiation of groundwater are often found near granite and phosphorus formations (OTA, 1984).

Radioactive wastes from the production and generation of nuclear fuel and radioactive materials can be categorized into five types: spent fuel, high level wastes, transuranic wastes, low level wastes and uranium mill tailings (OTA, 1984). Quantities of these waste types generated as of 1983 and projected quantities are presented in Table 3-8.

Spent nuclear fuel is irradiated fuel resulting from nuclear power plant operations and is highly radioactive (OTA, 1984). Most of these wastes are presently being stored in pools at reactor sites until underground repositories are found for permanent disposal (League of Women Voters, 1985).

High level wastes result from the initial processing of spent fuel and must be stored carefully in specially constructed facilities. Presently these wastes are being temporarily stored at four utility and defense sites awaiting permanent disposal (OTA, 1984). Possible sites for a high level repository are being considered in

Table 3-8   Current and Projected Quantities of Radioactive Waste and Spent Fuel (September 1983)

Type of Waste	Volume of Waste (1000 m³)		
	1983	2000	2010
High Level Waste			
Commercial	2.3	.436	3.17
Defense	304	294	269
Transuranic Waste			
Commercial	—	4.6	49.4
Defense	246	340	397
Spent Fuel[1]			
Commercial	4.626	19.378	33.258
Defense	0	0	0
Lower Level Waste			
Commercial	1,020	3,393	5,415
Defense	2,060	3,720	4,710
Inactive Uranium Mill Tailings[2]	14,314	23,700	23,700
Active Mill Tailings[3]	96,500	188,800	280,300

(DOE/Defense tailings are about 35% of these totals.)
1. No reprocessing.
2. Found primarily at inactive uranium mills located in western United States. DOE responsible for stabilization and control of mill tailings in safe and environmentally sound manner under the Uranium Mill Tailings Radiation Control Act of 1978. Anticipated cleanup completion in late 1980s.
3. Located at the 16 active licensed uranium mills operating in 1983, all in the western United States.

Source: Adapted from *Spent Fuel and Radioactive Waste Inventories, Projections and Characteristics.* U.S. Department of Energy, 1984.

79

Deaf Smith County, Texas, Yucca Mountain, Nevada, or Hanford, Washington. The site is scheduled to be complete by 1998 (Environmental Reporter, 1986).

Groundwater contamination by high level radionuclides has been reported in a few instances, largely due to the improper storage of wastes. For example, in the early 1950s, one hundred gallons of high level radioactive wastes leaked from a tank at the Savannah River defense storage site, contaminating some groundwater (Hileman, 1982). In 1973 a 115,000-gallon leak of high level radioactive wastes occurred at the Hanford Nuclear Reservation (Lindorff and Cartwright, 1977). No groundwater contamination resulted, although the soil beneath the site remains radioactive. Another extensive leak of high level radioactive wastes occurred in 1956, spilling 450,000 gallons of high level radioactive wastes in the ground at Hanford, Washington, but with no serious contamination resulting.

Transuranic wastes contain alpha-emitting radionuclides of atomic numbers greater than 92 (League of Women Voters, 1985). They result primarily from fuel reprocessing and the manufacture of plutonium containing products (OTA, 1984). These wastes take a long time to decay and require long-term isolation for proper disposal. Presently, there are a total of 7 sites where transuranic wastes may be deposited (OTA, 1984).

Low level wastes exist in all phases and consist of a wide range of waste material. Most of the radionuclides in low level wastes are short lived and have low radioactivity, but are produced in potentially hazardous amounts. These wastes are generated by hospitals, laboratories, industrial plants, nuclear power plants, and government and defense laboratories and reactors (League of Women Voters, 1985). In the past, low level wastes were either buried in trenches at government sites or dumped out at sea. Now all commercially produced low level wastes are permanently disposed of in one of the three commercial shallow nuclear burial facilities located in Beatty, Nevada, Barnwell, South Carolina, and Hanford, Washington, the major site being in Hanford (League of Women Voters, 1985). The Department of Energy and the Department of Defense also maintain a number of sites operated by private contractors for low level disposal (OTA, 1984). Shallow burial of low level wastes present a great potential for contamination. Groundwater may erode the structure of the burial site thus providing avenues for contamination to seep. A recent study found that water has come into contact with wastes in burial trenches at 6 of the 11 major low level governmental and commercial burial facilities (League of Women Voters, 1985).

Uranium mill tailings are earthen residues, often in the form of sand, left after uranium is extracted from ores. Disposal commonly takes place in shallow burial grounds located near refineries. Remedial action at inactive mill tailing sites are to be conducted by the Department of Energy under the Uranium Mill Tailings Radiation Control Act (OTA, 1984). According to the League of Women Voters (1985) twenty-four abandoned disposal sites are scheduled for clean up by DOE in Arizona, Colorado, New Mexico, North Dakota, Oregon, Pennsylvania, South Dakota, Texas, Utah, Washington, and Wyoming. Of the 28 active uranium mills licensed in the US, 21 reported some degree of groundwater contamination.

Remedial action has been taken at 16 of the sites so far (League of Women Voters, 1985).

Other Sources of Contamination

Accidental Leaks and Spills

Accidental leaks and spills can occur because of breaks in pipelines, leaks in underground storage tanks, spills from tanker trucks, and other accidents involving the transport and storage of materials. They commonly occur at airports, industrial sites, along highways, railroad sites, gas stations, and refineries.

81

Small hydrocarbon spills may be absorbed by the unsaturated soil zone, but large ones can reach the water table aquifer and float on the surface. Very small concentrations of petroleum products will render groundwater unfit to drink because of its objectionable odor and taste and also pose a threat to health.

Other chemicals have toxic properties. Ammonia, for example, increases the nitrification of groundwater, and acids speed up the solution of heavy metals and soil solids.

The U.S. General Accounting Office (1984) reported that leaks from underground storage tanks (USTs) have been reported in all 50 states; many states claim that this source is their leading cause of underground contamination (U.S. EPA, 1985b). There are an estimated 2.5 million underground storage tanks across the U.S., containing more than 25 billion gallons of non-hazardous liquid products, and an additional 2,031 hazardous storage and treatment tanks containing an estimated 13.8 billion gallons of substances (OTA, 1984).

Data was collected at the state level on the numbers of release incidents for EPA's (1986) report *Summary of State Reports on Releases for Underground Storage Tanks.* State files identified 12,444 release incidents that had occurred across the country as of 1984. Even though this data represents the most comprehensive currently available information on leaking USTs, these numbers cannot be regarded as representing all storage tank releases. Differences in state enforcement and reporting procedures have resulted in inherent biases in the data. The results of the study showed that the greatest number of incidents occurred in the Northeast, Mid-Atlantic, and Great Lake states (Figure 3-11). This is probably due to the more corrosive conditions of the soils, the higher tank population, and perhaps more efficient procedures for reporting and documenting leaks. Soils became contaminated in 68% of the cases, and groundwater contamination resulted from 45% of the releases (Figure 3-12). It was found that approximately 65% of the incidents originated from retail gasoline stations, and only 3% of the releases involved chemicals other than petroleum fuels. Ninty-five percent of the leaks were reported to have taken place from operating facilities. The mean tank-leakage age was found to be 17 years for all of the tanks. Eighty-one percent of the tanks were steel and the remaining 19% fiberglass. The most commonly reported cause for release was corrosion and structural failure (Figure 3-13).

82

The NAS estimated that 16,000 spills occur annually during transport operations (OTA, 1984). In 1981 alone, 10,072 spills from both oil and hazardous chemicals, totaling 19.6 million gallons were reported (OTA, 1984). The Department of Transportation reported that more than 9,000 spills of hazardous materials occurred during transportation in 1981 and 6,500 in 1982 (OTA, 1984). Emergency measures to deal with spills often consist of flushing the spilled compound away, thus permitting or even assisting in the contamination of groundwater (Harris, 1982). Flushing usually reduces the immediate threat of fire and explosion. Such practices are more likely to create surface-water pollution than groundwater pollution, as most storm sewers and ditches flow into surface water. If the spill is large enough and is not properly cleaned up, however, groundwater contamination can result. Under the Clean Water Act, Spill Prevention Countermeasure Control (SPCC) plans have been required of every facility since 1975 for spills to the ground. The National Contingency Plan, under CERCLA, effective December 1982, details cleanup procedures for spills of oil and hazardous materials on land, into surface water, or into groundwater. In addition, releases of quantities of oil or hazardous material in excess of stated amounts must be made known to the National Response Center. Many of the industrial trade associations have researched cleanup and containment and have published manuals—for example, the American Petroleum Institute manual for the cleanup of underground spills (API, 1980).

Pipelines are used to transport a variety of fluids, but the most common one is petroleum. In 1972 there were 174,000 miles of petroleum pipelines that transported a total of 8.5 billion barrels of petroleum products (U.S. Department of Commerce, 1974). In 1976 the same amount of pipeline transported 9.63 billion barrels of petroleum products (U.S. Department of Commerce, 1977). In 1971 there were 308 interstate pipeline accidents, involving a loss of about 245,000 barrels of liquids. Pipeline failure due to external corrosion caused 33% of these accidents, and damage to pipeline from excavating machines caused an additional 22% (Meyer, 1973). In 1980 there were 275 accidents; in 1981, 198 accidents. The decrease in incidents may be due to extensive regulations from the Department of Transportation regulating transportation of liquids by pipeline. Of the 4,112 accidents that occurred between 1968 and 1981, 1,372 were due to corrosion and 1,101 to pipeline ruptures. No information was given about the number of barrels involved in the incidents in 1980 and 1981 (U.S. Department of Commerce, 1982).

In some regions deliberate leakage and spillage of contaminants, so-called midnight dumping, has occurred. This source of contamination, which is a criminal offense, is considered to be a major problem in New England (D. Burmaster, consultant on surface and groundwater quality, personal communication).

Agricultural Activities

Irrigation return-flow; the use of pesticides, fertilizers, and manure; and changes in vegetative cover have all been known to cause changes in groundwater quality. Contamination of groundwater by nitrates is one of the largest sources of non-point pollution in the United States (Freeze and Cherry, 1979). The USGS has collected nitrate data over the past 25 years from approximately 124,000 wells nationwide. Its data (Table 3-9) revealed areas of groundwater that exceed EPA limits for nitrate in almost every state. The study suggests that a large contributing factor to elevated levels of nitrates are agricultural activities such as the use of fertilizers, irrigational practices, dryland farming, and livestock wastes. Other contributing factors mentioned are septic systems, land disposal of wastes, industrial wastes, and various natural sources. An EPA study of rural water conditions (1986), however, found that only 2.7% of the nations rural wells had nitrate levels exceeding EPA's limits. In the North Central region, 5.8% and in the Northeast region, 0.3% were reported in violation of standards (U.S. EPA, 1984c). Hallberg (1985) determined that since the 1960s, an increase in the use of nitrogen fertilizers has been paralleled by a similar increase of nitrates in groundwater. Fifty-four million tons of commercial fertilizer were used between 1980

Table 3-9. Summary of Nitrate-Nitrogen Concentrations in Groundwater, by State

State	No. of Wells Sampled	% of Wells with NO₃-N 10 mg/l
Alabama	244	0.0
Alaska	1,305	2.4
Arizona	4,164	13.9
Arkansas	2,436	3.9
California	2,732	10.1
Colorado	5,492	5.7
Connecticut	348	2.3
Delaware	165	9.1
Florida	3,140	2.0
Georgia	1,137	0.5
Hawaii	164	0.0
Idaho	1,806	1.7
Illinois	359	8.4
Indiana	650	1.4
Iowa	4,088	5.0
Kansas	1,140	20.0
Kentucky	3,227	4.2
Louisiana	3,177	0.6
Maine	147	2.0
Maryland	1,521	6.8
Massachusetts	414	1.2
Michigan	1,108	1.1
Minnesota	1,655	9.3
Mississippi	1,701	0.2
Missouri	2,165	2.1
Montana	2,821	3.8
Nebraska	2,326	9.3
Nevada	465	0.9
New Hampshire	69	1.4
New Jersey	1,385	1.4
New Mexico	4,685	2.9
New York	2,491	11.0
North Carolina	908	0.8
North Dakota	7,387	4.6
Ohio	339	2.6
Oklahoma	1,724	11.8
Oregon	685	1.2
Pennsylvania	4,326	5.9
Puerto Rico	79	2.5
Rhode Island	171	36.3
South Carolina	557	0.7

Table 3-9. Summary of Nitrate-Nitrogen Concentrations in Groundwater, by State (*continued*)

State	No. of Wells Sampled	% of Wells with NO₃-N 10 mg/l
South Dakota	1,996	6.7
Tennessee	109	0.9
Texas	36,196	9.4
Utah	3,301	2.0
Vermont	73	1.4
Virginia	762	0.8
Washington	1,158	4.3
West Virginia	954	0.5
Wisconsin	2,727	3.6
Wyoming	1,477	3.8
Total	123,656	6.4

Source: Madison and Brunett, 1985.

and 1981, 48.7 million tons between 1981 and 1982, and 42.3 million tons between 1982 and 1983 (OTA, 1984). Nitrogen accounts for 6.1% to 20.4% of these weights. Much of the applied nitrate is not utilized by crops and is subsequently lost through leaching (Madison and Brunett, 1985).

Less nitrate-nitrogen is available for leaching in the loamy fine-sand soils of subirrigation systems than in furrow and sprinkler systems. Thus, subirrigation systems provide a method for reducing solute concentrations in irrigation return-flow (Wendt et al., 1976). Irrigation only contributes further to nitrate leaching. Less than 1% of the total cropland in the Northeast is irrigated; as a result, nitrate contamination is not that much of a problem. In the West, however, due to the highly saline soils, irrigation often exceeds soil requirements. Dryland farming in the northern Great Plains does not utilize irrigation, and crop plants grown in the area often have lower evapotranspiration rates; therefore water may move down through the soil picking up salts and thus contaminating the groundwater. In some cases this salty water produces saline seeps.

In 1983, 12 states experienced groundwater contamination from pesticides applied to agricultural land. In 1984, a total of 12 pesticides were discovered in the groundwaters of 18 states. By 1985, 23 states discovered groundwater contamination from 17 different pesticides (Table 3-10) (U.S. EPA, 1986b). Although agricultural operations alone account for 69%–72% of all

Table 3-10. Pesticides Found in Groundwater as a Result of Normal Land Application†

Pesticide	Use*	State(s)	Typical Positive, ppb
Alachlor	H	MD, IA, NE, PA	0.1–10
Aldicarb (sulfoxide & sulfone)	I, N	AR, AZ, CA, FL, MA, ME, NC, NJ, NY, OR, RI, TX, VA, WA, WI	1–50
Atrazine	H	PA, IA, NE, WI, MD	0.3–3
Bromacil	H	FL	300
Carbofuran	I, N	NY, WI, MD	1–50
Cyanazine	H	IA, PA	0.1–1.0
DBCP	N	AZ, CA, HI, MD, SC	0.02–20
DCPA (and acid products)	H	NY	50–700
1,2-Dichloropropane	N	CA, MD, NY, WA	1–50
Dinoseb	H	NY	1–5
Dyfonate	I	IA	0.1
EDS	N	CA, FL, GA, SC, WA, AZ, MA, CT	0.05–20
Metolachlor	H	IA, PA	0.1–0.4
Metribuzin	H	IA	1.0–4.3
Oxamyl	I, N	NY, RI	5–65
Simazine	H	CA, PA, MD	0.2–3.0
1,2,3-Trichloropropane	N (impurity)	CA, HI	0.1–5.0

† Total of 17 different pesticides in a total of 23 different states.
* H = herbicide; I = insecticide; N = nematicide

Source: U.S. EPA, 1986b.

pesticide use, contamination by pesticides can also occur through spills, leaks, lawn and garden application, and disposal practices (OTA, 1984). EPA believes that with increased groundwater monitoring more cases of contamination involving additional agricultural chemicals will be discovered (EPA, 1986b). The pesticide dibromochloropropane (DBCP), now banned in the United States, adversely affects human health and causes reproductive dysfunction in laboratory animals when given at high concentrations. There is no direct evidence for adverse effects on human health at the concentrations found in contaminated groundwater (CEQ, 1981a). Although not really an agricultural activity, suburban lawn care can cause groundwater contamination (Miller et al., 1974).

Mining

The formation of acid mine-drainage water has been described above. In areas where coal mining has been carried out over long periods of time, in Appalachia and Pennsylvania, for example, it is estimated that due to its low pH, the groundwater would be unusable for decades after the cessation of mining activities. Metal-mine leachate has increased manganese in wells in Washington, caused arsenic poisoning in cattle in Idaho, and increased the radioactivity of groundwater in Wyoming (U.S. EPA, 1977).

Highway Deicing Salts

Many of the northern states use large quantities of salt in combination with abrasives such as sand on icy and snowy highways. The salt-solution runoff can percolate through adjacent soils or otherwise find its way into the groundwater via storm drains. Due to its low cost, salt is often stored in uncovered piles. Precipitation dissolves the stored salt, which may then infiltrate shallow aquifers. Large amounts of salt are used on northern highways each winter. Over 12 million tons were used in the winter of 1978–79 (Salt Institute, 1980). During 1982 and 1983 at least 9.35 million tons of dry salts and abrasives and an additional 1.78 million gallons of liquid salts were applied to highways (OTA, 1984). During a typical winter approximately 17.6 tons of salts are used per lane per mile (OTA, 1984). The amount of salt used depends, of course, upon the severity of the winter, upon the amount of abrasives used, and upon driver education. Nevertheless, salt usage in the Northeast has resulted in chloride contamination of many drinking-water wells adjacent to highways.

The dangers to public health of high salt concentrations in drinking water are not well known. Water is considered contaminated with chloride when the concentration of chloride exceeds 250 mg/liter; concentrations above that level leave a salty taste in the water (NAS, 1980b). High concentrations of sodium have been linked with hypertension and have been implicated in the exacerbation of certain liver and kidney problems. The American Heart Association has suggested a maximum level of 20 mg/liter in drinking water. High concentrations of salt can also be severely detrimental to plant life.

88 Atmospheric Contaminants and Acid Rain

Very little is known about the effects of atmospheric contaminants and acid rain on the quality of groundwater. In Michigan chromium in the dust from an industrial plant was leached into an aquifer and contaminated a well (Deutsch, 1963). Hubert and Canter (1980b) have reviewed the literature concerning the relation between acid rain and groundwater, and they conclude that acid rain may react with soils and surface waters in such a way as to increase the leaching of metals and nutrients. Thus, the potential for groundwater contamination is present and should be investigated further, especially the potential for lead and cadmium contamination via leaching because of their long environmental-residence time.

Infiltration of Surface Water

Aquifers are often in hydraulic connection with bodies of surface water. Under certain conditions, polluted surface water from lakes or streams can percolate into a water-table aquifer, thereby degrading the quality of the groundwater. Also, groundwater development near a body of surface water may draw contaminated water into the aquifer.

Development of Groundwater

Pumping of water wells may bring water of lesser quality into their zone of influence. In inland areas serious problems may be caused by the migration of water from saline aquifers through leaky aquitards into potable supplies. In coastal regions extensive groundwater development has led to saltwater intrusion. Saltwater encroachment is an important source of groundwater contamination in many populated coastal regions, particularly in parts of California, Texas, Louisiana, Delaware, Florida, and New York. Saltwater occurs naturally in the aquifers in coastal areas, and pumping fresh water from wells may cause saltwater intrusion into potable aquifers, rendering them unfit for use. In 1981 the USGS estimated that 21 billion gallons of water per day, or 26% of all groundwater withdrawn, is in excess of natural recharge capabilities (OTA, 1984). In 1985 the EPA reported 19 states having problems with saltwater intrusion from deep saline aquifers or from the ocean. Attempts to prevent such contamination include the injection of freshwater "barriers" to reverse the hydraulic gradient in the aquifer so that the flow is toward the sea rather than toward the pumping wells. This method has met with some success in the Los Angeles area. Land subsidence has occurred in some areas of the United States because of excess withdrawal of groundwaters. A dramatic example of this occurred over a period of 37 years in San Joaquin Valley, California, where an area of land subsided nearly 29 feet.

Improper Construction and Maintenance of Wells

Faulty, corroded, or ruptured well casings, the linking of two aquifers by a well screen, or open wells can cause contaminants to move from one aquifer to another.

Sources of contamination from the three main categories described above have been documented from all parts of the country. Whether they cause localized problems or whether the groundwater-contamination problems fall into larger regional patterns has been a matter of dispute. Our conclusions, based on a study of the documented cases of groundwater contamination available, are given in chapter 6 of this report, together with examples of known occurrences of groundwater contamination for selected states. It may be said, however, that groundwater contamination is not usually direct, and in most cases the contaminant must pass through the soil layer. Attenuation of contaminants has been thought to be greater in the unsaturated zone because of the greater possibility of aerobic degradation, adsorption, complexing, and ion exchange of organics and inorganics (Page, 1981). Recent investigations on anaerobic metabolism, however, have discovered that many polluting compounds can be significantly degraded under reduced conditions (Suflita, 1987). Dilution, buffering, precipitation, reduction or oxidation, mechanical filtration, volatilization, sorption, and other processes may operate within the aquifer to purify chemical wastes (Page, 1981).

Figure 3-1. Groundwater Pollution Problems (as Identified by Federal and State/Regional Study Teams)
Source: U.S. Water Resources Council, 1978a

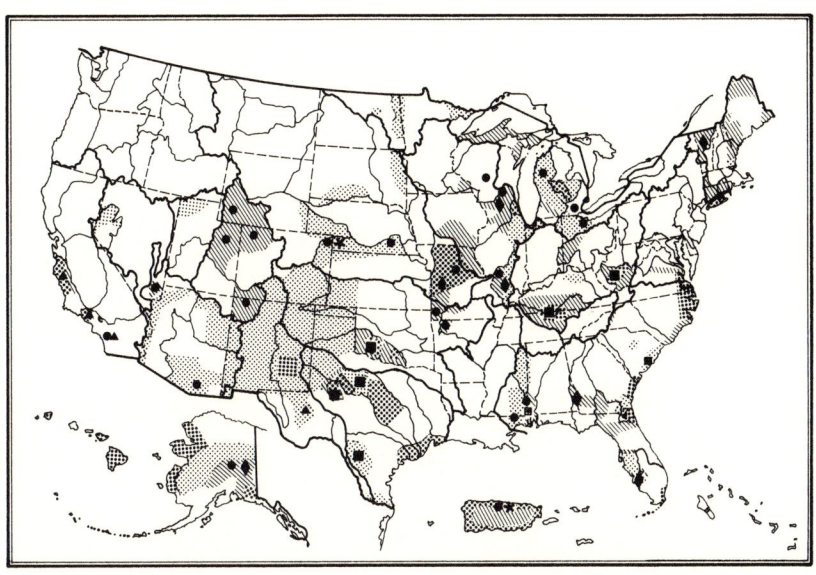

AREA PROBLEMS

▨ Significant groundwater pollution is occurring
▦ Salt-water intrusion or ground water is naturally salty
▩ High level of minerals or other dissolved
  solids in groundwater
☐ Unshaded area may not be problem-free,
  but problem was not considered major

BOUNDARIES

— Water resources region
— Subregion

SPECIFIC SOURCES OF POLLUTION

▤ Municipal and industrial wastes including
  wastes from oil and gas fields
● Toxic industrial wastes
◆ Landfill leachate
▲ Irrigation return waters
■ Wastes from well drilling, harbor dredging,
  and excavation for drainage systems
✶ Well injection of industrial waste liquids

Figure 3-2. Major Hydrochemical Processes in the Soil Zone of Recharge Areas
Source: R. Allan Freeze and John A. Cherry, *Groundwater* (Englewood Cliffs, N.J.: Prentice-Hall, Inc., 1979), p. 204

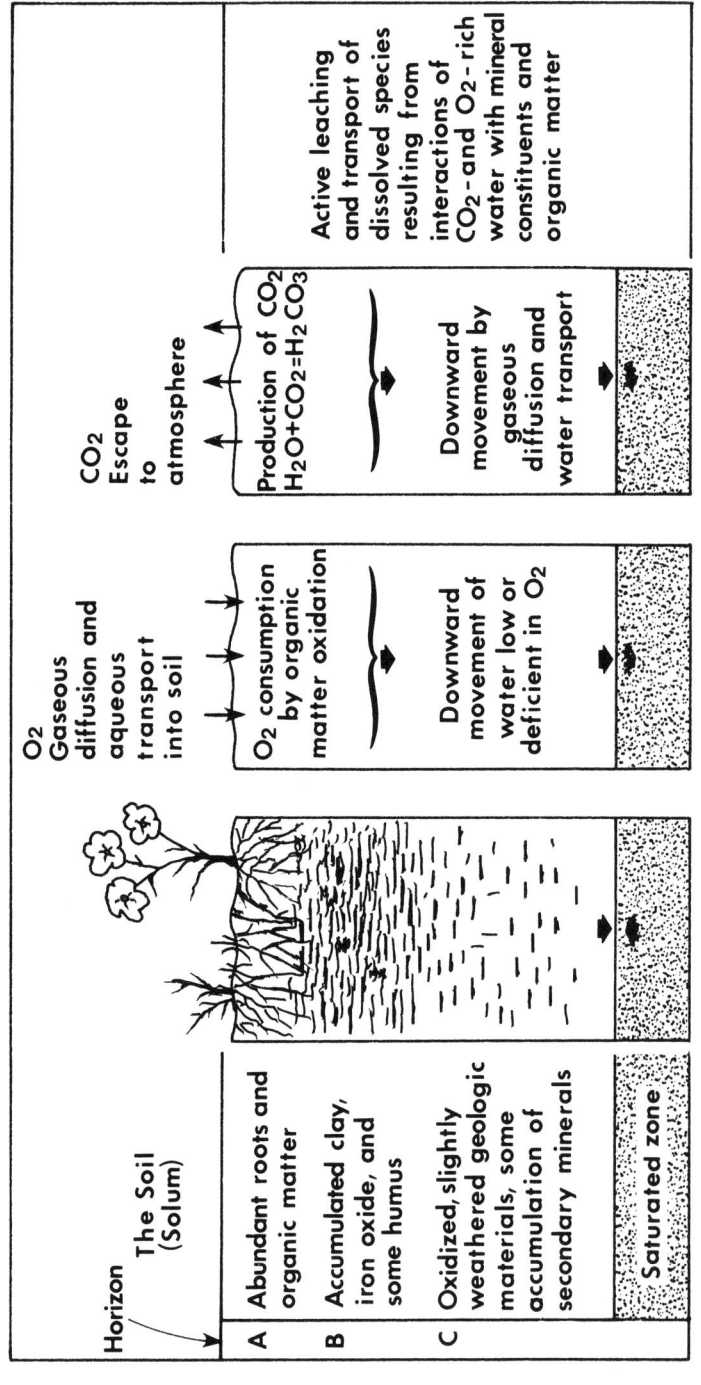

Figure 3-3. Changes in Plumes and Factors Causing the Changes
Source: U.S. EPA, 1977

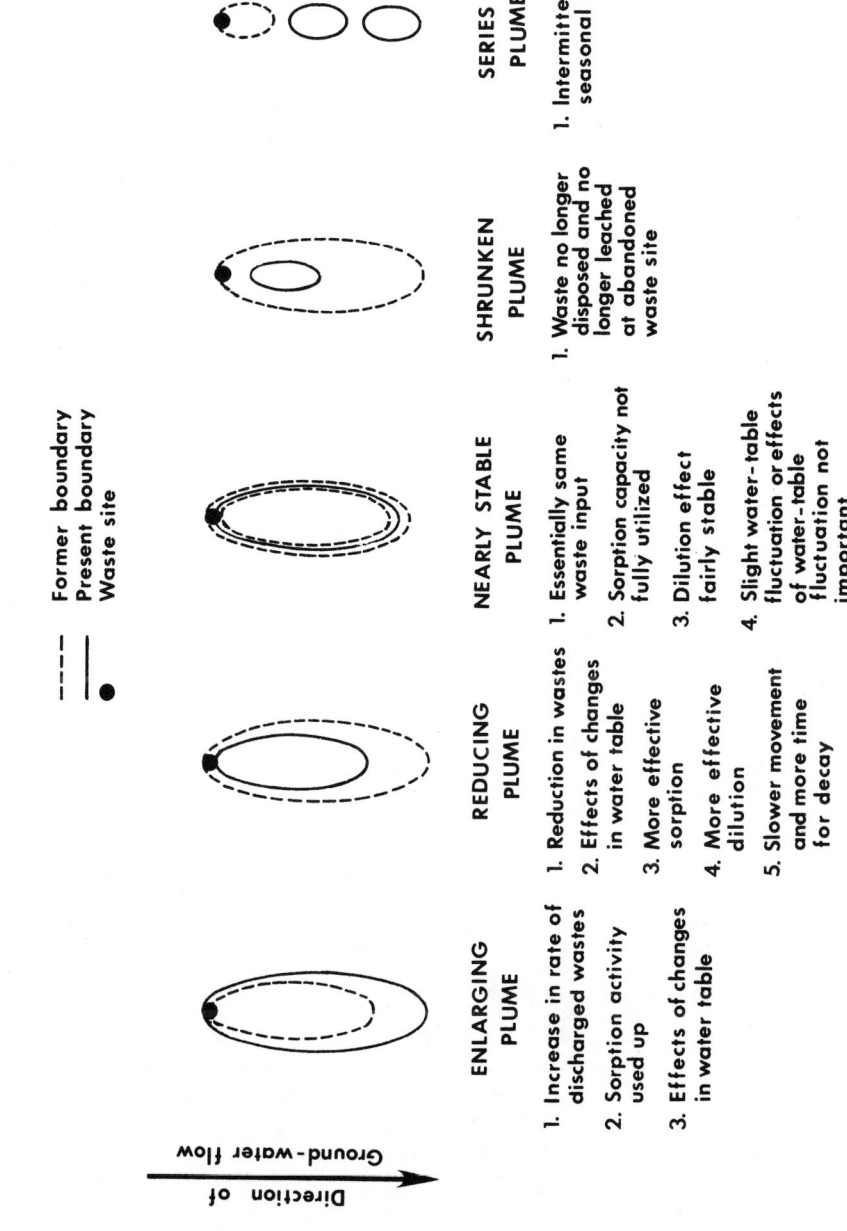

Direction of
Ground-water flow

- - - - Former boundary
———— Present boundary
● Waste site

ENLARGING PLUME

1. Increase in rate of discharged wastes
2. Sorption activity used up
3. Effects of changes in water table

REDUCING PLUME

1. Reduction in wastes
2. Effects of changes in water table
3. More effective sorption
4. More effective dilution
5. Slower movement and more time for decay

NEARLY STABLE PLUME

1. Essentially same waste input
2. Sorption capacity not fully utilized
3. Dilution effect fairly stable
4. Slight water-table fluctuation or effects of water-table fluctuation not important

SHRUNKEN PLUME

1. Waste no longer disposed and no longer leached at abandoned waste site

SERIES OF PLUMES

1. Intermittent or seasonal source

Figure 3-4. Waste Disposal Practices and Contamination of the Groundwater System
Source: U.S. EPA, 1977

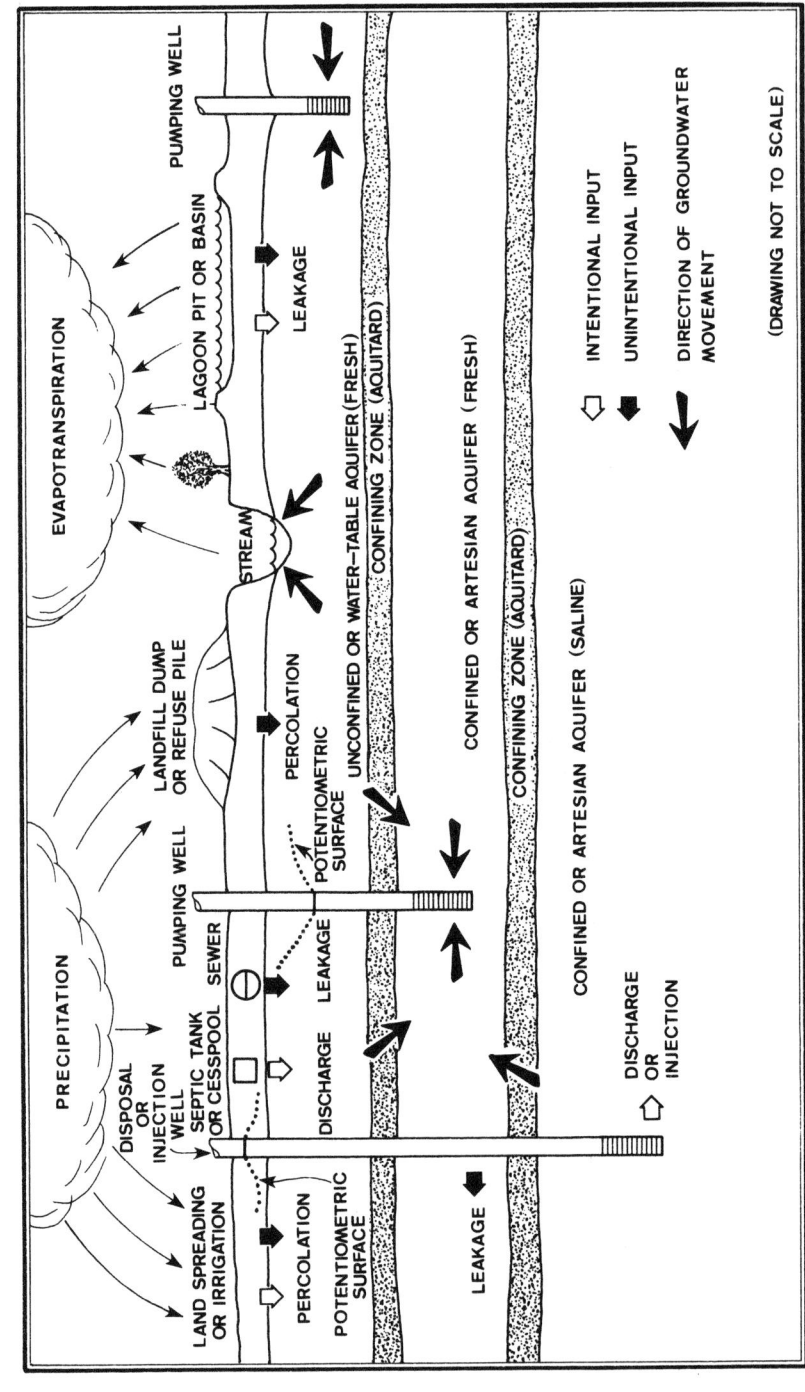

Figure 3-5. Density of Housing Units Using On-Site Domestic Waste Disposal Systems, by County

Source: U.S. EPA, 1977

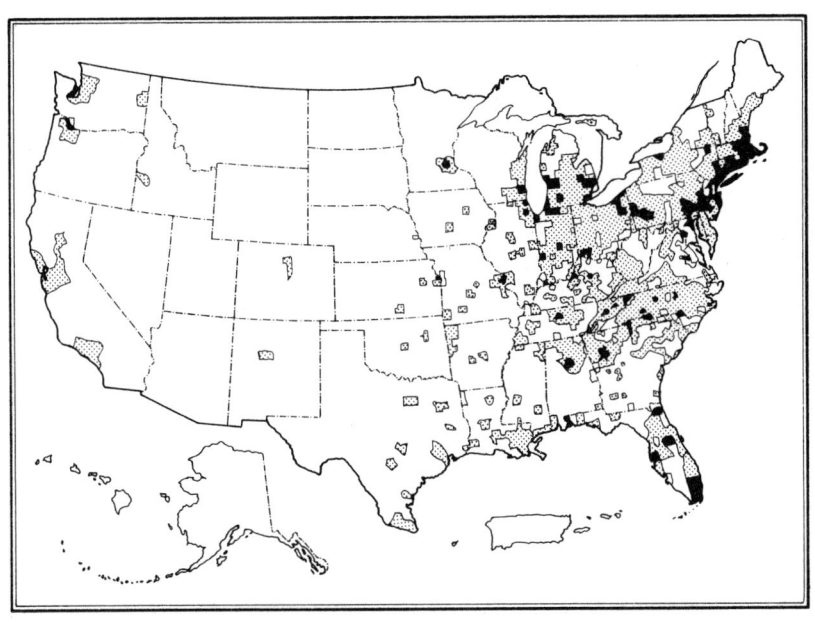

	UNITS/SQ.MI.	UNITS/SQ.KM.
□	< 10	< 3.8
▨	10 – 40	3.8 – 15.4
■	>40	>15.4

Figure 3-6. Density of Population Served by Municipal Sewage Treatment Facilities Discharging Effluent to Land, by County
Source: U.S. EPA, 1977

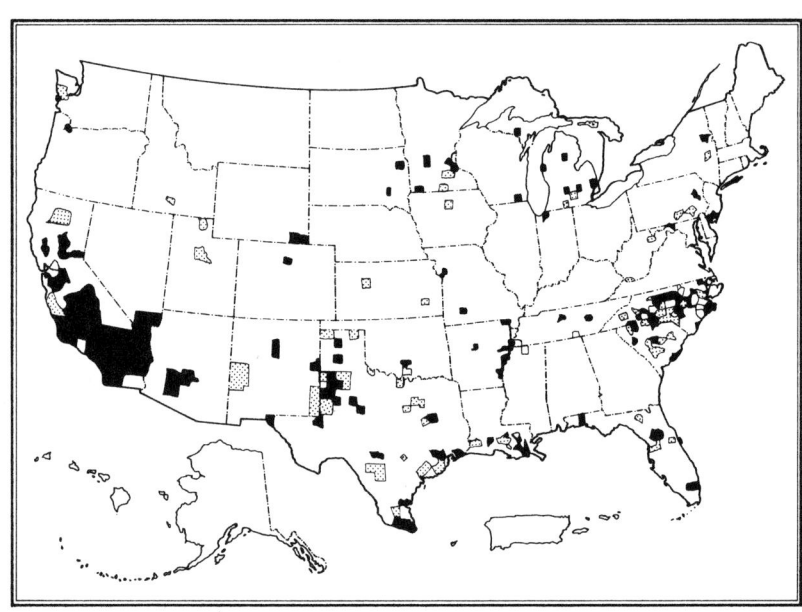

POPULATION
UNITS/SQ. MI.   UNITS/SQ. KM.

	< 6	< 2.3
	6 – 10	2.3 – 3.8
	> 10	> 3·8

Figure 3-7. Coal Fields of the Conterminous United States
Source: NAS, 1981b

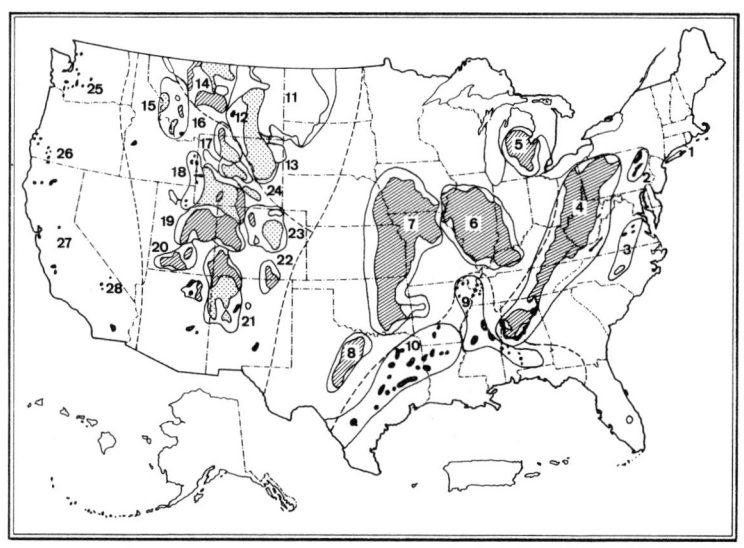

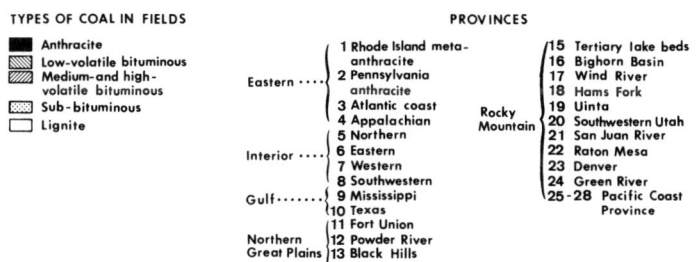

TYPES OF COAL IN FIELDS

■ Anthracite
▨ Low-volatile bituminous
▧ Medium-and high-
   volatile bituminous
▩ Sub-bituminous
☐ Lignite

PROVINCES

Eastern · · · ·
1 Rhode Island meta-
   anthracite
2 Pennsylvania
   anthracite
3 Atlantic coast
4 Appalachian

Interior · · · ·
5 Northern
6 Eastern
7 Western
8 Southwestern

Gulf · · · · · · ·
9 Mississippi
10 Texas

Northern
Great Plains
11 Fort Union
12 Powder River
13 Black Hills
14 North Central

Rocky
Mountain
15 Tertiary lake beds
16 Bighorn Basin
17 Wind River
18 Hams Fork
19 Uinta
20 Southwestern Utah
21 San Juan River
22 Raton Mesa
23 Denver
24 Green River
25-28 Pacific Coast
      Province

Figure 3-8. States in Which Significant Volumes of Wastewater Are Discharged from Mining and Ore Processing Operations, Excluding Coal and Petroleum, 1972
Source: U.S. EPA, 1977

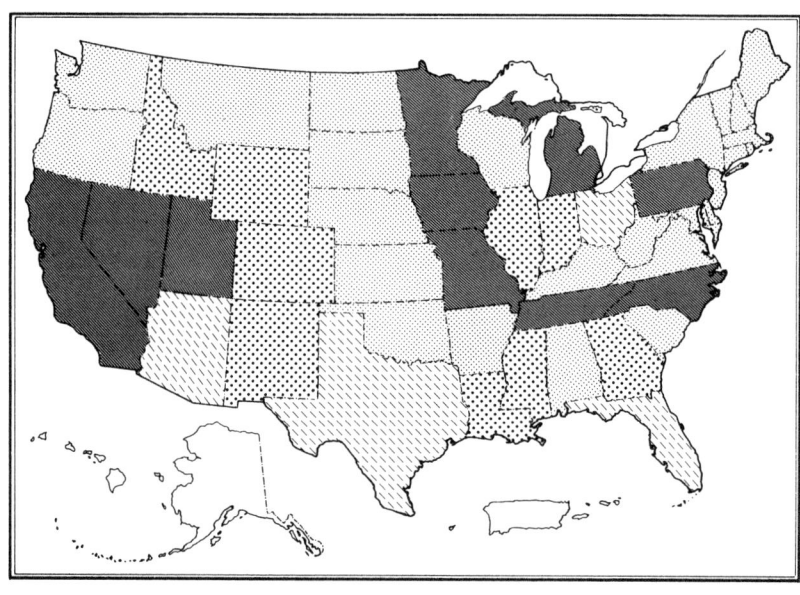

Reported as more than 20 billion gallons per year

Reported as 5-20 billion gallons per year

Volume undisclosed but assumed large

Reported as less than 5 billion gallons per year
or undisclosed but assumed small

Figure 3-9. States in Which Significant Volumes of Wastewater Are Discharged from Coal Mining and Processing Operations, 1972
Source: U.S. EPA, 1977

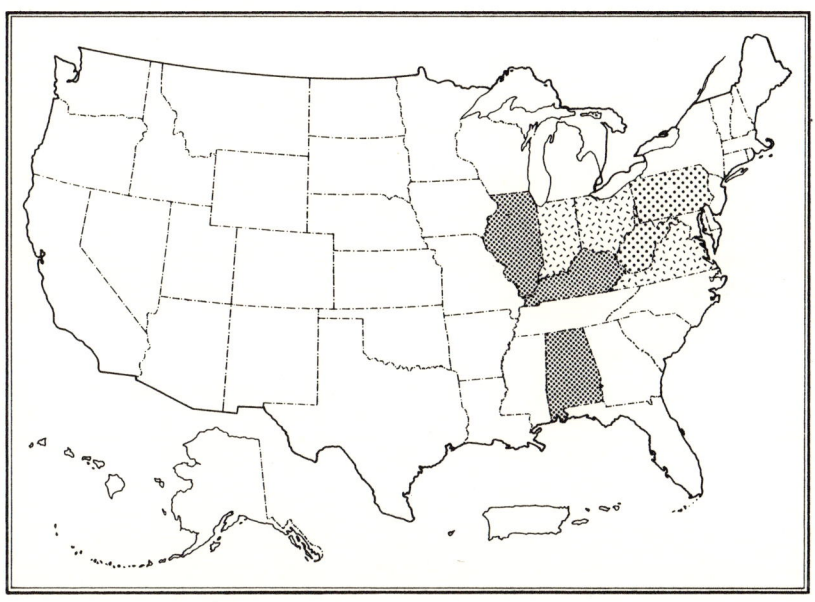

Reported as more than 20 billion gallons per year

Reported as 5-10 billion gallons per year

Volume undisclosed but assumed large

Reported as less than 5 billion gallons per year or undisclosed but assumed small

Figure 3-10. Distribution of Cattle Feeding Operations, by County
Source: U.S. EPA, 1977

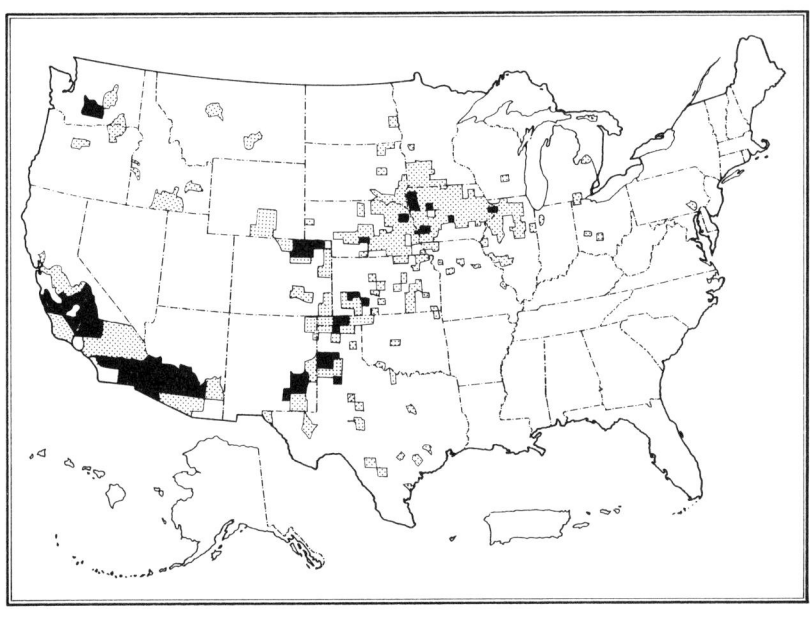

☐ 20,000 - 100,000 Head
■ >100,000 Head

Figure 3-11. Reported Underground Storage Tank Release Incidents, by State

Source: After U.S. EPA, 1986f

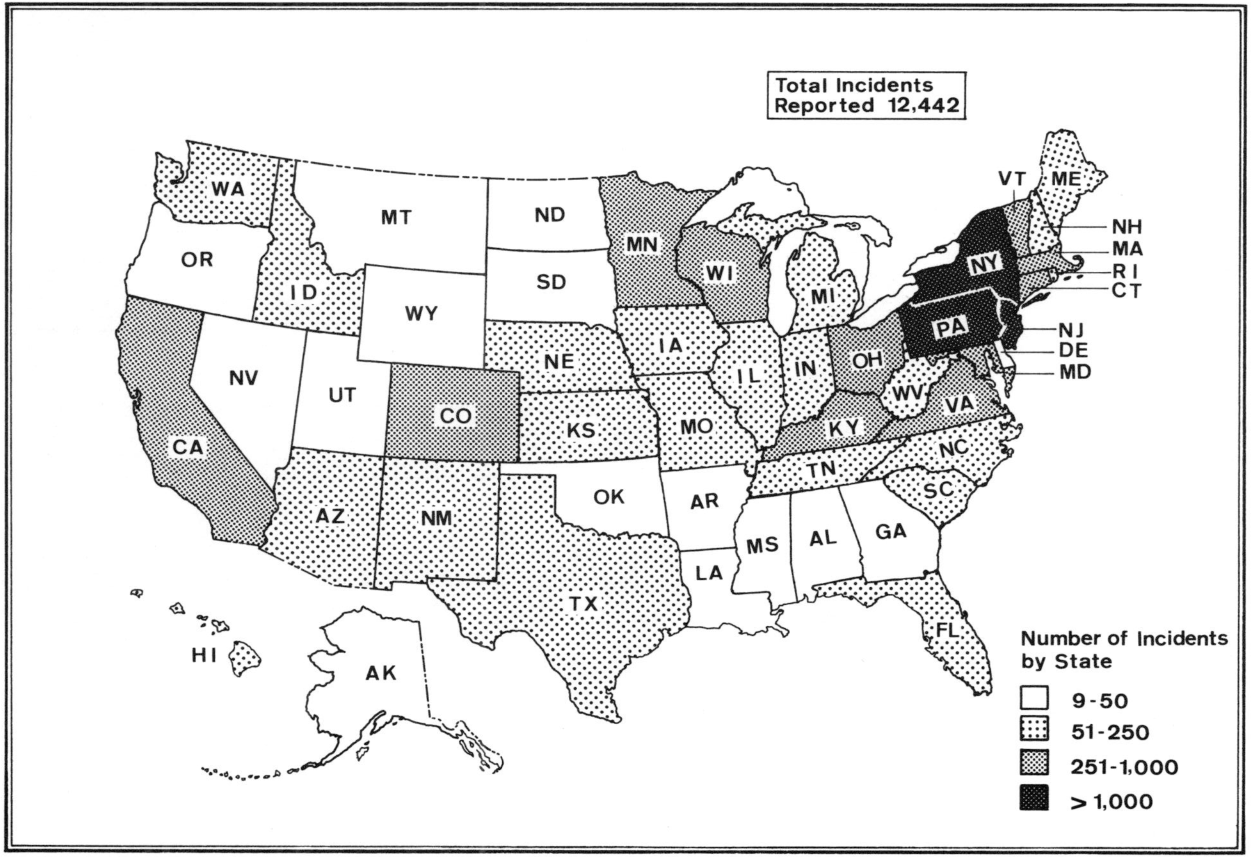

Total Incidents Reported 12,442

Number of Incidents by State

- 9 - 50
- 51-250
- 251-1,000
- >1,000

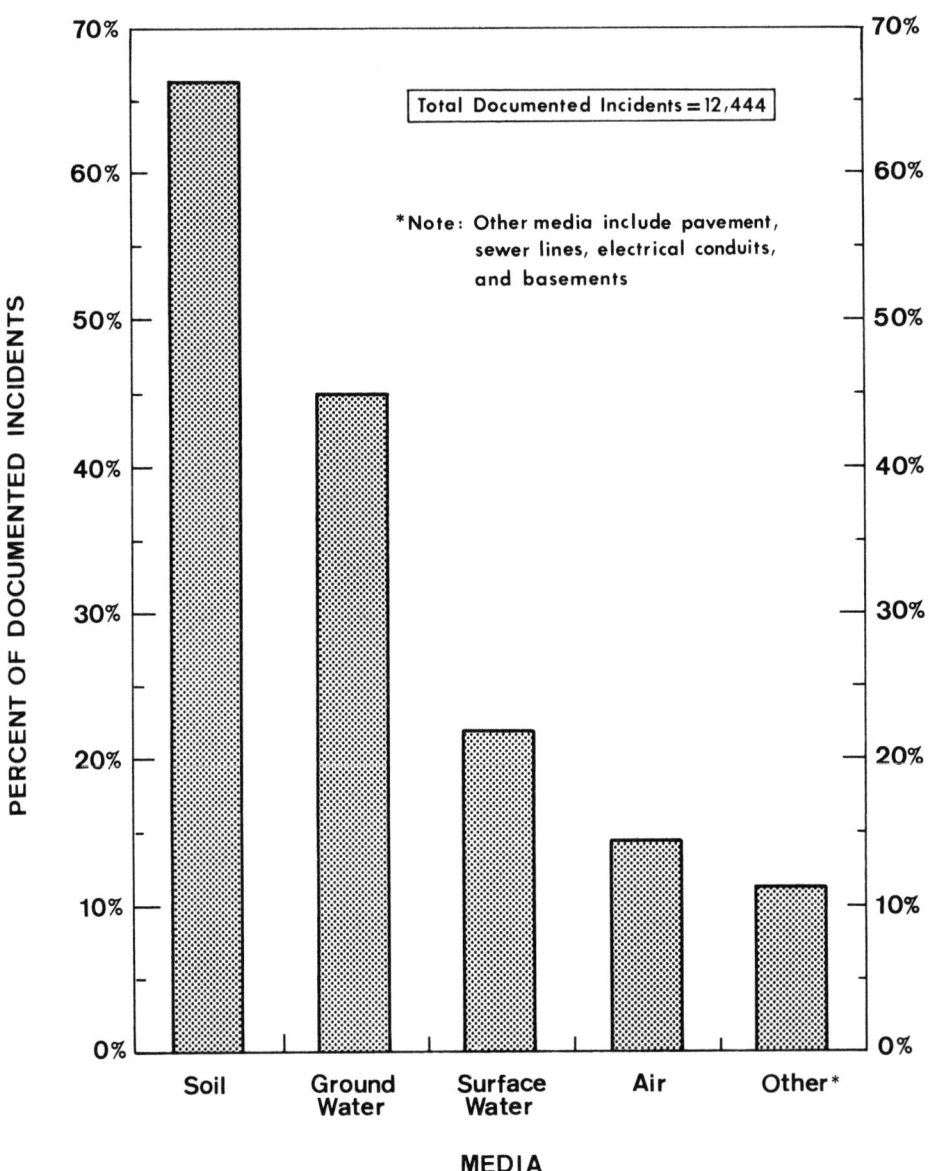

Figure 3-12. Documented Health/Environmental Impacts Reported for Each Medium due to Releases from Underground Storage Tanks
Source: After U.S. EPA, 1986f

Total Documented Incidents = 12,444

*Note: Other media include pavement, sewer lines, electrical conduits, and basements

PERCENT OF DOCUMENTED INCIDENTS

70%
60%
50%
40%
30%
20%
10%
0%

Soil   Ground   Surface   Air   Other*
        Water   Water

MEDIA

Figure 3-13. Causes of Underground Storage Tank Releases
Source: After U.S. EPA, 1986f

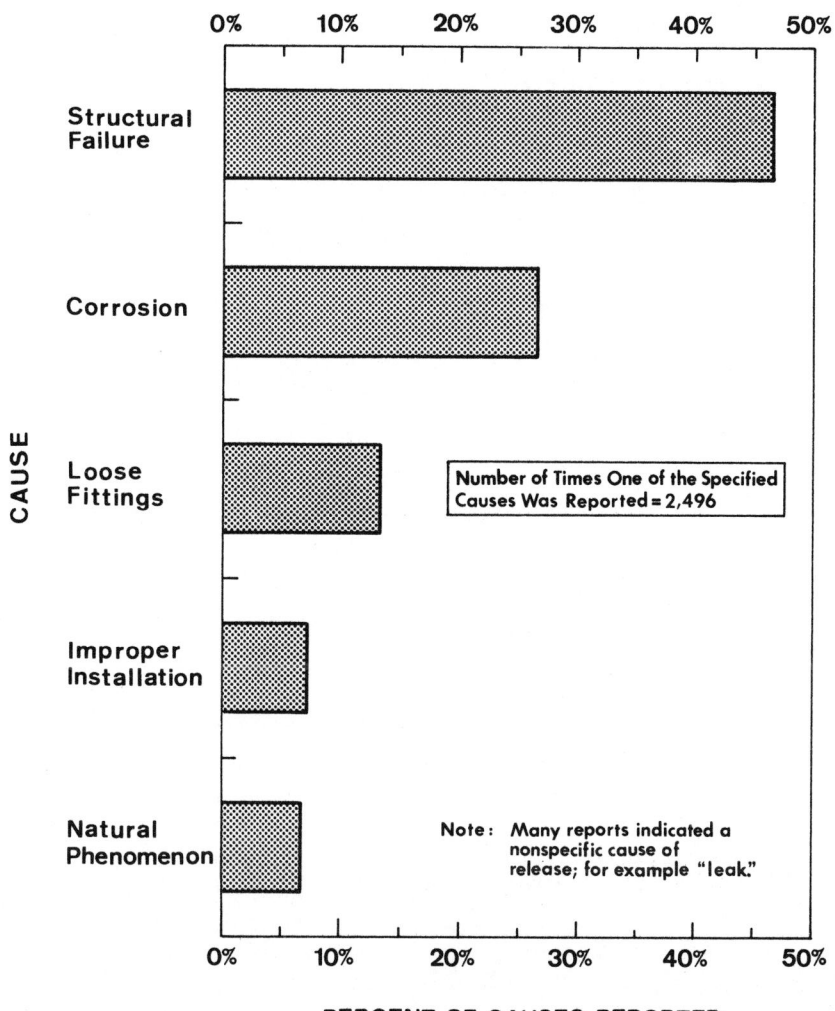

PERCENT OF CAUSES REPORTED

Four

The Severity of Groundwater Contamination

- Methods for assessing severity are varied. They include consideration of whether numerical quality standards for certain specific uses such as drinking water have been exceeded, calculating the number of people affected by well closings, estimating the percentage of usable aquifers that have been contaminated, determining the degree of hazard posed by contaminants, and determining the ease and cost of finding an alternative source of water.
- Estimates of the percentage of total contaminated groundwater have been made but vary and are inexact.

The question of the severity of groundwater contamination is a troublesome one in determining the overall magnitude of the national problem. The definition of severity can be approached in several ways.

1. If the contaminants in groundwater exceed the National Primary Drinking Water Standards (see Tables 4-1, 4-2, and 4-3), or other drinking-water standards set by individual states, then the problem could be said to be severe if the intended use is for drinking-water supplies. Consideration should be given to whether contamination renders the water unfit for its intended use, which could include industry, agriculture, drinking, or other purposes.
2. The number of persons affected by the contamination might

Table 4-1. National Primary Drinking-Water Regulations (Maximum Contaminant Levels)

Contaminant	MCL (mg/l unless otherwise specified)
Inorganic chemicals	
Arsenic	0.05
Barium	1
Cadmium	0.010
Chromium	0.05
Lead	0.05
Mercury	0.002
Nitrate (as N)	10
Selenium	0.01
Silver	0.05
Fluoride	1.4–2.4
Organic chemicals	
Chlorinated hydrocarbons	
Endrin	0.0002
Lindane	0.004
Methoxychlor	0.1
Toxaphene	0.005
Chlorophenoxys	
2,4-D	0.1
2,4,5-T, Silvex	0.01
Turbidity	1 unit
Microbiologic contaminants	1 coliform bacterium per 100 ml as the arithmetic mean of all samples per month
Radioactivity	
Combined radium-226 and radium-228	5 pCi/l
Gross alpha particle activity (including radium-226 but excluding radon and uranium)	15 pCi/l
Average annual concentration of beta particle and photon radioactivity not to produce annual dose equivalent greater than	4 millirem/year
Tritium	20,000 pCi/l
Strontium-90	8 pCi/l

Source: U.S. EPA, 1976

Table 4-2.   National Secondary Drinking-Water Regulations (Secondary Maximum Contaminant Levels)

Contaminant	MCL (mg/l unless otherwise specified)
Chloride	250
Color	15 color units
Copper	1
Corrosivity	Noncorrosive
Foaming agents	0.5
Hydrogen sulfide	0.5
Iron	0.3
Manganese	0.05
Odor	3 threshold odor number
pH	6.5–8.5 standard units
Sulfate	250 mg/l
TDS	500 ml/l
Zinc	5 mg/l

Source: J. M. Last in Maxcy–Rosenau, eds., *Public Health and Preventive Medicine.* 11th ed. (Norwalk, Conn.: Appleton-Century-Crofts, 1980), p. 976. Reprinted by permission of the publisher.

106

be taken into account. Thus, the contamination of an aquifer in the vicinity of a municipal well field that provides drinking water would be of more concern than contamination occurring in an isolated, sparsely populated area.

3. In the case of a single aquifer, severity of contamination may be related to the percentage of the aquifer that has been contaminated.
4. Nationwide, the severity of the problem may be indicated by the percentage of the aquifers affected.

Table 4-3.   Contaminants That May Be Included in the National Primary Drinking-Water Regulations

Volatile Organic Chemicals

Trichloroethylene	Chlorobenzene
Tetrachloroethylene	Dichlorobenzene(s)
Carbon tetrachloride	Trichlorobenzene(s)
1,1,1-Trichloroethane	1,1-Dichloroethylene
1,2-Dichloroethane	trans-1,2-Dichloroethylene
Vinyl chloride	cis-1,2-Dichloroethylene
Methylene chloride	
Benzene	

Table 4-3. Contaminants That May Be Included in the National Primary Drinking-Water Regulations (*continued*)

Microbiology and Turbidity

Total coliforms	Viruses
Turbidity	Standard plate count
Giardia lamblia	*Legionella*

Inorganics

Arsenic	Molybdenum
Barium	Asbestos
Cadmium	Sulfate
Chromium	Copper
Lead	Vanadium
Mercury	Sodium
Nitrate	Nickel
Selenium	Zinc
Silver	Thallium
Fluoride	Beryllium
Aluminum	Cyanide
Antimony	

107

Organics

Endrin	1,1,2-Trichloroethane
Lindane	Vydate
Methoxychlor	Simazine
Toxaphene	PAH's
2,4-D	PCB's Atrazine
2,4,5-TP	Phthalates
Aldicarb	Acrylamide
Chlordane	Dibromochloropropane (DBCP)
Dalapon	1,2-Dichloropropane
Diquat	Pentachlorophenol
Endothall	Pichloram
Glyphosate	Dinoseb
Carbofuran	Ethylene dibromide
Alachlor	Dibromomethane
Epichlorohydrin	Xylene
Toluene	Hexachlorocyclopentadiene
Adipates	
2,3,7,8-TCDD (Dioxin)	

Radionuclides

Radium-226 and -228	Uranium
Beta particle and photon radioactivity	Gross alpha particle activity
	Radon

Source: Safe Drinking Water Act Amendments of 1986 (S-124, PL 99-339)

5. A different measure of severity might be obtained if the volume of known and suspected contaminated plumes is expressed as a percentage of the known groundwater reserves nationwide.

6. The degree of hazard posed by the contaminants varies according to the volume discharged and the toxicity and concentration of the contaminant. Toxicity may be defined as the degree of hazard and may also depend on how contaminants move in the aquifer.

108

Thus, severity could depend on one or a combination of the following parameters: concentration and toxicity of the contaminants, the number of people affected if the contaminated aquifer is a source of drinking water, and the percentage of the available groundwater both locally and regionally affected by such contamination. Interwoven with each of these parameters is the economic cost of finding an alternate water source if the contamination renders the groundwater unfit for its prior or future uses or if treating the water before use is not possible.

Many of the numbers required for these various methods of assessing the severity of the contamination of groundwater in quantitative terms simply are not available. There have been no nationwide systematic surveys of the occurrence of groundwater contamination from all sources, although numerous reports and studies show that incidents of groundwater contamination have occurred in every state. The sort of information that is available from existing reports and studies is as follows:

- documentation of a large number of contamination incidents
- identification of the most important sources of contamination
- determination of the mechanisms of contamination
- in-depth studies of some contamination incidents
- nationwide surveys of the number of some types of potential contamination sources
- nationwide surveys of some contaminants focusing on either public drinking water wells or rural drinking water wells.

EPA (1980c) suggests two possible approaches for estimating the extent of groundwater contamination. The first is to conduct a systematic nationwide survey at randomly selected sites. EPA believes that if the sample is large enough, this method could provide a good estimate of both the extent and nature of the problem but not of the causes or possible solutions. Since the cost of the drilling required by such a project would be substantial and the time to accomplish it lengthy, EPA ruled out this approach, while recognizing its potential usefulness.

The second approach suggested (U.S. EPA, 1980c, see

appendices) is to estimate the number of sources of contamination and the amount of contamination per source, and from these to estimate the order of magnitude of the problem. Recognizing that the information available for this method is less precise, EPA points out that it can make use of existing information, both qualitative and quantitative. EPA used this method to develop a preliminary assessment, for which it was necessary to estimate both the sources of contamination that are the most important from a national perspective and the area of groundwater contaminated by these sources. This would be achieved by estimating the number of sites in each contaminating category and the area contaminating each site, calculating the total area estimated to be contaminated in each source category, and producing a final estimate of contamination from all categories of sources. In addition to these estimates, EPA (1980c) states that the probable extent of contamination from sources that could not be quantitatively measured should be discussed. A preliminary assessment of the problem nationwide would be based on a comparison of the area contaminated with the total area of usable aquifers. Naturally such a string of estimates is an irritant to the fastidious scientific mind, even though the object of the exercise, to achieve a preliminary assessment of the order of magnitude of the problem nationwide— a ball-park figure—is doubtless achieved. Such assessments do serve a useful purpose, but only when their inherent flaws are kept in mind.

OTA (1984) believes that a complete description of groundwater quality may never, for all practical purposes, be attainable. It contends that much of the current information available only reflects the nature of the investigation—what substances are being looked for and where. A complete nationwide survey would require detailed information on a site-by-site basis throughout the nation. Unfortunately, groundwater cannot be directly observed, and its behavior, its movement, and its chemical and biological processes are not well understood. In addition, OTA contends that attaining the needed information for a national survey requires a great deal of time and money.

Estimates of the Percentage of Groundwater That is Contaminated

The methods EPA used to estimate the area of contamination from impoundments and landfills are outlined in the Appendices of

EPA's Planning Workshops report (U.S. EPA, 1980c), but few details are given. For example, EPA estimated that the 25,000 active industrial impoundments identified in the 1978 Surface Impoundment Assessment are unlined and therefore could contaminate unconfined aquifers. Half the impoundments are known to be sited over usable aquifers on very permeable soil; thus 8,400 industrial impoundments may leak into usable aquifers. The extent of the plume of contamination caused by leaking impoundments is estimated from the volume and type of waste, the length of time the leakage occurs, and the hydrogeology of the area. EPA assumes that an impoundment is 10 acres and has been in operation for 10 years. Such an impoundment on permeable soil will produce 60 acres of contamination. If it is further assumed that none of the plumes overlaps or is intercepted by surface water, 790 square miles of contamination will be produced in total. Considering that the last two assumptions may not hold, EPA used a range from 400 to 800 square miles as the preliminary assessment of groundwater contamination due to surface impoundments.

Using a similar analysis to assess the contamination attributable to landfills, EPA came up with an estimate of 1,300 to 8,400 square miles. It should be pointed out, however, that no national survey of the number of landfill sites has been conducted, that leachate production is dependent on rainfall, which may be seasonal, and that the number of illegal dump sites is unknown.

The EPA combined estimate of contamination due to surface impoundments and landfills was thus from 1,700 to 9,200 square miles. The next stage was to assume that 60% of the 3.5 million square miles of U.S. land surface is underlain by usable aquifers; thus from 0.1% to 0.4% of the usable aquifers near the land surface are contaminated by industrial impoundments and landfill sites. EPA (1980c) cautions that the relatively small area of contamination should be viewed in relation to the fact that these two types of waste disposal sites are usually found in areas of significant industrial and domestic water use and that the problem could be exacerbated by the area's dependence on groundwater. Since groundwater pumping creates a zone of influence that affects plumes, the area over which water is rendered unusable may be larger than the area of the plumes.

EPA considered impoundments and landfills the most significant sources of contamination but also evaluated secondary sources, such as subsurface disposal systems and exploration and mining for petroleum. These sources were estimated to have contaminated

110

less than 1% and 0.1%, respectively, of the nation's usable aquifers. Thus EPA (1980c) concluded that by area up to 1% of the usable aquifers near the land surface in the United States may be contaminated at present by waste disposal practices and by petroleum exploration and mining and that the areas of contamination will increase with time. EPA did not include an estimate of the percentage of available groundwater contaminated by non-disposal, non-point-related activities.

Another independent assessment has been completed by Lehr **111** (1982). Lehr assumes that there are a total of 200,000 point sources, including septic tanks, landfills, pits, ponds and lagoons, and the like, twice the range of from 75,000 to 100,000 industrial sites often taken as a baseline. Lehr further assumes that each site creates a plume of pollution that is 1,000 feet wide and 100 feet thick, that it advances at the rate of 125 feet per year, and has been traveling for 40 years. Both the rate of travel and the distance traveled are faster and longer than most other assessments assume, and Lehr's is thus a worst-case estimate. Thus, each of the 200,000 plumes has moved a mile from its source and is polluting aquifers that are 25% porous. Assuming that each is a discrete plume, Lehr arrives at a figure of approximately 200 trillion gallons of polluted groundwater. Lehr assumes that 40% (a conservative figure) of all the available groundwater is found within one-third mile of the land's surface, and he arrives at a figure of 100 quadrillion gallons of available groundwater. Two hundred trillion gallons of polluted groundwater thus represents 0.2% of the available groundwater in the top one-third mile of the earth's crust in the United States. If one assumes a specific yield of groundwater closer to 5% than to 10%, a larger plume (1,500 feet wide, 1.5 miles long, 150 feet thick), then over 1% of the groundwater would be polluted. If there were 300,000 plumes, pollution would reach 2%. Lehr concludes that no matter how liberal or conservative the estimate, the fraction of groundwater polluted is very small.

Lehr (1982) outlines other calculations. He takes the EPA estimate of leakage of known pollution sources of 1.5 trillion gallons per year, doubles it to account for unknown sources, assumes a leakage time of 33.33 years to arrive at a figure of 100 trillion gallons of polluted groundwater, half that of his estimate outlined above. If the U.S. Geological Survey (USGS) estimate of groundwater availability (200 quadrillion gallons) is used, the estimated percentage of polluted groundwater is even lower. Lehr concedes that in certain areas of the country (industrial areas in New York, New Jersey, and New England) much more than 1% of the groundwater has been polluted, but there is still a considerable amount available that is not contaminated. Lehr estimates that nationwide, groundwater pollution is having an impact on less than 5% of the population. He predicts that within the next 10 years the percentages of groundwater polluted and population affected will both increase less than 10% and then decline during the following decade. By the year 2000, Lehr believes the nation will have reduced new emissions into groundwater by more than 90% of present levels (Lehr, 1985).

112

Population Affected by Groundwater Contamination

Unfortunately, very little information is readily available concerning the population affected by well closings due to groundwater contamination. In 1978 the New Jersey township of South Brunswick was obliged to close down public supply Well 11 due to contamination by 1,1,1-trichloroethane, an organic chemical often used as a degreasing agent and for cleaning septic tanks. The level of contamination of the pumped well water was usually in the 150 to 500 parts per billion (ppb) range but did, at times, exceed 1,000 ppb. Other contaminants subsequently found in the water from Well 11 or adjacent domestic and industrial wells include 1,1-dichloroethane, trichloroethylene, tetrachloroethylene, benzene, toluene, zinc, and arsenic. All contaminated domestic wells in the vicinity of Well 11 were taken out of service. Since 1978 the township has been dependent on Wells 12 and 13, which remain uncontaminated and are tested regularly (Geraghty and Miller, Inc., 1979).

New Castle County in Delaware recognized that the groundwater was being contaminated mainly due to solid-waste disposal, primarily at the Llangollen landfill. The estimated groundwater availability in New Castle County totals 52 million gallons per day (mgd) (Frick and Shaffer, n.d.), and the groundwater use in 1974 was 27.9 mgd, nearly 54% of that available. The total daily demand for both surface and groundwater in 1974 was 78 mgd and was projected to rise to 89 mgd in 1985 and to 105 mgd in 1995. In 1974 the population of New Castle County was 393,600, and this was projected to rise to 474,200 by 1995. It is estimated that in 1995 the amount of water resources that are rendered unusable because of pollution will be 6.7 mgd.

In Nassau and Suffolk Counties on Long Island, New York, the population of nearly 3 million is almost entirely dependent on groundwater as its only source of drinking water. More than 100 public community and non-community supply wells have been closed due to contamination by tetracholoethylene, trichloroethane, trichloroethylene, and other volatile synthetic organic compounds (Dennis Moran, Suffolk County Department of Health Services; and Donald Myott, Nassau County Department of Health; personal communication).

The most widespread area of groundwater contamination in **113** California is located in the San Joaquin Valley. An area extending over seven counties was found to have wells contaminated with pesticides, mainly DBCP. The total population of the counties affected is nearly 2 million. One hundred twenty-five large supply wells and 1,800 small water systems and private wells had concentrations of DBCP ranging from 0.05 to 46 μg/l. All large systems showing levels greater than 1 μg/l have been taken out of service. However, many small public systems and private wells continue to depend on these supplies (California Department of Health Services, 1986).

In Belleview, Florida, between 1979 and 1980 a release of 10,000 gallons of gasoline from an underground service station tank took place. Gasoline from this source was detected in all three of the city's underground drinking-water supplies two years later. The entire community of Belleview, comprising 2,500 people, was affected by this one leak and had to be provided with a new water supply (Schroeder, 1985).

In Bedford, Massachusetts, four municipal wells supplying 80% of the drinking water for the town were closed due to contamination by trichloroethylene and dioxane. Water had to be purchased from neighboring towns, one of which subsequently had to close its main well due to similar contamination.

Although these examples give an indication of the number of persons affected by well closings and possibly exposed to contaminants because of groundwater contamination, there are no national surveys on this. The volume and toxicity of wastes that may contaminate groundwater also affect our own perception of the severity of contamination. As Keeley (1977) points out, septic tanks are considered a major problem in some areas because of their numbers and distribution, but industrial lagoons, although less numerous, often contain a more hazardous waste from the point of view of toxicity.

Five

The Effects of Groundwater Contamination on Public Health

- The effects of groundwater contamination on public health may fall into one of three categories: (1) acute waterborne disease due to pathogenic organisms; (2) poisoning or other acute responses due to chemical contaminants; (3) chronic long-term effects.
- The impact of groundwater contamination on human health is difficult to quantify accurately.
- Pathogenic contaminants have caused outbreaks of disease.
- Chemical contaminants have caused acute public health effects such as poisoning.
- The correlation between cardiovascular disease and high mineralization of groundwater is uncertain.
- High concentrations of minerals in groundwater may be linked with various forms of cancer, but studies are not conclusive. It is important that these implications be defined.
- It is impossible to assess, in quantitative terms, the national risk of drinking contaminated groundwater due to the lack of comprehensive national surveys on the extent of groundwater contamination, the lack of knowledge of effects of many contaminants in low concentrations, and the paucity of testing of groundwater contaminants for carcinogenicity.

The effects of groundwater contamination on public health may fall into one of three categories: acute waterborne disease due to a variety of pathogenic organisms, poisoning or other acute responses to chemical contaminants, and chronic long-term disease. These effects are due to three distinct forms of groundwater contamination, namely, microbial, chemical, and radioactive. Acute illnesses are defined as those resulting from contact with high levels of a chemical over a short period of time (Conservation Foundation, 1987b). Chronic illnesses are defined as those that persist, either because the injury is persistent or progressive, or because the exposure is prolonged and the rate of new injury exceeds the rate of repair (NAS, 1977).

Human health can be affected by groundwater contaminants through the following routes of intake: by ingestion through drinking; by inhalation of contaminants that can volatilize when water is heated (as in a shower) or boiled; by absorption through the skin during washing, bathing, or handling of soil exposed to contaminated groundwater; and by ingestion through consumption of food derived from plants or animals exposed to groundwater (Environ Corp., 1983).

Every incidence of illness given in this chapter is attributable to contaminated groundwater. Some of the chemical causes of illness due to drinking water have been investigated for surface waters only. However, as groundwater may also contain some of these chemicals, the findings for surface waters are discussed where deemed appropriate.

Bunch and Jacobs (1979) estimated that in 1972 health hazards due to water pollution accounted for $4.4 billion of the total $188 billion estimated costs of all health hazards, but did not subdivide this total into separate costs attributable to groundwater and surface-water contamination (Table 5-1). These figures are somewhat out of date, but more recent ones are not available.

Risk Assessment and Toxicological Testing

The degree of understanding of health effects varies among the different types of contaminants. The effects of microorganisms and high to moderately high levels of radionuclides on public health have been well studied and are now fairly well understood. The effects of very low levels of radioactivity are not so well known, and many chemical effects are not understood at all. This lack of knowledge is due to the fact that only recently has the public

Table 5-1. Economic Costs of Environmental Health Hazards, 1972

Category of Hazard	Annual Health Costs 1972 Estimates (Billions of Dollars)
A. Health Costs Due to Environmental Hazards	
I. Direct Environmental Hazards	
Air pollution	8.6
Water pollution	4.4
Infectious diseases	
Hepatitis	0.73
Salmonellosis	0.0004
Influenza	4.9
Syphilis	0.007
Tuberculosis	0.7
Occupational hazards	4.4
Unsafe products	6.3
Motor vehicle accidents	10.1
Total All Direct (Total I)	40.1
II. Costs Resulting from Environmentally Induced Behavior	
Alcohol abuse	24.2
Drug abuse—heroin	3.7
Tobacco use	7.6
Total All Costs of Environmental Hazards (I and II)	75.6
B. Costs of All Illnesses	188.

Source: S. E. Bunch and P. Jacobs, *Journal of Environmental Health* vol. 11, no. 5 (1979), p. 268. Reprinted by permission of the National Environmental Health Association, Denver, Colorado.

become aware of exposure to this type of contamination (OTA, 1984). Assessing the risks of exposure to contaminants is, more often than not, very hard to do. The OTA (1984) points out that risk assessments require information about adverse effects, toxicity, and exposure. Epidemiological studies should be conducted when possible as well. Often, sufficient data for an assessment is not available, and especially not easily linked to the substances found in groundwater.

Toxicological testing falls into two main categories—in vivo and in vitro methods. In vivo testing involves the use of animals, can be of long or short duration, and is useful for determining either chronic or acute health effects. Most toxicological data is aquired by this type of testing (OTA, 1984). In vitro testing involves the use of isolated cells or microorganisms, usually in a culture, and the testing is usually of short duration. Epidemiological studies are of three kinds: the collection and evaluation of data resulting from observations on human populations followed over time (prospective studies), the comparison of death certificates of a control population (retrospective studies), and the study of the past history of patients (case-control studies). Hubert and Canter (1980a) note that epidemiological studies often produce inconclusive results because of the complexity of environmental factors that may be involved but not included in the studies. Most epidemiological evidence linking drinking-water contaminants with health effects has been based on surface-water sources.

117

Although the risk assessment of the ingestion of surface water cannot be directly applied to the ingestion of groundwater, because the two sources may contain quite different compounds, it nevertheless provides useful insights into the risks associated with some chemicals or classes of chemicals common to both. Absolute risks are very difficult to measure, usually apply only to the specific population for which they are determined, and are influenced by all the other pertinent characteristics of the specific population. Relative risks, however, are more likely to apply to other populations, although they too are influenced by the characteristics of the population exposed.

The maximum contaminant levels (MCLs) shown in Tables 4-1 and 4-2 were set by taking into account the human environmental exposure over a lifetime. Standards will be set for 83 additional chemicals by 1989 as a result of the 1986 Safe Drinking Water Act Amendments (Table 4-3). The federal government has provided additional guidance on selected substances through the establishment of non-enforceable health goals, called MCL goals. Standards for other chemicals that may be of concern have not been set due to lack of necessary data. The debate over whether or not there is a threshold below which no adverse response occurs complicates the task of setting standards for drinking water further (Okun, 1980). Synergistic effects of contaminants should also be taken into account when MCLs are set, because contamination of groundwater is usually from a group of chemicals. In addition to the factors listed above that hinder standard setting, the cost of determining a standard can be great. Page (1982) notes that estimates of the cost of establishing carcinogenicity in one compound can range from $500,000 to $1.25 million, and the procedure is time-consuming. Very few of the contaminants in water supplies have been so tested.

Until MCLs are established there are two possibilities for action that have been suggested. Either all potential carcinogens and toxic substances could be eliminated from the water supply in case they cause future harm, or such action could be delayed until its potential benefits are better elucidated. The long time lag between exposure to a carcinogen and the appearance of malignancy is yet another complicating issue. As it could take up to 30 years to establish MCLs for contaminants on the priority pollutant list, some prudent interim measure could be adopted if deemed necessary.

118

Incidence of Disease

Information concerning the effects of contaminated water on public health may be obtained from two main sources: the Health Effects Research Laboratory of the Environmental Protection Agency (EPA) in Cincinnati, and the Centers for Disease Control (CDC) in Atlanta. Each year the EPA laboratory collects information from state water-supply agencies on the occurrence of waterborne-disease outbreaks. These data are included in the annual summaries of water-related outbreaks of disease put together by the Centers for Disease Control.

The CDC defines a waterborne-disease outbreak as "an incident in which 1) 2 or more persons experience similar illness after consumption or use of water intended for drinking, and 2) epidemiologic evidence implicates the water as the source of illness. In addition, a single case of chemical poisoning constitutes an outbreak if laboratory studies indicate that the water was contaminated by the chemical. Only outbreaks associated with water intended for drinking are included." (CDC, 1985) Causes of outbreaks are subdivided by system deficiencies, namely, untreated surface water or untreated groundwater, treatment deficiencies, and distribution-system deficiencies. Waterborne-disease outbreaks are reported to CDC by state health departments on a standard reporting form. CDC believes that only a fraction of the outbreaks that actually occur are reported. The efficiency of reporting outbreaks to the health authorities varies tremendously from region to region. Once reported, the interest, diligence, and resources of the state or local health authority determine the quality of the investigation. Thus, the CDC reports cannot be considered a definitive record of the incidence of disease and etiology, a fact recognized by that agency. States with good reporting systems and a staff member interested in waterborne disease might appear to have a much higher incidence of disease than states with an equal or greater actual incidence but a poor reporting system.

The number of reported incidents of outbreaks across the country has been monitored and analyzed over the past decade by these two organizations. In the period between 1971 and 1982 there were 399 outbreaks of acute waterborne disease from both surface water and groundwater reported to the CDC and the EPA; these outbreaks were caused by all types of agents and affected nearly 86,100 persons. Of these, 204 of the outbreaks (51%) and 34,337 of the illnesses (40%) were caused by the use of untreated or inadequately treated groundwater (see Figure 5-1) (Craun, 1985). Most of the outbreaks occurred in non-community water systems, but the greatest number of people affected resulted from outbreaks in inadequately disinfected community water systems (Table 5-2). Craun (1985) observed that a large number of outbreaks were seasonal, affecting travelers, campers, visitors to recreational areas, and restaurant patrons during the summer months.

119

Of the 8558 illnesses caused by microbial contamination by the use of untreated well water, overflow or seepage through substrate of sewage from septic tanks and cesspools was responsible for the

Table 5-2. Waterborne Disease Outbreaks in Groundwater Systems, 1971–1982

Water System	Outbreaks		Cases of Illness	
	Wells	Springs	Wells	Springs
1. Untreated Groundwater				
A. Community	15	5	1,334	123
B. Non-community	73	8	7,029	1,116
C. Individual	22	2	195	29
Subtotal	110	15	8,558	1,268
2. Inadequately Disinfected Groundwater				
A. Community	26	4	17,125	1,519
B. Non-community	41	8	4,363	1,504
C. Individual	0	0	0	0
Subtotal	67	12	21,488	3,023
Totals	177	27	30,046	4,291

Source: Reprinted with permission from G. F. Craun, A Summary of Waterborne Illness Transmitted Through Contaminated Groundwater, *Journal of Environmental Health* vol. 48, no. 3 (1985), p. 123. Copyright © 1985. Published by the National Environmental Health Association, Denver, Colo.

greatest number (1626). Chemical contamination was responsible for only 157 of the illnesses. Surface runoff, flooding, and improper well construction caused another 11%. Thirty-five percent of the sources of outbreaks from untreated well water have remained unidentified (Craun, 1985). Of particular interest is that most of these acute illnesses were from organisms and not from chemical poisonings. It should be remembered, however, that these reported outbreaks represent only a fraction of the total actual occurrences.

120   Craun (1985) suggests that the apparent trend of increased outbreaks between the years 1971 and 1980 (Table 5-3) was most probably due to improved data collection and reporting systems. Evidence suggests, however, that in recent years increased surveillance of outbreaks and their origins has led to corrective measures, and thus the numbers of reported incidents are now decreasing. This apparent decrease, however, may also be a reflection of a decrease in the numbers of people obtaining water from untreated sources (CDC, 1985).

Table 5-4 lists some of the disease-causing pathogens and chemicals that are known to be responsible for outbreaks in the

Table 5-3. Waterborne Disease Outbreaks from Surface Water and Groundwater, 1971–1982

Year	Number of Outbreaks	Number of Illnesses
1971	20	5,182
1972	30	1,650
1973	25	1,762
1974	25	8,356
1975	24	10,879
1976	35	5,068
1977	34	3,860
1978	32	11,435
1979	42	9,750
1980	53	20,045
1981	35	4,450
1982	44	3,611
Totals	399	86,048

Source: Reprinted with permission from G. F. Craun, A Summary of Waterborne Illness Transmitted Through Contaminated Groundwater, *Journal of Environmental Health* vol. 48, no. 3 (1985), p. 122. Copyright © 1985. Published by the National Environmental Health Association, Denver, Colo.

Table 5-4. Guidelines for Confirmation of Waterborne Disease Outbreaks Associated With Both Surface Water and Groundwater Sources

Etiologic Agent	Clinical Syndrome	Laboratory and/or Epidemiologic Criteria
Bacteria		
1. *Escherichia coli*	a) Incubation period: 6–36 hours. b) Gastrointestinal syndrome—majority of cases with diarrhea.	a) Demonstration of organisms of same serotype in epidemiologically incriminated water and stool of ill individuals and not in stool of controls. *or* b) Isolation from stool of most ill individuals organisms of the same serotype that have been shown to be enterotoxigenic or invasive by special laboratory techniques.
2. *Salmonella*	a) Incubation period: 6–48 hours b) Gastrointestinal syndrome—majority of cases with diarrhea (often with fever), or extraintestinal infection.	a) Isolation of *Salmonella* organism from epidemiologically implicated water. *or* b) Isolation of *Salmonella* organism from stools or tissues of ill individuals.
3. *Shigella*	a) Incubation period: 12–48 hours. b) Gastrointestinal syndrome—majority of cases with diarrhea (often with fever and bloody diarrhea).	a) Isolation of *Shigella* organism from epidemiologically implicated water. b) Isolation of *Shigella* organism from stools of ill individuals. The recovery of *Shigella* is difficult and therefore failure to isolate the organism does not prove its absence (Kawata, personal communication).

Table 5-4. Guidelines for Confirmation of Waterborne Disease Outbreaks Associated With Both Surface Water and Groundwater Sources (continued)

Etiologic Agent	Clinical Syndrome	Laboratory and/or Epidemiologic Criteria
4. *Vibrio cholerae* 01	a) Incubation period: 1–5 days. b) Gastrointestinal syndrome—majority of cases with diarrhea and without fever, dehydration, cholera	a) Isolation of *V. cholerae* from epidemiologically incriminated water. *or* b) Isolation of organisms from stools or vomitus of ill individuals. *or* c) Significant rise in vibriocidal, bacterial agglutinating, or antitoxin antibodies in acute and early convalescent sera, or significant fall in vibriocidal antibodies in early and late convalescent sera in persons not recently immunized.
5. *Campylobacter fetus* ssp. *jejuni*	a) Incubation period: usually 2–5 days. b) Gastrointestinal syndrome—majority of cases with diarrhea, often with fever.	a) Isolation of *Campylobacter* organism from epidemiologically implicated water. *or* b) Isolation of *Campylobacter* organism from stools of ill individuals.
6. *Yersinia enterocolitica*	a) Incubation period: 3–7 days. b) Gastrointestinal syndrome—majority of cases with diarrhea.	a) Isolation of *Yersinia* organism from epidemiologically implicated water. b) Isolation of *Yersinia* organism from stools of ill individuals. c) Significant rise in bacterial agglutinating antibodies in acute and early convalescent sera.

7. *Leptospira*	a) Incubation period: 4–19 days. b) Protean group of diseases—headache, conjunctivitis, rash, meningitis, etc.	a) Isolation of leptospires from epidemiologically implicated water. b) Isolation of leptospires from blood of ill individuals. c) Significant rise in bacterial agglutinating or complement fixing antibodies in acute and early convalescent sera.
8. Others	Clinical data appraised in individual circumstances.	Laboratory data appraised in individual circumstances.
Chemical 1. Heavy metals Antimony Copper Iron Tin Zinc, etc.	a) Incubation period: 5 min.–8 hours (usually less than 1 hour). b) Clinical syndrome compatible with heavy metal poisoning—usually nausea, cramps, and metallic taste.	Demonstration of high concentration of metallic ion in epidemiologically incriminated water.
2. Fluoride	a) Incubation period: usually less than 1 hour. b) Gastrointestinal illness—usually nausea, vomiting, and abdominal pain.	Demonstration of high concentration of fluoride ion in epidemiologically incriminated water.
3. Other Chemicals	Clinical data appraised in individual circumstances.	Laboratory data appraised in individual circumstances.

Table 5-4. Guidelines for Confirmation of Waterborne Disease Outbreaks Associated With Both Surface Water and Groundwater Sources (*continued*)

Etiologic Agent	Clinical Syndrome	Laboratory and/or Epidemiologic Criteria
Parasitic		
1. *Giardia lamblia*	a) Incubation period: variable; 1–4 weeks. b) Gastrointestinal syndrome—chronic diarrhea, cramps, fatigue, weight loss, and bloating.	a) Demonstration of *Giardia* trophs in epidemiologically incriminated water. *or* b) Demonstration of *Giardia* trophs in stools or duodenal aspirates of ill individuals.
2. *Entamoeba histolytica*	a) Incubation period: variable; usually 2–4 weeks. b) Variable gastrointestinal syndrome—from acute fulminating dysentery with fever, chills, and bloody stools to mild abdominal discomfort with diarrhea.	a) Demonstration of *Entamoeba histolytica* cysts in epidemiologically implicated water. *or* b) Demonstration of *Entamoeba histolytica* cysts in stools of affected individuals.
3. Others	Clinical data appraised in individual circumstances.	Laboratory data appraised in individual circumstances.
Viral		
1. Hepatitis A	a) Incubation period: 2–6 weeks. b) Clinical syndrome compatible with hepatitis—usually including jaundice, GI symptoms, dark urine, anorexia.	Liver function tests compatible with hepatitis in affected persons who consumed the epidemiologically incriminated food.

2. Parvovirus-like agents (Norwalk, Hawaii, Miami)	a) Incubation period: 4–77 hours. b) Gastrointestinal syndrome—vomiting, watery diarrhea, abdominal cramps.	a) Demonstration of virus particles in stool or ill individuals by immune electron microscopy. *or* b) Significant rise in antiviral antibody in acute and convalescent sera.
3. Rotavirus	a) Incubation period: 24–72 hours. b) Gastrointestintal syndrome—vomiting, watery diarrhea, abdominal cramps often with significant dehydration.	a) Demonstration of the virus in the stool of ill individuals. b) Significant rise in antiviral antibody in paired sera.
4. Enterovirus	a) Incubation period: variable. b) Syndrome—variable; poliomyelitis, aseptic meningitis, herpangina, etc.	a) Isolation of virus from epidemiologically implicated water. *or* b) Isolation of virus from ill individuals.
5. Others	Clinical evidence appraised in individual circumstances.	Laboratory evidence appraised in individual circumstances.

Source: After CDC, 1985; Harris, 1986.

United States, as well as the clinical symptoms or laboratory or epidemiological criteria required for their confirmation (CDC, 1985). Table 5-5 lists toxic substances that have been found in groundwater, and their effects; radioactive substances, however, are not included (Environ Corp., 1983).

Acute Illness Due to Contaminated Groundwater

126

Fortunately, most acute waterborne poisonings cause mild, self-limited illnesses (Harris, 1986). The symptoms vary with each substance ingested. However, commonly reported symptoms suggestive of acute illness include gastrointestinitis, nausea, vomiting, and skin rashes; these effects usually occur soon after exposure (Conservation Foundation, 1987b).

Acute Illness Due to Pathogenic Contaminants

Five categories of pathogens are found in water: bacteria, viruses, protozoa, worms, and fungi. The most significant waterborne diseases worldwide are typhoid and cholera, which were a scourge in the industrialized nations, including the United States, up to the late nineteenth century (Okun, 1980). Gastrointestinal illnesses constitute the major category of illnesses resulting from groundwater contamination (Hubert and Canter, 1980a).

Bacteria

Of the more than 400 types of bacteria present in human and animal feces, *Escherichia coli,* an organism that is not usually pathogenic and is always found in the gastrointestinal tracts of humans and other warm-blooded animals, has been used for most of this century as a monitor of the safety of water for human consumption (NAS, 1977). The potential for human infection by pathogenic bacteria depends upon the bacterial survival rate and the contamination potential, which, in turn, is affected by the type of soil, the rate and duration of bacterial loading, the groundwater flow, and the integrity of the media surrounding the aquifer (Hubert and Canter, 1980a). Survival times may vary from less than one day for *Salmonella typhi, Vibrio cholerae,* and *Entamoeba histolytica* cysts on vegetables to several months for *Salmonellae* in soil and *Shigellae* in water containing humus (Burge and Marsh, 1978). The viability of bacteria in porous media under favorable conditions in the tropics may be five years, although 60 to 100 days might be the

Table 5-5. Summary of Toxic Effects of Known Groundwater Contaminants (Chemicals Listed by Structural Class)

Key: The numbers after each chemical correspond to the numbers that precede each toxic effect listed below.

1. Eye irritation
2. Skin irritation
3. Allergic sensitization
4. Upper respiratory tract irritation
5. Lung/respiratory effects
6. Liver damage
7. Kidney damage
8. Pancreatic damage
9. CNS effects
10. Peripheral nervous system effects
11. Blood cell disorders
12. Immunological effects
13. Cardiovascular effects
14. Gastrointestinal effects
15. Cholinesterase inhibition
16. Methemoglobinemia
17. Skin damage
18. Vision damage
19. Endocrine effects
20. Reproductive effects
21. Embryotoxicity
22. Teratogenicity
23. Mutagenicity
24. Carcinogenicity

I. Organic Chemicals Other than Pesticides	Toxic Effects
A. Aliphatic Hydrocarbons	
Cyclohexane	6, 7, 9
Propane	13
Dicyclopentadiene	1, 2, 4, 5, 7, 9
(Note: many petroleum products are in this class; see III below)	
B. Aromatic Hydrocarbons	
Benzene	1, 2, 3, 9, 10, 11, 12, 14, 24
Naphthalene	1, 2, 4, 11, 18
C. Polynuclear Aromatic Hydrocarbons	
Acenaphthalene	2, 4
Anthracene	3, 23, 24
Chrysene	24
Dibenz[a.h.]anthracene	24
Phenanthracene	3, 23, 24
Pyrene	2, 5, 6, 11, 14, 22, 23
D. Alkyl-Substituted Aromatic Hydrocarbons	
Ethylbenzene	1, 2, 4, 6, 7, 9
n-Propylbenzene	10
Styrene (unsaturated side chain)	1, 2, 4, 6, 7, 9, 14, 21, 23

Table 5-5. Summary of Toxic Effects of Known Groundwater Contaminants (Chemicals Listed by Structural Class) (*continued*)

I. Organic Chemicals Other than Pesticides	Toxic Effects
Toluene	1, 2, 4, 6, 7, 9, 13
1,2,4-Trimethylbenzene	2, 5, 9
Xylenes	1, 4, 6, 7, 9, 13
E. Aliphatic Alcohols	
Benzyl alcohol	2
Methanol	1, 6, 9, 18
F. Phenols	
p-Tert-butylphenol	2
G. Ethers	
Tetrahydrofuran	2, 4, 6, 7, 9
Dioxane	1, 4, 6, 7, 24
Diethyl ether	1, 2, 4, 9
H. Ketones	
Acetone	1, 4, 9
Benzophenone	3
I. Carboxylic Acids	
Acetic acid	1, 2, 4, 5
Formic acid	1, 2, 4, 9
J. Halogenated Aliphatic Hydrocarbons	
Bromochloromethane	5, 6, 7, 9
Bromoform	1, 4, 6, 9
Carbon tetrachloride	6, 7, 9, 20, 24
Chloroform	1, 2, 6, 7, 9, 14, 21, 24
Chloromethane	5, 6, 7, 9
Dibromochloromethane	23
1,1-Dichloroethane	6, 7, 9, 21
1,2-Dichloroethane	1, 2, 5, 6, 7, 9, 13, 14, 19, 20, 23, 24
1,1-Dichloroethylene	1, 2, 4, 5, 6, 7, 9, 13, 19, 21, 23, 24
Dichloromethane	1, 2, 6, 7, 9, 21
1,2-Dichloropropane	6, 7, 23
Fluoroform	23
Hexachlorocyclopentadiene	5, 6, 7, 13, 14, 19
Hexachloroethane	6, 7, 9, 21, 24
1,1,1,2-Tetrachloroethane	6, 21, 23
1,1,2,2,-Tetrachloroethane	4, 6, 7, 9, 14, 21, 22, 23, 24

Table 5-5. Summary of Toxic Effects of Known Groundwater Contaminants (Chemicals Listed by Structural Class) (*continued*)

I. Organic Chemicals Other than Pesticides	Toxic Effects
Tetrachloroethylene	1, 4, 5, 7, 9, 21, 24
1,1,1-Trichloroethane	1, 2, 5, 6, 7, 9, 13, 23, 24
1,1,2-Trichloroethane	1, 2, 4, 5, 6, 7, 9, 23, 24
Trichloroethylene	2, 5, 6, 7, 9, 11, 13, 24
Trichlorofluoromethane	9, 13
Vinyl chloride	2, 6, 7, 9, 11, 23, 24

K. Halogenated Aromatic Hydrocarbons

Bromobenzene	1, 2, 5, 6, 8, 23
Chlorobenzene	1, 4, 5, 6, 7, 9, 20, 23
2-Chloro-naphthalene	2, 6
o-Chlorotoluene	1, 2, 9
o-Dichlorobenzene	1, 2, 4, 5, 6, 7, 9, 11
p-Dichlorobenzene	1, 4, 5, 6, 7, 9, 11
Polybrominated biphenyls	6, 12, 21, 22
Polychlorinated biphenyls	5, 6, 12, 13, 14, 20, 21, 22, 23, 24

L. Aromatic Amines

Aniline	9, 16, 24
Benzidine	24

M. Esters

Bis-(2-ethylhexyl)phthalate	1, 2, 6, 20, 21, 22, 24
Dibutyl phthalate	1, 4, 9
Diethyl phthalate	2, 4

N. Chlorophenols

Pentachlorophenol	1, 2, 4, 6, 9, 21
2,4,6-Trichlorophenol	6, 23, 24

O. Multifunctional and Miscellaneous Structural Types

Alkyl benzene sulfonates	1, 2, 3, 14
Alkyl sulfonates	1, 2, 3
Benzoyl chloride	2, 4
Chlorinated dibenzo-p-dioxins	6, 9, 10, 17, 20, 22, 23, 24
4,4'-Methylene-bis-2-chloroaniline (MOCA)	1, 4, 6, 7, 16, 23, 24
Dichlorobenzidine	1, 4, 5, 14, 24
Dimethyl disulfide	22
4-Dinitrosodiphenylamine	24
Dodecylmercaptan	1, 2, 3, 9, 23
o-Nitroaniline	16, 23
Nitrobenzene	6, 7, 9, 11, 16, 19, 20, 21
4-Nitrophenol	5, 16
N-Nitrosodiphenylamine	6, 24

Table 5-5. Summary of Toxic Effects of Known Groundwater Contaminants (Chemicals Listed by Structural Class) (*continued*)

II. Pesticides	Toxic Effects
A. Chlorinated Hydrocarbons	
Aldrin	2, 6, 7, 9, 20, 24
1,2-Dibromo-3-chloropropane (DBCP)	1, 2, 5, 6, 7, 20, 23, 24
Chlordane	4, 6, 9, 13, 24
DDT	5, 6, 9, 11, 20, 23, 24
DDD	19, 24
Dieldrin	6, 9, 20, 24
Endosulfan	6, 7, 9, 21, 23
Endrin	6, 9
Kepone	6, 7, 20, 21, 24
Methoxychlor	7, 20
Toxaphene	6, 7, 9, 21, 23, 24
B. Carbamates	
Aldicarb	15
Carbofuran	6, 15
C. Organic Phosphates	
Chlorpyrifos	6, 15
Dichlofenthion	15
Parathion	15, 20, 21
Phorate	15
D. Phenoxy Herbicides	
2,4-Dichlorophenoxyacetic acid (2,4-D)	9, 10, 22, 23
2,4,5-Trichlorophenoxyacetic acid (2,4,5-T)	6, 7, 9, 22, 23
Silvex	6, 7, 9
E. Other Herbicides	
Atrazine	2, 5, 9, 11, 21, 23
Bromacil	6, 9
Simazine	9, 23
F. Miscellaneous	
2,4-Dinitrophenol	1, 2, 7, 9, 20

130

Table 5-5. Summary of Toxic Effects of Known Groundwater Contaminants (Chemicals Listed by Structural Class) (*continued*)

III. Petroleum Mixtures	Toxic Effects
Gasolines	2, 4, 9, 14
Fuel oils and diesel	2, 5, 7, 9, 14
Jet fuels	6, 7, 9, 10

IV. Inorganic Chemicals	Toxic Effects
Ammonia	1, 2, 4, 5, 6, 7, 9
Cyanide	1, 2, 3, 4, 9
Fluoride	7, 9
Aluminum	14
Antimony	2, 5, 6, 7, 11, 13
Arsenic	5, 10, 13, 17, 20, 22, 23, 24
Barium	2, 4, 5, 7, 10, 19
Beryllium	1, 2, 3, 4, 5, 6, 24
Cadmium	4, 5, 7, 9, 20, 21, 22, 24
Chromium	2, 3, 4, 5, 6, 21, 22, 23, 24
Cobalt	1, 2, 3, 5, 11, 13, 14
Copper	6, 7, 11
Iron	5, 6, 14
Lead	9, 11, 20
Lithium	7, 9, 13, 14
Magnesium	1, 9
Manganese	5, 6, 9
Mercury	7, 9, 18, 22
Nickel	1, 3, 4, 5, 9, 14, 20, 24
Palladium	7, 11
Selenium	1, 4, 5, 6, 9, 14
Silver	5, 7, 11, 14, 17, 18
Thallium	6, 7, 9, 13, 14, 20, 22
Vanadium	3, 5, 6, 7, 9, 11
Zinc	1, 2, 11, 19, 20

131

Note: The data presented here are limited: 1) not all of the substances listed have been tested for all impacts; 2) documentation is not available for cases in which specific impacts were not observed during studies of specific chemicals; 3) chemicals that dominate the list are the ones that have been most thoroughly studied; 4) the data were obtained from secondary sources.

Source: Environ. Corp., *Approaches to the Assessment of Health Impacts of Groundwater Contaminants.* Prepared for the Office of Technology Assessment. Washington, D.C., 1983.

maximum in temperate regions (Romero, 1972). Bacteria in water flowing through porous media are removed by filtration, die-off, and adsorption (Freeze and Cherry, 1979). In medium-grained sand and finer materials, pathogenic coliform organisms only penetrate a few yards (Krone et al., 1958), whereas in sand or gravel aquifers such transport may be in the order of hundreds of yards (Krone et al., 1957). Transport in fractured rock, where relatively high groundwater velocities occur (upwards of several feet per day), permits transport in the order of miles for a microorganism whose viability may range from days to months (Freeze and Cherry, 1979).

132

EPA-approved testing for bacterial contamination is standardized throughout the United States (Hubert and Canter, 1980a) and consists of three phases: a presumptive test, a confirmed test, and a completed test. These procedures are summarized by NAS (1977).

The information provided by the EPA laboratory in Cincinnati and by the CDC annual summaries shows that between 1945 and 1984 nationwide there were 104 groundwater-related outbreaks of disease of bacterial origin causing 26,763 cases of illness.

Viruses

More than 100 different enteric viruses have been identified in human fecal material (WHO, 1979), and these may survive for several months in wastewater, tapwater, soil, and shellfish. In addition, the normal practice of wastewater chlorination may not be sufficiently effective to inactivate viruses. The diseases caused by enteric viruses range from rashes, diarrhea, and fever to meningitis, paralysis, encephalitis and myocarditis (WHO, 1979) and are summarized in Table 5-6.

The sources of infection include contaminated water and food pathways. The minimum infective dose of ingested viruses depends on the type of virus involved, the route of entry, and the susceptibility of the host. Doses as low as a single infectious unit may cause disease in man. The 1979 WHO report concludes that current epidemiological techniques are not sensitive enough to detect low-level transmission of viral diseases through water for two reasons. Firstly, most enteric viruses cause such a broad spectrum of symptoms that the identification of the single entiological agent causing the illness is often too difficult to identify. Secondly, viral agents causing infections are difficult to recognize as being waterborne. Nearly 60% of the recently reported cases of waterborne disease in the United States were caused by unknown or unrecognized agents (WHO, 1979). Infectious hepatitis can often

Table 5-6. Human Enteric Viruses That May Be Present in Water

Virus Group	No. of Types	Disease Caused
Enteroviruses:		
Poliovirus	3	Paralysis, meningitis, fever
Echovirus	34	Meningitis, respiratory disease, rash, diarrhea, fever
Coxsackievirus A	24	Herpangina, respiratory disease, meningitis, fever
Coxsackievirus B	6	Myocarditis, congenital heart anomalies, rash, fever, meningitis, respiratory disease, pleurodynia
New enteroviruses	4	Meningitis, encephalitis, respiratory disease, acute hemorrhagic conjunctivitis, fever
Hepatitis type A (probably an enterovirus)	1	Infectious hepatitis
Gastroenteritis virus (Norwalk type agents)	2	Epidemic vomiting and diarrhea, fever
Rotavirus (Reoviridae family)	?	Epidemic vomiting and diarrhea, chiefly of children
Reovirus	3	Not clearly established
Adenovirus	>30	Respiratory disease, eye infections
Parvovirus (adeno-associated virus)	3	Associated with respiratory disease in children, but etiology not clearly established

Note: Other viruses which because of their stability, might contaminate water are the following:

(1) SV40-like papoviruses, which appear in the urine. The JC subtype is associated with progressive multifocal leukoencephalopathy.

(2) Creutzfeld–Jakob (CJ) disease virus. Like scrapie virus, the CJ virus resists heat and formaldyhyde; it causes a spongiform encephalopathy, characterized by severe progressive dementia and ataxia.

Source: WHO, 1979.

be attributed to the waterborne category, and 84 outbreaks between 1946 and 1977 were documented in the United States (Hubert and Canter, 1980a). The CDC and EPA reported only 19 cases of infectious hepatitis-A between 1971 and 1982 and an additional 165 cases between 1983 and 1984.

The principal sources of viral contamination of groundwater contain human and animal fecal material. They include sewage effluent disposed of in injection wells, land-waste disposal, animal feedlots and dairies, septic tanks, and cesspools (Hubert and Canter, 1980a).

Land application of wastewater for irrigation and for disposal may also lead to viral contamination of the groundwater. The movement of viruses in the soil depends upon the rate of application, the composition and structure of the soil, its pH, organic content, and strength of the effluent (WHO, 1979). Viruses are readily absorbed onto clays under suitable conditions and are also removed by sandy loams and soils with a high organic content. Fractured rock permits the viruses to travel great distances unimpeded.

Current methods of monitoring viral contamination are summarized in Table 5-7 and require special facilities: thus they are not routinely performed on drinking-water samples (Hubert and Canter 1980a). Sewage-treatment processes are not likely to remove all viruses present. Sedimentation may remove 50%; the activated-sludge process, 60%–99%; trickling filters and stabilization ponds, 80%–95%. Chemical coagulation with lime achieves a 90%–99% reduction (WHO, 1979).

Information from the EPA laboratory in Cincinnati and the CDC's annual reports show that nationwide between 1945 and 1984 there were 64 outbreaks of disease due to groundwater-related viruses. These resulted in 4,365 cases of illness. The majority of the outbreaks were due to the hepatitis virus, which caused 51 outbreaks and affected 1,919 people. The total number includes 9 outbreaks of the Norwalk-type viral agents, affecting 1,476 persons; 3 outbreaks due to parvovirus, affecting 937 people; and 1 outbreak of polio virus affecting 16 persons (CDC, 1982–85; U.S. EPA, 1981a).

Parasites

This category covers protozoa, worms, and fungi. Hubert and Canter (1980a) report no recorded incidents of the outbreak of disease attributable to the contamination of groundwater by a parasitic organism. Outbreaks due to contamination of surface water by *Giardia, Entamoeba histolytica,* and *Ascaris lumbricoides*

Table 5-7. Methods for the Detection of Viruses in Different Types of Water and Sludges

Method	Type of Sample
Direct inoculation without concentration	Wastewater, sludge
Swab sampling	Wastewater samples (method sensitive but not quantitative)
Filter adsorption elution methods	0.2–5.0-liter samples where more
Adsorption and precipitation methods employing polyvalent cation salts	than 1 infectious unit of virus per liter is expected
Use of preformed alum, aluminum hydroxide, iron (III) oxide (50), iron (III) chloride, or lime flocs	
Flocculation by added salts	
Hydroextraction and aqueous polymer two-phase separation techniques	
Soluble alginate	
Precipitation by low pH	
Flat membranes, hollow fibre	Large-volume samples, of 5–400 liters or more. In general, single-stage procedures may be used for samples of up to 20 liters— i.e., for samples containing relatively large amounts of virus such as are likely to occur in sewage, treated sewage effluents, and polluted surface water
Flow-through filter adsorption of acidified samples followed by elution at high pH	
Single or multistage procedures	
Filtration through pleated filters followed by elution	Groundwaters, less polluted surface waters, and highly treated wastewater (multistage procedures)
Organic flocculation followed by elution	
Filtration through adsorptive filters with positive charge followed by elution	
Adsorption to glass powder followed by elution	

Source: WHO, 1979.

were reported, and Hubert and Canter concluded that the absence of outbreaks due to the contamination of groundwater is probably attributable to the size of the organisms involved. The EPA, however, in a printout of the incidence of disease due to groundwater contamination from 1945 to 1980, reported that 9 outbreaks of protozoan parasitic disease were reported between 1953 and 1980. An additional 9 outbreaks were reported by the CDC between 1981 and 1984. These 18 included major outbreaks of illness due to *Giardia* in Arizona, Colorado, Florida, Virginia, Wisconsin, Vermont, and Pennsylvania, and to *Entamoeba histolytica* in Arizona, Indiana, Missouri, Mississippi, Oklahoma, and South Carolina.

136

Giardia lamblia is the most commonly reported protozoan pathogen that causes outbreaks of disease from water supplies (Harris, 1986; Spofford, 1986). Between 1965 and 1982 there were 54 outbreaks caused by *Giardia,* affecting approximately 20,000 people (Spofford, 1986). However, one must remember that in only about one-half of all of the cases reported is the cause of disease ever established; therefore, both of these numbers may be gross underestimates. Of the *Giardia* outbreaks reported, 72% of the cases involved contaminated surface water, 13% involved groundwater, and in 15% of the cases the water source was not reported. Most of the cases were a result of untreated drinking-water supplies or deficiencies in the distribution or treatment processes (Spofford, 1986).

Giardia passes through four stages in its life cycle. (Spofford, 1986). The third stage, the cyst, is the most infective. In this stage the parasite has great resistivity to chlorine and can survive in water for more than two months. The virus can be contracted through contact with an infected person or animal or through consumption of water contaminated with the pathogen. Although boiling destroys the cyst, the most effective treatment appears to be chemical pretreatment (coagulation) followed by filtration and chlorination (Spofford, 1986).

Recently, federal regulation for controlling the incidence has been established under the Safe Drinking Water Act Amendments of 1986. An MCL goal of 0 for *Giardia* was proposed by EPA and is now included in the list of 83 contaminants being considered for MCLs.

An EPA-sponsored conference on the Microbial Health Considerations of Soil Disposal of Domestic Wastewaters was held by the National Center for Ground Water Research (NCGWR) in 1981. The aim of the conference was to identify important health-related research needs associated with the deposit of domestic wastes on land via septic tanks, cesspools, and land-application systems (NCGWR, 1981). The papers included reviews of the current practices of disposal, the transport and fate of various classes of pathogens, outbreaks of waterborne disease, risk evaluations, and possible control measures. Pathogen removal by soils in the land application of wastes can be improved by changing the infiltration rate (Lance, 1981). Increased pollution is often traced to the contaminated recharge water from partially treated wastes that comes in contact with groundwater (Hagedorn, 1981). Sobsey (1981) concludes that the information on the transport and fate of viruses is inadequate, especially on the mechanisms of microbially mediated antiviral activity in soils. Gerba (1981) states that the current body of information concerning viruses in groundwater is not sufficient to develop guidelines for estimating safe distances between waste sources and drinking-water wells and proposes that this gap be filled by further field studies. Cabelli (1981) describes the criticism levelled at the microbial-indicator systems most frequently used to monitor drinking water in the United States, the total-coliforms and the standard-plate count. It is thought that coliform organisms do not survive environmental stress, including disinfection or residence in wastewaters, as well as do some of the viral enteric pathogens. He suggests that alternative indicators, such as enterococci, *Clostridium perfringens* spores, acid-fast bacilli, yeasts, and coliphage, would better reflect the survival characteristics of viral pathogens and noted that two of these are used along with *E. coli* as indicators in Europe. Deficiencies in the indicator system could mean that the risk of disease is understated. An evaluation of the risk to public health of various methods of disposing of wastewater explored by Cooper and Olivieri (1981) uses a probability-matrix technique that requires data from two groups of experts. The first group delineates the health problems associated with a given process and estimates the probability of their occurrence, and the second group judges the severity of the problems delineated by the first group. Boyle (1981) describes the major causes of subsurface-disposal failure leading to contamination and suggests technical and managerial remedies, particularly for septic-tank failures. Despite its aim, the conference does not appear to have produced a list, in order of priority, of the health-related research needs.

137

When groundwater is used for drinking water by municipal suppliers, pretreatment, such as chlorination, minimizes the possibility of outbreaks of disease caused by pathogens. Deficiencies in treatment cause problems. In domestic wells with no pretreatment of the water, prevention of contamination is of paramount importance.

Acute Illness Due to the Chemical Content of Water

Acute illness due to chemical poisoning is reported in the EPA computer printout of the incidence of illness attributable to the ingestion of contaminated groundwater. Hubert and Canter (1980a) listed both the known cases of chemical contamination affecting public health and their accompanying symptoms. In Craun's *Summary of Waterborne Illness Transmitted through Contaminated Groundwater*, 45 (11%) of the 399 outbreaks that occurred between 1971 and 1980 were due to chemical poisonings. These chemicals, natural and man-made, organic and inorganic, included petroleum products (lubricating oil, fuel oil, gasoline), nitrate, polychlorinated biphenyls, trichloroethylene, phenols, benzene, arsenic, and selenium. Only 157 persons were affected by these outbreaks (Craun, 1985).

Harris (1986) reports that many of the acute water poisonings that have occurred in the past have been due to fluoride equipment failure, contaminating drinking-water sources with excess amounts of flouride, and heavy metal poisoning, occurring mainly as a result of corrosion of plumbing materials by acidic waters. Organic chemical poisonings have resulted from many factors, but mainly from accidental spills.

Recent investigations have shown widespread contamination of groundwater by toxic organic chemicals (for a review, see CEQ, 1981; OTA, 1984). In 1984 the National Academy of Sciences (NAS) reported that there is virtually no toxicity information in existence for approximately 70% of the estimated 66,000 chemicals currently in commercial use. The NAS itself has studied approximately 21 organic chemicals now found in drinking water (Conservation Foundation, 1987b).

Estimates for the number of chemicals in existence today vary greatly. Weimar (1980) noted that approximately 1,000 new chemicals enter the environment each year, in addition to an estimated 30,000 that already exist there. Estimates range as high as 75,000 (Henderson, et al., 1984), but not all of these chemicals would find their way into landfills or impoundments, nor would all of them prove harmful to human health.

Many of the organic chemicals recently found in groundwater originate from industrial waste. However, a common inorganic contaminant, nitrate, comes mainly from natural sources, sewage disposal, animal feedlots, and the use of agricultural fertilizer. The Maximum Contaminant Level (MCL) established by EPA for nitrate was set at a level to prevent infantile methemoglobinemia. This disease occurs almost exclusively in infants less than three months old who have ingested high levels of nitrate. Before or after ingestion, nitrate is bacterially converted to nitrite. In the body this nitrite oxidizes the ferrous iron in hemoglobin to ferric iron, converting hemoglobin to methemoglobin. Methemoglobin is incapable of binding molecular oxygen, and thus cyanosis, respiratory distress, or suffocation of the infant may result (Craun, 1986). Death may occur from the ingestion of water containing 50–100 mg/l NO_3 (Hubert and Canter, 1980a). Cases of blue-baby syndrome have not been well documented, and national statistics are not available. Hallberg (1986), however, reports that four acute cases have been reported in western Iowa, eastern Nebraska, and southeastern South Dakota in recent years. Only one incident of death due to high nitrate levels from public water supplies has been reported since 1960, and few cases due to concentrations from individual supplies have been reported in the past 10 years (Craun, 1986). A chronic health hazard associated with the ingestion of nitrites is their interaction with amines or amides within the body to form nitroso compounds, nitrosamines, and nitrosamides. These substances have been shown to produce cancer in laboratory animals (Craun, 1986; Madison and Brunett, 1985).

139

Local deposits of arsenic, selenium, and other naturally occurring inorganic compounds may contaminate groundwater and render it unfit for human consumption. Selenium and other chemicals, which are toxic in high concentrations and even carcinogenic, are also required nutrients.

In general, very little is known about the acute effects caused by the ingestion of the chemicals listed above; however, many of the effects they produce through inhalation and skin absorption have been summarized. These effects, however, may differ somewhat from the effects caused by ingestion. In conclusion, it may be said that relatively few cases of acute effects on public health have been documented in response to chemical contamination. This is in part caused by our lack of knowledge of the effects on human health of many chemicals.

Chronic Effects of Contaminated Groundwater on Public Health

140   Incidents suggestive of the chronic effects on public health of groundwater contamination have been reported for cancer, malformations, miscarriage, central nervous system disorders, and cardiovascular disease, although there have been few carefully controlled epidemiologic investigations that confirm these associations (Harris, 1982). The controversy surrounding the question of whether contamination can have one or a variety of long-term effects may be divided into two parts, namely, the effects associated with the mineral characteristics of the water supply and those that may be associated with its organic chemical contamination. Each will be dealt with in turn.

Chronic Disease Due to the Mineral Content of Groundwater

Early studies of the relation between the mineral content of water and apoplexy in Japan and cardiovascular disease in the United States raised the possibility that a so-called water factor was involved (Neri et al., 1974; Sharrett, 1979; Comstock, 1979). Hubert and Canter (1980a) believe that this controversy will continue even though the American Medical Association has stated that adverse effects on health are probably due to life-style rather than to environmental factors.

Hubert and Canter (1980a) give an extensive summary of all the literature available on the relation between the hardness of water and coronary disease, arteriosclerosis, ischemic heart disease, hypertension, and stroke. In addition, they review the inherent weaknesses of the studies to date.

Hubert and Canter (1980a) summarize the conclusions to be drawn from the available literature as follows:

- The relation between water hardness and cardiovascular disease and mortality is uncertain.
- Cardiovascular disease and mortality is known to be greatly influenced by factors varying from diet to stress, and the

situation is further complicated by the mineral content of potable groundwater.

- Although many studies show a statistical correlation between water hardness and cardiovascular disease, the responsible element or compound in water has not been conclusively identified.

Because of the uncertainty surrounding the water factor, Hubert and Canter (1980a) conclude that it is impossible to design a water treatment for potable groundwater to remove allegedly harmful constituents related to cardiovascular disease.

141

The effects on health of inorganic solutes, organic solutes, and radioactive substances in drinking water have been admirably summarized in the study on drinking water and health carried out by the National Research Council of the National Academy of Sciences (NAS, 1977, 1981a). In a later continuation of this study the epidemiological studies of water hardness and cardiovascular disease were also assessed (NAS, 1980a, 1980b). The panel noted that current knowledge is derived mainly from ecological epidemiological studies, those in which individual exposures and risk factors were not considered. A prior study, *The Geochemistry of Water in Relation to Cardiovascular Disease* (NAS, 1979), came to the following conclusions:

- Studies encompassing large geographical areas generally showed a correlation between hard water and low cardiovascular-disease rates. This correlation breaks down for smaller study areas or if the population studies are grouped by altitude or by proximity to a seacoast. The association of some noncardiovascular illnesses with soft-water areas raises the possibility that soft water may be a concomitant of some more basic risk factor.
- Most studies reported correlation coefficients rather than risk estimates as a function of exposure. Upper estimates of risk ratios for soft compared with hard water, from those studies with sufficient data to be useful in such an assessment, average approximately 1 to 25 for cardiovascular disease and 1 to 2 for arteriosclerotic, stroke, and hypertensive disease.
- Autopsy studies have reported low magnesium levels in the tissue of the heart, diaphragm, and pectoral muscle of persons who die from myocardial infarction as compared with those who die from accidental causes. Similar deficits were observed in persons living in soft-water areas as compared with hard-water areas. These findings are in line with those of

Sharrett (1979), Comstock (1979), and Hubert and Canter (1980a).

It is recognized that while certain trace metals are essential to human nutrition, higher concentrations may be harmful. Trace metals may be introduced into groundwater as a result of the natural local occurrence of specific minerals in the aquifer or the disposal of certain types of industrial waste. Berg and Burbank (1972) point out that present-day occurrences of cancer should be correlated with trace-metal levels of 20 to 30 years ago, but such data are not available. They compared mortalities for 34 types of cancer with the incidence of eight metals in surface water in the United States. The positive correlations of 28 were double the number expected. No significant correlations were found for iron, cobalt, or chromium. The correlations for lead were positive for kidney cancer, all lymphoma and leukemia, and intestinal cancer mortalities, which correspond to the known biological activities of lead. Beryllium was significantly correlated with all cancers, and this, the authors considered, also reflected its biological activity. Arsenic and nickel are thought to be human carcinogens. It should be mentioned that arsenic compounds have not been reliably demonstrated to produce tumors in laboratory animals, but epidemiological studies show that the incidence of certain cancers and precancerous conditions increases in humans who are chronically exposed to arsenic compounds by oral or respiratory routes (NAS, 1981a). The results are shown in Table 5-8. Several milligrams per liter of arsenic have been found naturally in the groundwaters of Indiana, Utah, Alaska, Oregon, and California (Craun, 1986). Cadmium showed the strongest correlations with the most types of cancer, and although dose-response relationships are uncertain, the authors suggest further studies.

In the 1970s asbestos fibers were discovered in drinking water and beverages throughout the country, resulting in an increased concern over the possible health effects associated with ingestion. Asbestos is a group of naturally occurring silicate minerals that have a fibrous morphology and are found in drinking water as a result of both natural and industrial processes. Recent studies have concluded that lung cancer and mesothelioma has occurred in experimental animals after inhalation of asbestos. However, similar effects were not found in these same animals after ingestion of the same amounts of asbestos (Craun, 1986). It has also been reported that epidemiologic studies conducted in the United States and Canada have yielded no definitive conclusions regarding adverse effects of ingested asbestos. Likewise, none of the epidemiologic

142

Table 5-8. Statistically Significant Positive Correlations Between Metal Concentrations and Cancer Death Rates

Metal	Cancer (ICD)	Probability of Positive Association*
Arsenic	Larynx (161)	0.024
	Eye (192)	0.009
	Myeloid leukemia (204.1)	0.042
Beryllium	Breast (women) (170)	0.040
	Uterine cervix (171)	0.016
	Other uterus (172–174)	0.006
	Bone (196)	0.024
	All cancers (140/205)	0.016
Cadmium	Mouth and pharynx (140/149)	0.003
	Esophagus (150)	0.00004
	Large intestine (153/154)	0.00001
	Larynx (161)	0.004
	Lung (162/163)	0.001
	Breast (women) (170)	0.003
	Bladder (181)	0.009
	Myeloma (203)	0.008
	All lymphomas (200/203, 205)	0.016
	All cancers (140/205)	0.0005
Chromium	None	
Cobalt	None	
Iron	None	
Lead	Stomach (151)	0.0026
	Small intestine (152)	0.038
	Large intestine (153/154)	0.009
	Ovary (175)	0.02
	Kidney (180)	0.008
	Myeloma (203)	0.042
	All lymphomas (200/203, 205)	0.0005
	All leukemias (204)	0.006
Nickel	Mouth and pharynx	0.044
	Large bowel	0.031

* Expressed in percentage form 0.024 = 2.4%

Source: J. W. Berg and F. Burbank, Correlations Between Carcinogenic Trace Metals in Water Supplies and Cancer Mortality, *Annals of the New York Academy of Sciences* 199 (1972), p. 252. Reprinted by permission of New York Academy of Sciences.

studies of human exposure through ingestion have been able to define levels that pose a risk. Despite the inconsistent results of these asbestos studies, EPA is currently considering regulating asbestos levels in public drinking waters (Craun, 1986).

Trace metals may cause additional chronic health problems other than cancer. The Environmental Protection Agency reported that the accumulation of lead in bone and tissue may have deleterious effects on the nervous system and kidneys, and may also cause circulation problems (Conservation Foundation, 1987b). In some cases high doses of lead in children have been shown to affect intelligence, behavior, and school performance (Conservation Foundation, 1987b). Effects of mercury poisonings include central nervous system dysfunctions, eye damage, urinary tract damage, embryo injury, physiological disturbances, and even death. High doses of cadmium accumulate in the liver and kidneys and may eventually cause kidney disease in old age. There is some indication that cadmium may cause birth defects as well. Elements such as arsenic, selenium, and nitrates may also create health problems at high levels. Arsenic has been reported to damage human gastrointestinal tracts and cause heart abnormalities. Selenium poisoning may cause damage to the nervous system and the gastrointestinal tract, and high levels have been known to induce dental decay. Nitrate at high levels, as mentioned earlier, may cause methemoglobinemia and cancer. In addition, it has been reported to cause nervous system impairments and birth defects (Conservation Foundation, 1987b).

Some evidence indicates that low levels of certain minerals may have a protective effect against cancer and other pathologies. Pories et al. (1972), stating that iodine is the best known inhibitor of experimental neoplastic growth, also found that such inhibitory properties were shown by arsenic, copper, platinum, selenium, and zinc. Inclusion of arsenic in the diet decreased the incidence of all tumors in animals and also stunted carcass growth. Copper is known to potentiate the biological activity of certain anti-cancer agents. This could be due to the fact that it is a potent free radical inhibitor or that it shields specific sites on liver protein molecules. It is known that there is a high frequency of preexisting goiter, a condition caused by iodine deficiency, in patients with thyroid cancer. Platinum is known to halt cell division, and although the neoplastic inhibitory mechanism is unknown, it may be due to a platinum-induced release of viral genomes initiating an antibody reaction against the tumor. Dietary selenium of 1 ppm reduced the incidence of animal tumors. It is known that cancer death rates are lower in areas with drinking water containing high levels of selenium from natural sources. These seleniferous areas are found in Wyoming, Colorado, and South Dakota. In the past, concerns that selenium may cause cancer have been expressed; however, the NAS (1977, 1980) does not believe that scientific evidence is able to establish that selenium is indeed carcinogenic (Craun, 1986). Zinc deficiency decreases tumor growth in animals, but its role in human cancer is not well understood, and the data are contradictory. Concentrations of fluoride in drinking water between **145** 0.7 and 1.2 mg/l are thought to be beneficial in preventing dental caries. However, with the ingestion of excess concentrations, mottling of teeth and dental fluorosis may result in children, and increased density and calcification of the bone may occur in adults.

Thus, although trace minerals are necessary to health, higher concentrations may be harmful. Excessive concentrations of certain minerals have been implicated in the various chronic forms of cardiovascular disease and cancer, but the studies to date are not conclusive. It is possible that water containing certain heavy metals may also protect against a broad spectrum of human tumors. Although the evidence is fragmentary, it deserves further investigation (Pories et al., 1972).

Chronic Disease Due to Contamination by Toxic Organic Chemicals

Burmaster's comprehensive report. *Contamination of Ground Water by Toxic Organic Chemicals,* was published by the Council on Environmental Quality (CEQ, 1981a). He reviewed the sources of toxic organic contamination, methods of prevention, incidents involving drinking-water contamination, and estimates of the health risks due to toxic organic chemicals. The Safe Drinking Water Committee of the National Academy of Sciences has also reviewed the effects of organic solutes on human health (NAS, 1977) and the toxicity of selected organic contaminants (NAS, 1981a, 1983, 1986). The Environ Corporation conducted an extensive study of the effects of toxic chemical contamination for the Office of Technology Assessment (Environ Corp., 1983).

Many toxic organic chemicals are colorless, tasteless, and odorless in the concentrations at which they occur in drinking water. Often, chemicals found in groundwater are in concentrations orders of magnitude higher than those found in surface water. This phenomenon is a reflection of the lack of mixing and the minimal dispersion that occurs in contaminated plumes of groundwater in comparison with surface water.

Page (1981), using data collected by the New Jersey Department of Environmental Protection, compared groundwater and surface water for patterns and levels of contamination by toxic substances. Samples from 1,000 wells and 600 surface-water supplies in New Jersey were analyzed for 56 toxic substances: 27 light chlorinated hydrocarbons, 20 heavy chlorinated hydrocarbons, and 9 heavy metals. The probabilities of detecting them are shown in Table 5-9. For 64% of the toxics, the highest concentration was found in groundwater, and Page considered this a potentially significant threat to some persons consuming groundwater. Similarly high concentrations of contaminants in groundwater were also found in other states and reported to CEQ (1981a). These are shown in Table 5-10. Page (1981) states that in New Jersey, since groundwater is at least as contaminated as surface water and the patterns of contamination are similar for the two, the control of toxic chemicals in groundwater should receive attention equal to that given their control in surface water.

Greenberg et al. (1981a), also using data collected by the New Jersey Department of Environmental Protection, examined population exposure to toxic substances in New Jersey. The State of New Jersey organized a research program in response to the news that its residents showed the highest white male cancer rate in the United States (Mason and McKay, 1974). Non-white male and female and white female rates were also among the highest in the country. These findings have been criticized for lack of proper controls for critical variables (Demopoulos and Gutman, 1980). Analyses for 45 substances were carried out on 408 groundwater samples, representing a cross section of locations and land and water uses. The results showed that pesticide contamination tends to occur in agricultural and forest areas, and light chlorinated hydrocarbon pollution in urban areas.

The contamination of groundwater associated with hazardous chemicals has come to light only in the last decade, mainly because these chemicals have not been routinely monitored in the past. Through a number of surveys conducted in the past 10 years by the states and the federal government, several organic chemicals have been frequently detected in groundwater-supplied drinking water (OTA, 1984). The OTA and the EPA have both

146

Table 5-9.   Probabilities of Detecting Toxics in New Jersey Groundwater and Surface Water*

	Probability Detectable in Groundwater	Probability Detectable in Surface Water	Significantly Different at 0.05 Level
Fluoroform	0.03	0.08	Yes
Methyl chloride	0.01[a]	0.04	Yes
Vinyl chloride	0.01[a]	0.03	Yes
Methylene chloride	0.23	0.45	Yes
Chloroform	0.64	0.64	no
1,2-Dichloroethane	0.10	0.12	no
1,1,1-Trichloroethane	0.78	0.79	no
Carbon tetrachloride	0.64	0.68	no
1,1,2-Trichloroethylene	0.58	0.56	no
Dichlorobromoethane	0.34	0.43	yes
1,1,2-Trichloroethane	0.07	0.09	no
Dibromochloromethane	0.14	0.18	yes
1,2-Dibromoethane	0.08	0.06	no
1,1,2,2-Tetrachloroethylene	0.43	0.88	yes
Bromoform	0.22	0.33	yes
1,1,2,2-Tetrachloroethane	0.06	0.11	yes
Diiodomethane	0.06	0.02	yes
Total dichlorobenzene	0.03	0.07	yes
m-Dichlorobenzene	0.02	0.04	yes
p-Dichlorobenzene	0.03	0.06	yes
o-Dichlorobenzene	0.03	0.03	no
Aroclor 1242	0.11	0.08	no
Aroclor 1248	0.06	0.14	yes
Aroclor 1254	0.03	0.14	yes
gem-Dichloroethylene	0.44	0.65	yes
Dibromomethane	0.12	0.28	yes
trans-Dichloroethylene	0.51	0.63	yes
Bromodichloroethane	0.18	0.06	no
BHC-alpha	0.16	0.39	yes
Lindane	0.21	0.34	yes
BHC-beta	0.50	0.60	yes
Heptachlor	0.21	0.21	no
Aldrin	0.26	0.24	no
Heptachlor epoxide	0.26	0.40	yes
Chlordane	0.40	0.56	yes
o,p'-DDE	0.19	0.44	yes
Dieldrin	0.17	0.39	yes
Endrin	0.11	0.14	no

Table 5-9. Probabilities of Detecting Toxics in New Jersey Groundwater and Surface Water* (*continued*)

	Probability Detectable in Groundwater	Probability Detectable in Surface Water	Significantly Different at 0.05 Level
o,p'-DDT	0.09	0.18	yes
p,p'-DDD	0.10	0.27	yes
p,p'-DDT	0.08	0.17	yes
Arsenic	1	1	no
Beryllium	1	1	no
Cadmium	1	1	no
Copper	1	1	no
Chromium	1	1	no
Nickel	1	1	no
Lead	1	1	no
Selenium	1	1	no
Zinc	1	1	no

* Probability of detecting difference in groundwater concentration from that of surface water
[a] Probability is less than 0.01

Source: G. W. Page. *Environmental Science and Technology* vol. 15, no. 12, (1981), p. 1425. Reprinted by permission of the American Chemical Society.

Table 5-10. Organics in Drinking-Water Wells

Chemical	Highest Concentration (ppb)	State
Trichloroethane	27,300	PA
Toluene	6,400	NJ
1,1,1-Trichloroethane	5,440	ME
Acetone	3,000	NJ
Methylene chloride	3,000	NJ
Dioxane	2,100	MA
Ethyl benzene	2,000	NJ
Tetrachloroethylene	1,500	NJ

Source: After R. A. Saar, Behavior and Movement of Contaminants. In *The Fundamentals of Ground-Water Contamination*. Presented by Geraghty and Miller, Inc., and American Ecology Services, Inc. Syosset, N.Y., 1985.

reported that serious contamination of groundwater by associated toxic chemicals has occurred in at least 34 states.

Synthetic organic chemicals, including most chlorinated hydrocarbons, can cause health problems ranging from acute effects from high doses, such as nausea, dizziness, tremors, and blindness to skin eruptions and central nervous system impairment at lower concentrations. CEQ (1981a) suggested that ingestion of low concentrations of synthetic organic chemicals over an extended period of years may prove fatal. Studies of occupational medicine, cancer studies using laboratory animals, and epidemiological studies were reviewed by CEQ to determine the possible long-term effects of such exposure.

149

Studies of occupational exposure have shown that some chemicals have adverse effects on human reproductive, central nervous, and other systems, but there is no direct human evidence to indicate whether exposure to such chemicals, at the highest concentrations found in groundwater, would represent a significant threat to human reproduction (CEQ, 1981a). DBCP, vinyl chloride, EDB, benzene, toluene, xylene, selected chlorinated ethanes and phthalate esters, PCBs, and the chlorinated dibenzo-p-dioxins are a few substances suspected of damaging the reproductive system or causing birth defects in humans (OTA, 1984).

Known or suspected carcinogens found in groundwater have been summarized and are presented in Table 5-11 (Environ Corp., 1983). Thirty-two of the listed substances are organic chemicals and five are heavy metals. The Office of Technology Assessment (OTA) claims there is very little doubt that benzene, benzidine, inorganic arsenic, vinyl chloride, chromium, and nickel are human carcinogens. However, the latter two are not likely to be present in groundwater in their carcinogenic forms (hexavalent chromium, nickel and nickel salts) or at carcinogenic levels. Evidence from animal studies has also been reviewed, and the following substances are confirmed animal carcinogens: chlorinated aliphatic hydrocarbons (carbon tetrachloride, chloroform, TCE, PCE, and others) and chlorinated hydrocarbon pesticides (aldrin, chlordane, DDT, dieldrin, heptachlor, toxaphene, and others) (OTA, 1984).

Many epidemiological studies have been conducted on the health effects resulting from ingestion of chlorinated water. The effects of chlorination are likely to be limited to their use in surface waters, however, because groundwater generally has a low humic-acid content, and it is these acids that react with chlorine during the chlorination process to produce trihalogenated methanes.

Table 5-11.   Known or Suspected Carcinogens Identified in Groundwater

Only substances classified as carcinogens by the International Agency for Research on Cancer, the Department of Health and Human Services, and the Environmental Protection Agency are listed. In all cases, the evidence has been subjected to peer review.

I. Organic Chemicals Other Than Pesticides

150

Aromatic Hydrocarbons
  Benzene
Polynuclear Aromatic Hydrocarbons
  Anthracene
  Phenanthracene
Ethers
  Dioxane
Halogenated Aliphatic Hydrocarbons
  Carbon tetrachloride
  Chloroform
  1,2-Dichloroethane
  1,1-Dichloroethane
  Hexachloroethane
  1,1,2,2-Tetrachloroethane
  Tetrachloroethylene
  1,1,1-Trichloroethane
  1,1,2-Trichloroethane
  Trichloroethylene
  Vinyl chloride
Halogenated Aromatic Hydrocarbons
  Polychlorinated biphenyls (PCBs)
Aromatic Amines
  Aniline
  Benzidine
Chlorophenols
  1,4,6-Trichlorophenol
Esters
  bis-(2-ethylhexyl)phthalate
Multifunctional or Miscellaneous Structural Types
  Dichlorobenzidine
  4,4-Methylene-bis-2-chloroaniline (MOCA)
  2,3,7,8-Tetrachlorodibenzo-p-dioxin

II. Pesticides

Chlorinated Hydrocarbons
  Aldrin
  Chlordane
  DDD

Table 5-11.   Known or Suspected Carcinogens Identified in Groundwater (continued)

DDT
1,2-Dibromo-3-chloropropane (DBCP)
Dieldrin
Heptachlor
Kepone
Toxaphene

III. Inorganic Chemicals

Arsenic
Beryllium*
Cadmium
Chromium*
Nickel*

* May be carcinogenic only through inhalation and in forms found in occupational settings.

Source: After Environ Corp., *Approaches to the Assessment of Health Impacts of Groundwater Contaminants.* Prepared for the Office of Technology Assessment. Washington, D.C., 1983.

Furthermore, groundwater from domestic wells, as opposed to municipal wells, usually does not undergo chlorination prior to use (G. W. Comstock, Johns Hopkins University, personal communication). The ecological epidemiological studies of surface-water ingestion suggest an association between either chlorination or the presence of trihalogenated methanes and cancer mortality in several areas of the United States. In five case studies—in New York, Illinois, Wisconsin, Louisiana, and North Carolina—the risk of rectal cancer was found to be higher for populations supplied by chlorinated water than for those supplied by unchlorinated water. Colon cancer risks were statistically significantly raised in three of the studies (New York, Wisconsin, and North Carolina), and bladder cancer risks in two (New York and North Carolina). No clear trend was evident to determine whether cancer risks increased with increasing exposure to organic contaminants (CEQ, 1981a). The associations of certain types of cancer with the ingestion of chlorinated water are thought by many investigators to be non-specific and probably associated with something else beside water exposures (G. W. Comstock, Johns Hopkins University, personal communication).

A new area of study in which very little is yet known is that of the synergistic effects of contaminated groundwater. Synergistic effects in groundwater would involve interactions between chemicals and biological processes. OTA considers synergism as potentially one of the most important health issues concerning groundwater thus far. In addition, very little is known about the metabolism of many chemicals. It is known that some chemicals, such as dichloroethylene and dichloroethane, pass through reactive intermediate stages in which they have a potential for damaging living tissue. The actual damage depends on the particular chemical, the rate of metabolism, and the ability of the organism to tolerate the presence of reactive intermediates.

152

Health Effects Due to Radionuclides in Groundwater

The effects on public health of high levels of radionuclides have been studied thoroughly for some time (OTA, 1984). Effects from low levels, however, are often more difficult to identify. Trace amounts of radioactive substances and their degradation products occur naturally in most groundwater; concentrations of these substances vary from area to area depending on the materials through which the groundwater has flowed. Man-made sources of radionuclides also contribute to the potential of groundwater contamination. Such sources include nuclear waste disposal sites, nuclear fallout, nuclear testing, nuclear accidents, waste tailings and piles, and mine drainage (OTA, 1984). Radium-226 is the most significant radionuclide found in groundwater; its presence is most often due to natural geologic conditions (Craun, 1984).

According to the National Academy of Sciences (1977) the risk of cancer from total average background radiation is estimated to be 4.5–45 fatalities per million persons per year (based on certain assumptions). Less than 1% of this total can be attributed to drinking contaminated water (OTA, 1984). Under average conditions, no measurable developmental or teratogenetic effects due to radiation are expected to occur from drinking groundwater containing radionuclides. However, in some areas in the Midwest—particularly Iowa, Illinois, Wisconsin, and Missouri—approximately 1 million people have been exposed to significant levels of radiation through groundwater (Craun, 1984). In Illinois and Iowa alone, approximately 500,000 people have water-well supplies containing radium-226 levels of 3–6 pCi/l, some 300,000 with levels of 6–9 pCi/l, and some 120,000 with levels of 9–80 pCi/l. Continuous consumption of water containing radium-226 or radium-228 at 5 pCi/l may cause between 0.3 and 3 incidents of cancers per year per million exposed persons (Craun, 1984).

The lack of comprehensive nationwide surveys of the extent and severity of groundwater contamination and the paucity of groundwater contaminants that have actually been tested for carcinogenicity make it impossible to assess the national risk of drinking groundwater (CEQ, 1981a; Hubert and Canter, 1980a). We do know, however, that in their pure form, at least some of the chemical contaminants are carcinogenic, that at high concentrations and in pure form some cause reproductive problems or affect the central nervous system or the cardiovascular system, and that in high concentrations some compounds cause poisoning and other problems such as skin rashes and diarrhea. We also know that pathogens probably cause more outbreaks of disease than are actually reported (Hubert and Canter, 1980a). Thus, adverse effects on human health have resulted from exposure to contaminated groundwater, but more research is required on the actual incidence and causal agents of disease, and a more rigorous inventorying of outbreaks of water-related disease would be very useful. Long-term monitoring of well water and the health of the population dependent on it might elucidate the cancer, cardiovascular, and other health risks and should be undertaken in the near future.

153

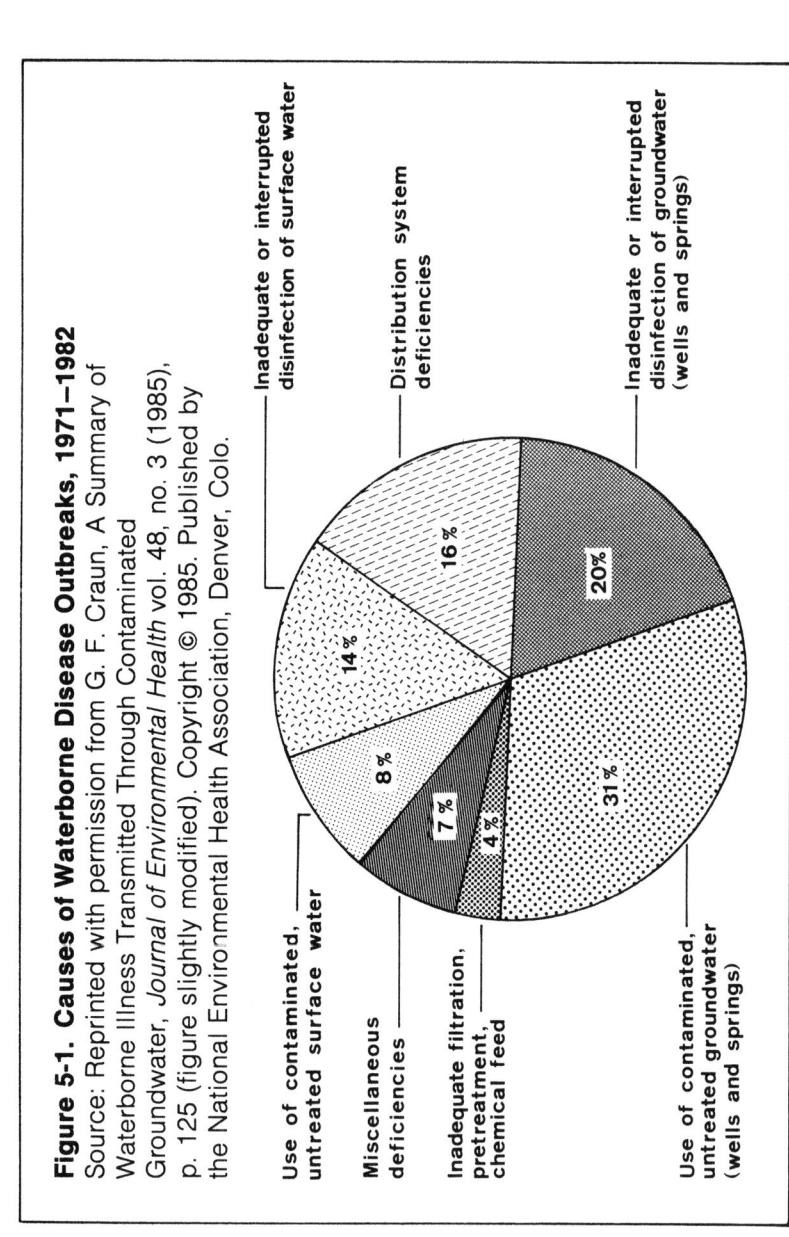

Figure 5-1. Causes of Waterborne Disease Outbreaks, 1971–1982

Source: Reprinted with permission from G. F. Craun, A Summary of Waterborne Illness Transmitted Through Contaminated Groundwater, *Journal of Environmental Health* vol. 48, no. 3 (1985), p. 125 (figure slightly modified). Copyright © 1985. Published by the National Environmental Health Association, Denver, Colo.

Inadequate or interrupted disinfection of surface water

Distribution system deficiencies

Inadequate or interrupted disinfection of groundwater (wells and springs)

Use of contaminated, untreated surface water

Miscellaneous deficiencies

Inadequate filtration, pretreatment, chemical feed

Use of contaminated, untreated groundwater (wells and springs)

16%

20%

14%

8%

7%

4%

31%

Six

The Geographical Extent of Groundwater Contamination

- Contamination in each region or state reflects the hydrogeology, climatic conditions, population density, and degree of industrial and/or agricultural activity.
- The U.S. EPA commissioned five regional assessment reports summarizing geology, major aquifers, natural groundwater quality, and major pollution problems for the southwest, south-central, southeast, northeast, and northwest regions of the United States.
- Anecdotal data of case histories of contamination for nineteen states in different regions of the country have been compiled for this report: Arizona, California, Connecticut, Florida, Idaho, Illinois, Massachusetts, Montana, Nebraska, New Jersey, New Mexico, North Dakota, Oregon, Pennsylvania, Rhode Island, South Carolina, Texas, Vermont, and Washington.
- Major types of contamination across the country revealed by these studies are organisms, organic chemicals, chlorides, nitrates, heavy metals, and hydrocarbons.
- New studies might well reveal new sources of contamination.

The following chapter includes assessments of groundwater contamination in the United States compiled by the U.S. Environmental Protection Agency, as well as state data collected for this report. The EPA compiled state groundwater information for its 1985 report *State Ground-Water Program Summaries*. These summaries cover groundwater uses, sources of groundwater contamination, and basic elements of groundwater programs for each state. For our purposes, the report lists and ranks the major groundwater pollution problems for each of the states; the findings are shown in Tables 6-1 and 6-2. It should be noted that some states did not provide EPA with needed information. The charts, then, should not be viewed as a comprehensive summary of all state groundwater problems, but as a general overview of significant problems that have been reported in the past. In addition to EPA's assessment, State Water Quality Inventory Reports were reviewed for this report. State Water Quality Inventory Reports are biennial reports prepared pursuant to section 305(b) of the federal Clean Water Act (PL 95-217). The major sources of contamination and overall groundwater quality as listed in each report are presented in Table 6-3. In instances where 305(b) reports were not available, each state provided us with the needed information through telephone or written communication.

156

The listings of contaminants presented in Tables 6-1, 6-2, and 6-3 enable one to judge the relative severity of the various sources in the states listed. This data reflects the hydrogeology, climatic conditions, population density, and degree of industrial or agricultural activity. Natural pollution is considered an important source of contamination in the arid south-central and southwestern states. The extensive agricultural irrigation that has transformed parts of the Southwest into a major crop-producing area causes problems from irrigation return-flow. The slow rate of recharge of many of the confined and unconfined aquifers in this area, coupled with extensive withdrawals, has led to an overdraft of groundwater and to concomitant changes in its quality. The south-central region, being the major area of petroleum production, has a major problem with oil-field brines and associated petroleum-production problems. Road salts are a major source of non-point groundwater pollution in many northern states. Saltwater intrusion has contaminated many wells in coastal states. The main cause for concern in the Northeast, Northwest, and Southeast is considered to be industrial development. However, with the growing concern with the occurrence of toxic organic chemicals in groundwater, the rest of the country is becoming aware of the contamination of their own groundwaters by these substances. Contamination from underground storage tanks is a major concern across the country, particularly in densely populated areas. Septic-tank effluent seems to be a nationwide problem, as well. In general, areas with high

Table 6-1. Sources of Groundwater Pollution in the United States, Compiled by the U.S. EPA*

States	Septic Tanks	Municipal Landfills	On-site[a] Industrial Landfills	Other Landfills	Surface[b] Impoundments	Oil and Gas Brine Pits	Underground Storage Tanks	Injection Wells	Abandoned Hazardous Waste Sites	Regulated Hazardous Waste Sites
Alabama		x–3	x–4	x	x–2	x	x–1	x	x	x
Alaska	x–1	x–3					x–2			
Arizona		x								
Arkansas				x		x				
California		x	x	x	x		x		x	x
Colorado	x–2	x–3	x	x	x–3	x	x		x–4	
Connecticut		x–2	x–1				x–4			
Delaware		x–4	x–1	x–1	x–3		x–2			
Florida	x–3			x	x		x–1	x–2	x–1	
Georgia	x–4				x–3			x		
Hawaii	x–3							x–2		
Idaho	x–1				x–2		x–3	x		
Illinois	x		x				x	x		
Indiana			x				x	x	x	x
Iowa	x–7	x–3			x–2		x–4	x–6	x–5	
Kansas						x				x
Kentucky		x	x	x	x			x	x	x
Louisiana				x	x	x		x		x
Maine	x	x–4	x		x		x–1		x–2	
Maryland	x–3	x–1	x–1	x–1	x–1	x–2	x–2		x–1	x–2
Massachusetts	x–4	x–1	x		x–3		x		x–2	
Michigan		x–4	x–3	x	x–2	x	x–1		x	
Minnesota	x	x–1	x–1	x–2	x–2		x	x–4	x–1	
Mississippi	x		x		x–4	x–1	x–2	x–3		x–3
Missouri	x	x		x	x			x		
Montana	x					x	x			x

Table 6-1. Sources of Groundwater Pollution in the United States, Compiled by the U.S. EPA* (cont.)

States	Septic Tanks	Municipal Landfills	On-site[a] Industrial Landfills	Other Landfills	Surface[b] Impoundments	Oil and Gas Brine Pits	Underground Storage Tanks	Injection Wells	Abandoned Hazardous Waste Sites	Regulated Hazardous Waste Sites
Nebraska	x	x			x	x	x			
Nevada	x-1				x-2		x-3		x	
New Hampshire	x	x-2	x-1		x		x-3			
New Jersey	x-4	x	x-3	x	x-2	x	x-1	x		x
New Mexico	x				x	x	x	x	x	
New York	x	x	x	x			x-1		x-4	
North Carolina		x-3	x-3		x-2		x-1			
North Dakota	x	x	x	x-3	x-1	x		x-4	x	
Ohio	x-1		x-4	x-4	x-3	x-2	x-2			x
Oklahoma							x			
Oregon	x-2	x-3	x-1				x	x	x	x
Pennsylvania	x-3	x-2	x-2		x-2	x-2		x-3	x	
Puerto Rico	x	x		x-1			x			x
Rhode Island	x	x-3	x-4		x	x	x-2		x-1	
South Carolina		x-3	x	x	x		x-2		x	x-1
South Dakota	x-3	x	x-4	x	x	x-2	x			
Tennessee	x		x-3		x-4	x	x	x	x-1	x-2
Texas	x	x	x	x	x		x		x	
Utah	x			x		x	x			
Vermont	x	x-4	x		x		x-2		x	
Virgin Islands	x									
Virginia	x-1				x		x			
Washington	x-5	x-1			x-3		x	x-7	x-2	
West Virginia			x		x	x				
Wisconsin	x						x		x	
Wyoming	x					x	x			

Table 6-1. Sources of Groundwater Pollution in the United States, Compiled by the U.S. EPA*
(continued)

States	Saltwater Intrusion	Agricultural	Road Salting	Land Application of Waste	Spills & Leaks	Mining	Improper Well Construction/Plugged Wells	Effluent from Wastewater Treatment	Other
Alabama	x	x							
Alaska		x-4							
Arizona	x								
Arkansas	x				x				
California	x	x-5							
Colorado		x-2				x-6			A-1
Connecticut		x-1	x						
Delaware	x-1	x-1							
Florida	x-2	x-1							
Georgia	x-1	x-2							
Hawaii		x-1							
Idaho		x		x-4					
Illinois		x			x		x		B
Indiana					x				C
Iowa		x-1		x-8					
Kansas			x-3						
Kentucky			x-3						
Louisiana		x	x				x		D
Maine	x	x	x-3						
Maryland	x-2	x-3	x-3	x-3					
Massachusetts		x	x						
Michigan		x	x	x	x				
Minnesota		x-2	x					x	
Mississippi	x								
Missouri		x			x	x	x		

Table 6-1. Sources of Groundwater Pollution in the United States, Compiled by the U.S. EPA* (cont.)

States	Septic Tanks	Municipal Landfills	On-site[a] Industrial Landfills	Other Landfills	Surface[b] Impoundments	Oil and Gas Brine Pits	Underground Storage Tanks	Injection Wells	Abandoned Hazardous Waste Sites	Regulated Hazardous Waste Sites
Montana	x					x	x			
Nebraska			x							
Nevada							x			
New Hampshire				x–4						
New Jersey	x		x	x						
New Mexico			x				x			E
New York			x			x–2				
North Carolina	x					x–4				
North Dakota										
Ohio				x						C
Oklahoma	x									
Oregon			x							
Pennsylvania							x–2		x–3	
Puerto Rico	x		x							
Rhode Island			x	x						
South Carolina	x		x		x–4					
South Dakota			x					x		F–1
Tennessee			x				x			
Texas	x		x				x	x		
Utah										
Vermont		x–3	x–3	x–1						
Virgin Islands	x									
Virginia										
Washington	x–6	x								G–4
West Virginia										

		C
Wisconsin	x	x
Wyoming		x

* x indicates the presence of a source; numbers indicate severity with 1 being most severe.
a Excluding surface pits, lagoons, surface impoundments
b Excluding oil and gas brine pits

Other
A Colorado Rocky Mountain Arsenal
B Inter-aquifer exchange
C Disposal of waste into abandoned wells
D Contaminated water piping
E Milling
F Petroleum products storage
G Disposal sites (more specific information needed)

Source: U.S. EPA, 1985 c and e.

161

Table 6-2. Major Types of Groundwater Contamination, by State

States	Organic Chemicals		Inorganic Chemicals						Radioactive Materials	Pesticides	Other
	Volatile	Synthetic	Nitrates	Fluorides	Arsenic	Brine/ Salt	Other	Metals			
Alabama	X	X	X	X[a]		X		X	X	X	
Alaska	X				X[a]						1
Arizona		X	X[a]	X[a]						X	
Arkansas		X						X	X[a]		3
California	X	X	X[b,a]	X	X	X		X		X	
Colorado		X	X	X[a]	X	X		X			
Connecticut	X	X					X			X	
Delaware			X								
Florida	X	X	X			X		X		X	2
Georgia	X	X				X				X	
Hawaii		X								X	
Idaho	X		X	X	X		X	X			1
Illinois	X		X	X	X		X[c,d]	X	X[a]		
Indiana		X	X			X				X	
Iowa		X	X[b]	X[a]					X[a]	X	
Kansas						X					
Kentucky	X	X				X		X	X		
Louisiana		X		X[a]		X	X[d]				3
Maine	X		X				X	X		X	
Maryland	X		X								
Massachusetts	X	X	X					X		X	
Michigan	X	X	X	X	X	X	X	X		X	2
Minnesota	X	X	X		X			X		X	
Mississippi	X	X				X		X			
Missouri		X									
Montana	X		X[b]		X	X	X[c]	X		X	4
Nebraska	X		X[b]			X		X		X[b]	2

Geographical Extent of Groundwater Contamination

State											
Nevada		x	x[a]	x[a]	x[a]	x[a]		x	x	x	
New Hampshire											
New Jersey	x	x	x	x	x	x	x	x	x	x	1
New Mexico	x	x	x[c,a]			x	x	x		x	2
New York	x	x	x			x	x			x	
North Carolina	x				x	x				x	1
North Dakota	x	x				x		x		x	1
Ohio	x	x				x		x			
Oklahoma	x		x			x		x			
Oregon	x	x	x	x	x	x		x			1
Pennsylvania			x								
Puerto Rico			x								
Rhode Island	x		x			x		x	x	x	
South Carolina	x	x	x	x		x		x	x	x	
South Dakota	x		x	x		x		x	x	x	2
Tennessee	x	x	x		x	x		x			
Texas	x	x				x		x	x	x	
Utah	x	x			x	x		x			
Vermont	x	x	x		x	x		x	x	x	1
Virgin Islands	x		x		x	x		x		x	
Virginia								x			
Washington	x	x	x	x	x	x	x[d]	x	x	x	1, 2
West Virginia	x	x								x	
Wisconsin	x		x			x			x	x	1
Wyoming						x		x			
Total	**33**	**30**	**34**	**16**	**15**	**28**	**10**	**28**	**13**	**25**	**18**

a Natural mineral deposits.
b Agricultural activities.
c Sulfates.
† Chlorides (other than salt water).
Key: 1 = bacteria; 2 = petroleum products; 3 = sodium; 4 = acids.

Source: U.S. EPA, 1985c

163

Table 6-3. Sources of Groundwater Contamination in the United States—Collected from Water Quality-305(b) Reports

	Overall Quality	Manufacturing- & Industrial- Waste Disposal*	Manufacturing & Industrial Landfills**	Underground Storage Tanks	Injection Wells	Oil & Gas Products	Hazardous- Waste Transport & Storage	Toxic- Waste Sites
				Point Source Contamination				
Northeast								
Connecticut	G/M	X	X	X		X	X	
Delaware	G		X	X			X	
Maine	G	X	X	X		X		X
Maryland	G			X				
Massachusetts	M	X	X	X			X	
New Hampshire	E	X	X	X				X
New Jersey	G	X	X	X				X
New York	M/G	X	X	X				
Pennsylvania	M/G			X		X		
Rhode Island	G/E	X	X	X			X	X
Vermont	G			X			X	
West Virginia	M					X		
Southeast								
Alabama	G	X	X	X				X
Florida	G			X				
Georgia								
Kentucky	G/M							

State	Code							
Mississippi	G			X		X	X	X
North Carolina	G	X				X	X	X
South Carolina	E	X	X		X	X	X	X
Tennessee	G		X	X		X		
Virginia	G							
Southwest								
Arizona	M/G						X	
California	G							X
Nevada	M				X			
Utah	M							
South-Central								
Arkansas	G	X	X			X	X	
Louisiana				X		X	X	X
New Mexico				X		X		
Oklahoma		X		X		X	X	X
Texas	G/P		X	X		X	X	
Northwest								
Colorado	E	X						
Idaho	G					X	X	
Montana	G		X			X		X
Oregon		X				X		X
Washington		X				X	X	X
Wyoming		X				X	X	X

Table 6-3. Sources of Groundwater Contamination in the United States—Collected from Water Quality-305(b) Reports (continued)

		Point Source Contamination						
	Overall Quality	Manufacturing-& Industrial-Waste Disposal*	Manufacturing & Industrial Landfills**	Underground Storage Tanks	Injection Wells	Oil & Gas Products	Hazardous-Waste Transport & Storage	Toxic-Waste Sites
North-Central								
Iowa	G	X	X	X			X	
Kansas	VG	X	X			X		
Missouri	G	X		X				X
Nebraska	G	X	X	X			X	
North Dakota	G							
South Dakota	G				X			
Great Lakes								
Illinois	G	X	X	X		X		
Indiana	M		X	X		X	X	
Michigan	E			X				X
Minnesota	G	X	X	X			X	
Ohio			X	X				
Wisconsin	G	X	X	X				
Non-Contiguous United States								
Alaska	E			X		X		
Hawaii	G				X			

Geographical Extent of Groundwater Contamination

	Quantity			Non-Point Source Contamination						
	Land Subsidence	Over-draft	Saltwater Intrusion	Agri-culture	Septic Tanks	Mining	Road Salts	Animal Wastes	Urban Runoff	Natural Elements
Northeast										
Connecticut				X			X			
Delaware			X	X						X
Maine				X	X		X			
Maryland			X		X					
Massachusetts			X	X	X		X			
New Hampshire				X						X
New Jersey			X	X	X					X
New York			X	X	X				X	
Pennsylvania				X	X	X	X			X
Rhode Island				X			X			
Vermont							X			
West Virginia			X		X	X		X		X
Southeast										
Alabama				X		X				
Florida			X	X	X					
Georgia			X							
Kentucky										X
Mississippi		X		X						
North Carolina										
South Carolina				X						
Tennessee										
Virginia			X							

Table 6-3. Sources of Groundwater Contamination in the United States—Collected from Water Quality-305(b) Reports (continued)

	Quantity			Non-Point Source Contamination						
	Land Subsidence	Over-draft	Saltwater Intrusion	Agri-culture	Septic Tanks	Mining	Road Salts	Animal Wastes	Urban Runoff	Natural Elements
Southwest										
Arizona				X						
California			X	X		X		X	X	
Nevada				X	X	X		X		
Utah				X			X	X		X
South-Central										
Arkansas		X	X							
Louisiana								X		X
New Mexico						X				
Oklahoma		X		X						
Texas	X	X	X	X	X			X		
Northwest										
Colorado				X	X	X				
Idaho	X			X	X	X				
Montana				X		X		X		X
Oregon					X					X
Washington				X						X
Wyoming				X		X				
North-Central										
Iowa				X	X					
Kansas				X						

Missouri			X					X	
Nebraska	X	X	X					X	
North Dakota		X	X			X		X	
South Dakota									
Great Lakes									
Illinois		X	X	X	X		X	X	
Indiana				X	X			X	
Michigan									
Minnesota		X	X						
Ohio		X	X	X					
Wisconsin		X						X	
Non-Contiguous United States									
Alaska									
Hawaii	X								

Legend
E - Excellent
VG - Very good
G - Good
M - Moderate
P - Poor

* Manufacturing- & Industrial-Waste Disposal—Lagoons, Pits, Surface Impoundments
** Manufacturing & Industrial Landfills—Solid Waste

Sources: 1) State 1984 & 1986 305(b)—Water Quality Reports. 2) Personal Communication: John Robertson, Arizona Department of Health Services; David Knuth, Alaska Department of Environmental Conservation; Greg Brand, Colorado Health Dept.; Bill McClemore, Georgia Geologic Survey; Connie Cousins, Iowa Dept. of Water, Air & Waste Management; Vic Robbins, Kansas Bureau of Water Protection; Dan Gross, Nevada Division Groundwater Protection; New Mexico Groundwater & Hazardous Waste Bureau; Tom Allen, Ohio Environmental Protection Agency; Oklahoma Groundwater Division of Water Resources Board; Tony Barrett, Washington Department of Ecology; Tom Williams, Wyoming Department of Environmental Quality.

septic-tank densities have a greater potential for polluted groundwater.

State Summaries Compiled From 1981–1986 For This Report

170 Surveys mandated by the federal and state governments have produced an abundance of information concerning the number of potential and actual sites of groundwater contamination. In addition, many individual states have undertaken inventories of case histories of their groundwater contamination. No systematic national sampling survey has yet been carried out. The data presented here are from known incidents of contamination. They give an indication of the most commonly reported groundwater-contamination problems in the states studied. They should not be generalized on a nationwide basis. For an accurate nationwide assessment, a systematic survey is indispensable. For the purpose of this report, each state was asked to provide such information if it was available. The information received varied in its usefulness to this study. Some states have completed their inventories, while others have only just started or are still in the process of documentation. One or two states report that, for them, groundwater contamination is not a major problem. We have taken a sample of the states to investigate further whether the additional information available over the last few years could cast more light on the severity and patterns of occurrence of groundwater problems. The states discussed in this report were chosen mainly because they had the most extensive case histories of groundwater contamination, but they also serve as examples of differing levels of industrialization, agricultural activity, population density, dependence on groundwater, and climatic conditions. These characteristics are summarized in Tables 6-4 and 6-5. The states we have chosen to look at in more detail are the following: Arizona, California, Connecticut, Florida, Idaho, Illinois, Nebraska, New Jersey, New Mexico, and South Carolina. Additional states examined for this revised edition include Massachusetts, Montana, North Dakota, Oregon, Pennsylvania, Rhode Island, Texas, Vermont, and Washington. For each of these, summaries of the groundwater usage, the major aquifers and their characteristics, the natural groundwater quality, and major sources of contamination are given.

Table 6-4. Economic Activities That Could Contribute to Groundwater Contamination

Arizona	*manufacturing*—electrical, communications, aeronautical equipment. *mining*—produces over half the nation's copper.
California	*manufacturing*—transportation equipment, machinery, electronic equipment. *principal natural resources*—petroleum, cement, natural gas.
Connecticut	*manufacturing*—weapons, sewing machines, jet engines, helicopters, motors, hardware, tools, cutlery, clocks, ballbearings, submarines. *agriculture*—grows, per acre, the nation's most valuable tobacco crop.
Florida	*agriculture*—oranges and grapefruits; also, sugarcane, tomatoes, beans, celery, potatoes, field corn, watermelons, limes, mangos. *natural resources*—produces 80% of nation's phosphate.
Idaho	*mining*—produces more than ⅓ of all silver mined in U.S.; also, antimony, lead, garnet, cobalt, phosphate rock, vanadium, zinc, mercury. *agriculture*—produces ¼ of nation's potatoes.
Illinois	*agriculture*—nation's biggest exporter of agricultural products. *livestock*—second largest producer of hogs. *manufacturing*—iron and steel production in Chicago area.
Maine	*agriculture*—produces 9.4% of the national potato crop and 95% of nation's low-bush blueberry crop. *silvaculture*—90% of land is forested. *manufacturing*—fifth largest producer of boots and shoes.
Massachusetts	*manufacturing*—electronics and communications equipment industries. *agriculture*—nation's largest cranberry crop.
Montana	*mining*—once supplied ½ of U.S. copper. *agriculture*—wheat, barley, rye, oats, flaxseed, sugar beets, potatoes; sheep and cattle raising are important to the state economy as well.
Nebraska	*agriculture*—leading grain producer; bumper crops of rye, corn, wheat; Omaha—nation's largest meatpacking center; second largest cattle market in world.
New Jersey	*industry*—chemicals; also, pharmaceuticals, instruments, machinery, electrical goods, apparel; one of world's foremost research centers. *agriculture*—garden vegetables, poultry farming, dairying.

171

Table 6-4. Economic Activities That Could Contribute to Groundwater Contamination (*continued*)

New Mexico	*mining*—leader in output of uranium, potassium salts; also, petroleum, natural gas, copper, gold, silver, zinc, lead, molybdenum. *industry*—leader in energy research and development—nuclear, geothermal.
New York	*manufacturing*—second largest manufacturing center in country—apparel, printing, publishing. *agriculture*—leading wine producer; also, dairying, poultry.
North Carolina	*manufacturing*—largest producer of furniture, brick, textiles; also, metalworking, chemicals, paper. *mining*—leading producer of mica and lithium. *agriculture*—largest producer of tobacco.
North Dakota	*agriculture*—farms cover more than 90% of the state; principal crops are barley, rye, oats, flaxseed, sugar beets and hay. *manufacturing*—food processing and farm equipment. *natural resources*—natural gas, lignite, salt, clay, sand, gravel.
Oregon	*manufacturing*—5 billion dollar wood processing industry. *mining*—source of all nickel produced in the U.S. *agriculture*—leading state in the growing of peppermint, winter pears, plums, blackberries, boysenberries, filberts, Blue Lake beans, coverseed crops.
Pennsylvania	*manufacturing*—produces 23% of all American pig iron steel, ranks first among states in steel wire and structural metal production. *mining*—produces almost all the nation's anthracite coal.
Rhode Island	*manufacturing*—largest manufacturer of jewelry and silverware; other industries include metal processing and metal products, machinery, rubber and plastics, food processing, chemicals, transportation equipment, and electronics. *agriculture*—dairy, poultry, and truck farming.
South Carolina	*manufacturing*—asbestos, wood, pulp, steel products, chemicals, machinery, apparel. *agriculture*—second largest grower of peaches; fourth in tobacco.
Texas	*natural resources*—largest mineral producer due to the abundance of sulfur, salt, helium, asphalt, graphite, bromine, natural gas, clay. *manufacturing*—chemicals, oil refining, food processing, machinery, transportation equipment.

Table 6-4.   Economic Activities That Could Contribute to Groundwater Contamination (*continued*)

	agriculture—leads nation in cattle, sheep, cotton production.
Vermont	*natural resources*—leads nation in production of monument granite, marble, maple syrup, asbestos, talc. *agriculture*—more dairy cows per capita than any other state. *manufacturing*—machine tools, computer components, stone and clay products, lumber, furniture, paper.
Washington	*silvaculture*—leading lumber producer. *agriculture*—leading producer of apples, blueberries, hops, red raspberries. Commercial *fishing* makes a significant contribution to economy. *manufacturing*—aircraft and missile industry, food processing, metals and metal products, chemicals, machinery.
Wyoming	*mining*—most important industry; oil and natural gas; leads nation in production of sodium carbonate and bentonite; second in production of uranium. *agriculture*—wheat, oats, sugar beets, corn, potatoes, barley, alfalfa; second largest producer of wool.

173

Source: Dolmatch, 1982.

The format used for the charts reporting contamination is a modification of that used by Lindorff and Cartwright (1977). An explanation of the categories follows. Note that not all categories appear in every table.

Contamination Incidents. The contaminants are divided by source rather than constituents as the latter are not always specified. *Industrial* and manufacturing products and wastes may be liquid or solid. Wood processing plants were included in this category. Where the wastes were specified as petroleum or its derivatives the case histories were included under that category. *Landfill* leachate is derived from solid, semisolid, or liquid waste of either municipal or industrial origin. *Petroleum* products include, among many others, home heating oil and aircraft fuel. *Chlorides* usually are designated as originating from highway deicing salts, agricultural return flow, oil-field brines, or saltwater intrusion. *Organic* wastes are those derived from plant, animal, or human wastes and include those from feedlots, dairy barns, fruit and

Table 6-5. General Land-Use Profiles for States Appearing in the "State Summaries" and "Aquifer Classification" Sections

State	Arizona	California	Connecticut	Florida	Idaho	Illinois	Maine
Groundwater as % of total water use, 1980	58	39	11	52	35	5	9
Population, 1980 (millions)	2.718	23.667	3.107	9.746	0.943	11.426	1.124
% Change in population from 1970	53.1	18.5	2.5	43.5	32.4	2.8	13.2
% of population that is rural, 1980*	16.2	8.7	21.2	15.7	46.0	16.7	52.5
Total area (millions of acres)	72.96	101.57	3.21	37.54	53.48	36.06	21.29
Total land area (millions of acres)	72.65	100.03	3.11	34.66	52.74	35.61	20.48
Rank in size	6	3	48	22	13	24	39
% of total acreage used for cropland, 1982	2.0	11.3	7.2	11.8	12.3	69.5	3.0
% of cropland that is harvested farmland, 1982	70.8	77.9	76.1	64.6	75.4	93	74.8
% of cropland that is pastureland, 1982	8.7	11.9	19.0	26.2	11.8	4.3	14.2
% of cropland that is used for cover crops, 1982**	0.6	0.6	1.6	1.2	1.0	0.8	3.4
# of manufacturing establishments, 1982	3,407	47,625	6,693	13,723	1,404	18,618	2,009
Value added by manufacture, 1982 (millions of dollars)	6,162	94,388	16,373	18,112	2,077	47,705	4,038

State	Massachusetts	Montana	Nebraska	New Jersey	New Mexico	New York	North Carolina
Groundwater as % of total water use, 1980	13	2	59	25	47	12	10
Population, 1980 (millions)	5.737	0.786	1.569	7.364	1.302	17.558	5.881
% Change in population from 1970	0.8	13.3	5.7	2.7	28.1	-3.7	15.7
% of population that is rural, 1980*	16.2	47.1	37.1	11.0	27.9	15.4	52.0
Total area (millions of acres)	5.3	94.11	49.51	4.98	77.82	31.43	33.71
Total land area (millions of acres)	5.0	93.05	49.05	4.78	76.65	30.32	31.26
Rank in size	45	4	15	46	5	30	28
% of total acreage used for cropland, 1982	5.3	17.7	45.7	14.0	2.9	18.8	19.0
% of cropland that is harvested farmland, 1982	74.4	56.9	76.1	85.1	58.0	77.8	78.3
% of cropland that is pastureland, 1982	78.5	6.8	10.7	9.6	20.1	15.6	13.6
% of cropland that is used for cover crops, 1982**	1.8	0.9	1.0	?	2.6	10.0	2.0
# of manufacturing establishments, 1982	11,017	1,090	1,928	15,126	1,223	32,651	10,133
Value added by manufacture 1982 (millions of dollars)	25,968	718	4,445	31,656	1,398	62,907	28,492

Table 6-5. General Land-Use Profiles for States Appearing in the "State Summaries" and "Aquifer Classification" Sections (continued)

State	North Dakota	Oregon	Pennsylvania	Rhode Island	South Carolina	Texas	Vermont	Washington	Wyoming
Groundwater as % of total water use, 1980	11	17	6	22	4	61	13	9	10
Population, 1980 (millions)	0.653	2.633	11.863	0.947	3.121	14.229	0.511	4.132	0.469
% Change in population from 1970	5.7	25.9	0.5	−0.3	20.5	27.1	15.0	21.1	41.3
% of population that is rural, 1980*	51	32.1	30.7	13.0	45.9	20.4	66.2	27	37.3
Total area (millions of acres)	46.25	62.13	29.0	0.76	19.91	170.76	6.15	43.61	62.60
Total land area (millions of acres)	44.35	61.56	28.73	0.68	19.33	167.69	5.93	42.57	62.07
Rank in size	17	10	33	50	40	2	43	20	9
% of total acreage used for cropland, 1982	63.4%	8.5	19.3	4.1	16.4	21.8	13.0	19.2%	4.4
% of cropland that is harvested farmland, 1982	72.2%	63.1	78.7	75.4	77.8	56.7	71.0	64.4%	66.2
% of cropland that is pastureland, 1982	5.5%	16.4	15.5	16.4	15.2	27.4	26.6	7.5%	16.8

% of cropland that is used for cover crops, 1982**	1.3%	0.9	1.4	1.9	1.0	1.9	0.5	0.8%	0.5
# of manufacturing establishments, 1982	587	5,659	17,666	2,855	4,206	20,288	1,104	6,788	511
Value added by manufacture, 1982 (millions of dollars)	652	7,973	44,824	3,793	12,219	53,358	2,037	12,596	408

* Rural areas are those with less than 1,000 people/sq. mile.
** Cover crops are those planted to protect the land from freezing or erosion and are not harvested.

Sources: U.S. Dept. of Commerce, 1982, 1986, USGS, 1985.

vegetable processing plants, and sewage. Remaining categories are *Pesticides and Fertilizers, Radioactive* wastes, *Mine* wastes, and *Miscellaneous* (e.g., residue from water softeners).

The numbers of incidents reported are tabulated by the nature of the contaminant and by the actual or potential effects. The latter ·include: affected or threatened public water supplies (*Aff., Thr.*), actual or threatened *Fire or Explosion,* and documented or potential public health effects (*Doc., Pot.*).

178 *Means of Detection.* Not all case histories specify how the contamination was detected. Where provided, the information is included.

Remedial Action. The discovery of groundwater contamination is not an end in itself. Many remedial actions can be taken to alleviate or reduce the problem, even though these are often costly. Where such actions have been taken they are listed under the categories *Direct* and *Indirect.* A discussion of possible remedial actions and their costs is given in chapter 9.

Finally, it should be emphasized again that the case histories listed here are not part of a national, random-sampling survey; rather, they are so-called anecdotal data based on reported individual cases of groundwater contamination. As such, they only indicate the potential magnitude of the problem rather than document its actual magnitude. Because of the nature of the charts, the information contained therein provides an indication of the range and importance of some point sources, but not of non-point sources. Where the latter are of importance—for example, in Nebraska—a separate analysis is given where information is available.

Arizona

In 1980, Arizona used 4,200 million gallons per day (mgd) of groundwater, representing 58% of its total water use (USGS, 1985). The state is divided into five groundwater basins: the Upper and Lower Santa Cruz, White Mountain, Salt River Valley, and Upper Salt River basins (Arizona Division of Health Services, 1979). The Upper and Lower Santa Cruz basins and the Salt River Valley comprise an area known as the Basin and Range Lowlands Province, where the two major population centers of Arizona— Tucson and Phoenix—are located. This area uses the most groundwater in the state and has the highest potential for contamination. Greater than 95% of all groundwater pumped in Arizona is consumed in this province. Sixty percent of the water consumed here is groundwater. The Upper Santa Cruz Basin is 100% dependent on groundwater. Eighty-five percent of all industrial activity and 60% of all mining activity occurs in the Basin and Range area. A majority of its aquifers are highly permeable, unconsolidated alluvial material that produces high well yields. The northeastern part of the state has very limited groundwater potential (Fuhriman and Barton, 1971). Groundwater withdrawals exceed recharge throughout most of the state.

179

Very few cases of groundwater contamination were known to have occurred as of 1983; only a total of 23 incidents were reported (see Table 6-6), almost all of which threatened or affected the water supplies. The most common source of contamination was industrial wastes at 30%, or 7 cases, closely followed by landfill leachate at 26%, or 6 cases, and human and animal wastes at 26%, or 6 cases. Each of the incidents involving human and animal wastes led to outbreaks of disease. The incidents were discovered, for the most part, by investigation, and in only 26% of the cases was some sort of remedial action taken. It should be noted that, although only 2 incidents of natural pollution were reported Fuhriman and Barton (1971) consider natural pollution to be significant in Arizona. Arizona is considered to have areas of hard water (see Figure 2-7). Natural leaching of the soil by percolating waters causes natural accumulations of minerals that may limit the uses to which the groundwater may be put (Fuhriman and Barton, 1971). The presence of phreatophytes also increases the mineral content of the water and Unfortunately, updated information was not made available for the revision of this table. Therefore, it should be noted that this information is outdated and may misrepresent the present condition of Arizona's groundwater.

California

California uses more groundwater than any other state in the nation—approximately 14 bgd (California State Water Resources Control Board, 1981). Groundwater supplies about 40% of the state's water needs. The largest groundwater reservoir underlies the Central Valley Region, which occupies 10% of California's land area and includes the San Joaquin Valley (Thomas and Phoenix, 1976).

Table 6-6. Arizona: Summary of Information from Documented Case Histories

Contaminant	Total	Water Supplies Aff.	Water Supplies Thr.	Public Health Effects
Industrial[1]	7(30%)	3	4	
Landfill	6(26%)		6	
Organic	6(26%)	6		6
Pesticides[2]	1			
Mine[3]	1	1		
Natural				
Lead	1		1	
Hexavalent chromium	1		1	
Totals	23	10(43%)	12(52%)	6

Means of Detection

Well Cont.	7
Investigation	11
None mentioned	5

Remedial Action

Indirect

Extent determined	2
Remedial action	1
No action	1
Landfill closed	2
Well abandoned	3
None mentioned	16(70%)

[1] Industrial Wastes—Trichloroethylene (TCE) contamination in Tucson area from Hughes Aircraft dumping during WW II. Counted as one case.
[2] Pesticides—Dibromochloropropane (DBCP), pesticide for nematodes, is a major contaminant of wells in Yuma County. 33 of 76 wells sampled are contaminated. Counted as one case.
[3] Mine Wastes—Tailings ponds from copper mining are a major contaminant in the Globe-Miami area. Counted as one case.

Sources: Arizona Division of Health Services, 1979; Fuhriman and Barton, 1971; Hadeed, 1979; Lemmon, 1980; Robertson, 1975; Schmidt, 1972; Schmidt, 1973; U.S. EPA, 1981a.

180

Most of California's groundwater occurs in alluvial sediments in the valleys and plains flanking mountain ranges. These sediments have mainly been deposited by existing streams. Groundwater basins underlie approximately 40% of the state's land area (Figure 6-1) and have a total storage capacity for water of about 424 trillion gallons (California State Water Resources Control Board, 1984). The usable portion of this total, that is, "the reservoir capacity that can be shown to be economically capable of being dewatered during periods of deficient surface supply and capable of being resaturated, either naturally or artificially, during periods of excess surface supply" (Fuhriman and Barton 1971), is estimated to be 47 trillion gallons (California Department of Water Resources, 1975). The proportion of usable storage that is actually occupied varies from region to region. In northern California, the annual draft on groundwater is replenished by recharge. Central and southern California, on the other hand, experience serious overdrafts of groundwater due to low-precipitation and high-evaporation rates. Precipitation falling on valley floors in the southern half of the state generally remains within the depth of the soil penetrated by plants (California Department of Water Resources, 1975). Only in years with exceptionally heavy precipitation is there enough moisture to penetrate below the root zone. Annual groundwater pumping exceeds recharge in several basins, resulting in an overdraft of 651.7 billion gallons annually. In 1972 about 489 billion gallons of this annual overdraft was in the San Joaquin Valley. The California Water Plan of 1957 was designed to answer overdraft problems by proposing facilities for the storage and transport of surface water from places of surplus in order to satisfy water demands in the areas of overdraft. Today California has an elaborate aqueduct system to recharge needy aquifers artificially (Figure 6-2). Overall groundwater quality in the state is considered to be good to fair (California State Water Resources Control Board, 1984).

181

There is no readily accessible complete inventory of case histories of groundwater contamination in California, but some information on general and persistent contamination problems in the state will be summarized. As the data often do not refer to specific cases of contamination but to generalized occurrences, it has not been possible to compile a chart similar to those of the other states.

The California State Water Resources Control Board identified six general statewide groundwater problems of present or potential concern (California State Water Resources Control Board, 1980):

1. Nitrate concentrations from various sources are increasing,

causing a current problem in some areas, a potential problem in other areas. Animal wastes are one potential source of nitrate contamination. In 1968 beef and dairy cattle numbered almost 1,900,000 head, most being fed on the open range (Fuhriman and Barton, 1971). Poultry numbered 260 million, and the hog population was at 150,000 head.

Estimated Wastes Generated per Year

Solid Wastes	23,000,000 tons/yr.
BOD	810,000 tons/yr.
Nitrogen	220,000 tons/yr.
Phosphorus	44,000 tons/yr.

2. The overdraft of groundwater has resulted in seawater intrusion of the 262 coastal groundwater basins (Fuhriman and Barton, 1971), mineralization due to recirculation or percolation of used water, and induced connate water migration. The most serious seawater intrusion has occurred in the following areas (Fuhriman and Barton, 1971):
- The West Coast of Los Angeles County
- East coastal plain pressure area of Orange County
- Petaluma Valley in Sonoma County
- Napa-Sonoma Valley in Napa and Sonoma Counties
- Santa Clara Valley in the San Francisco Bay area
- Pajaro Valley in Monterey and Santa Cruz Counties
- Salinas Valley area in Monterey County
- Oxnard Plain Basin in Ventura County
- Mission Basin in San Diego County

In the Los Angeles area, three barriers have been constructed against seawater encroachment. Overdrafting has also resulted in land subsidence, which has been most severe in the San Joaquin Valley, where subsidence in excess of 20 feet has occurred in some areas (Fuhriman and Barton, 1971). Land subsidence may be responsible for cases of arsenic contamination in this region. It is thought that arsenic adheres to clay particles, and as the pressure is increased in the soil due to the subsidence of the land, the arsenic goes into solution.

3. In the absence of officially designated hazardous-waste dumpsites, there has been a significant amount of illegal dumping. Many rubbish sites do not conform to regulation. Fuhriman and Barton (1971) identified 207 legal sites that had inadequate control of surface drainage.

4. Percolation ponds for handling industrial and military wastes

are often inadequate for the types of wastes being disposed. Oil-field brines and brines from water-softener regeneration plants are particularly troublesome (California Department of Water Resources, 1975).

5. Crop-dusting has introduced numerous pesticides that have the potential to reach groundwater.

6. The design of proper monitoring wells may be the biggest roadblock to the process of establishing waste-discharge requirements.

183

Contamination problems have also been identified on a regional basis. The U.S. Geological Survey has divided California into nine hydrologic regions (Figure 6-1), each of which has its own Water Quality Control Board. All of California's groundwater and surface water quality programs function at this level.

Region One, North Coast

This area consists of remote wilderness and redwood stands (California State Water Resources Control Board, 1980). Tourism and timber-harvesting are important to the economy. Water quality is generally good to excellent. However, contamination problems do exist. Groundwater contamination has been experienced in this region along the Russian River, where wood-processing wastes have leached into groundwater from unlined ponds. On the Smith River Plain, the use of agricultural fumigants has been reported to have contaminated groundwater. A build-up of nitrates in groundwater has occurred where percolation ponds overlie recharge areas along rivers and coastal terraces (California State Water Resources Control Board, 1981 and 1984).

Region Two, San Francisco Bay

This region is highly urbanized. Agriculture is the major activity in the Napa, Petaluma, Sonoma, and lower Santa Clara Valleys, and in eastern Contra Costa County (California State Water Resources Control Board, 1980). Contamination sources of major concern include the poultry ranches in the Petaluma area, which have increased the level of nitrates, sodium, chlorides, boron, and dissolved solids in local groundwaters (California State Water Resources Control Board, 1981 and 1984). In the Santa Clara Valley, elevated levels of trace metals from unknown sources were discovered in 1981 (California State Water Resources Control Board, 1980). In addition, it was reported in 1984 that TCE concentrations exceeding EPA's criteria were found in some industrial areas of the valley (California State Water Resources

Control Board, 1984). Another ongoing problem found in the valley is saline-water intrusion due to extensive groundwater pumping (California State Water Resources Control Board, 1984).

Region Three, Central Coast

Santa Cruz and Monterey Peninsula are major urban areas where agriculture and food processing are the major industries. The economy is also supported by oil production, tourism, and manufacturing. The quality of groundwater in this region varies from excellent to very poor. Coastal groundwaters plagued with natural saltwater intrusion, such as those in Santa Ynez, Lumpa, Santa Maria Valley, and others, have been an ongoing problem that has worsened in recent years (California State Water Resources Control Board, 1981, 1984). Waste discharge to land, the use of water softeners, and certain agricultural practices have contributed to the increasing salinity in this area. In order to alleviate this problem, salt limits have been placed on municipal-waste discharges (California State Water Resources Control Board, 1984). Other concerns include problems that recently have been discovered through hazardous-waste inspections. Most of these cases have been linked to agricultural chemical companies in Monterey County and to an electronics firm in Santa Cruz County (California State Water Control Board, 1984). Lastly, serious overdrafts have been reported within the Paso Robles groundwater basin (California State Water Control Board, 1981, 1984).

Region Four, Los Angeles

Half the water supply in this region is imported, and the other half comes from local groundwater supplies. The southern part of this region is densely populated and industrial. There is intensive urban development in the coastal plain, San Fernando Valley, San Gabriel Valley, and the adjoining foothills. The northern Ventura-Santa Clara basin is rugged and mountainous, with large undeveloped areas. Contamination problems of major concern involve the lack of hazardous-waste disposal sites and high nitrate concentrations in San Gabriel Valley. TCE concentrations exceeding limits suggested by the EPA have been found in a number of wells in Los Angeles County. Investigations carried out since 1980 have sought to determine the source of contamination (California State Water Resources Control Board, 1981, 1984). Mineralization of groundwater and saltwater intrusion continue to be a problem in the Santa Clara River Basin due to overpumping (California State Water Resources Control Board, 1981, 1984).

184

Region Five, Central Valley

This area contains 40% of the state lands and the majority of its streams. Rainfall averages over 100 inches per year in the north and less than 5 inches per year in Kern County. The Sacramento and San Joaquin Rivers dominate the stream system. Stream water is used for irrigation and municipal water supply. More than 40 groundwater pollution problems were in various phases of investigation and cleanup in 1984 (California State Water Resources Control Board, 1984). The majority of these cases involved industrial, military, railroad, and wood-treatment facilities. Industrial and military discharges and illegal dumpings of contaminants have been found at the Aerojet-General, Occidental, Mather Air Force Base, McClellan Air Force Base, Sacramento Army Depot, and Southern Pacific Yard (California State Water Resources Control Board, 1984). Aero-General, one of the nation's largest producers of liquid- and solid-propellant rocket engines, created a major groundwater contamination problem when groundwater beneath the corporation's land became polluted with inorganic chemicals, a herbicide, arsenic, phenols, perchlorate, trichloroethylene, and several other suspected carcinogens. A cleanup project began in mid-1981 and was still continuing in 1984 (California State Water Resources Control Board, 1981, 1984).

185

The following sources have contaminated this region to a lesser degree. The leaching of wood-processing wastes into groundwater has been an ongoing problem in the Upper Sacramento Valley and other foothill areas. The improper disposal of oil-field brines and cut water was reported in the Central Valley, where brines were known to have percolated into the groundwater near Tulare Lake in 1980. In the past, the pesticide dibromochloropropane (DBCP) and methyl bromide have polluted groundwaters of both the Telare and San Joaquin basins. Nitrate from many different sources often has been the cause of problems in the foothills of the Sacramento and San Joaquin Valleys (California State Water Resources Control Board, 1981, 1984).

Region Six, Lahontan

This region is divided into northern and southern sub-basins. Agriculture and cattle grazing dominate the economy in the north. Lake Tahoe and the Truckee River are the major tourist highlights of the north basin. The southern basin is primarily desert. Agriculture and military and space installations support the economy. Most contamination in this area has been associated with development, though natural contamination from high concentrations of dissolved solids, boron, fluoride, sodium, high temperatures, and extremely high mineral content are not uncommon in some portions of this region (California State Water Resources Control Board, 1981, 1984). Nitrate contamination from inadequate sewage-disposal systems has occurred in the Madeline Plains, Bridgeport Valley, Mustang/Mesa area, Sierra Valley, and along the Truckee River (California State Water Resources Control Board, 1981, 1984). An increase in agricultural return-flow and dairy runoff in past years has increased nitrate concentrations, as well. In the Mojave River Basin municipal and industrial wastewater discharges, which occurred 20 to 50 years ago, created a plume of organic chemicals in Barstow's aquifer known as the "Barstow Slug." The disposal of toxic substances at Edwards Air Force Base and geothermal exploration in Coso Hot Springs have also had a great impact on this region (California State Water Resources Control Board, 1984).

186

Region Seven, Colorado River Basin
This region includes the irrigated agricultural lands of the Coachella, Imperial (valued for winter crops), Palo Verde, and Bard Valleys. Excellent-quality groundwater is available for domestic use. Colorado River water is imported for irrigation and groundwater recharge. The contamination problems of major concern include elevated concentrations of nitrates originating from sewage systems and agricultural operations, which are found in most of the groundwater basins, particularly in the upper Coachella Valley. Serious overdraft has also been a problem in these areas and has resulted in a need for artificial recharge with imported water. In 1980 it was reported that waste disposal sites in Imperial Valley had to be relocated, since they were subject to washout and flooding (California State Water Resources Control Board, 1981). Leaking underground storage tanks have recently been detected as an abundant source of contamination. The reinjection of geothermal brines into groundwater zones has degraded otherwise usable groundwaters. The build-up of pesticides in soils is continuing to lead to groundwater problems (California State Water Resources Control Board, 1981). Lastly, high levels of fluoride and arsenic have been identified as natural pollutants in many springs throughout this region (California State Water Resources Control Board, 1984).

Region Eight, Santa Ana
This area includes 2,800 square miles of valley floor and mountains. Valleys overlie extensive groundwater basins divided

into the upper and lower Santa Ana and San Jacinto River watersheds. Upper basin groundwaters are somewhat overdrawn but of good quality. The lower Santa Ana River watershed is highly urbanized and covers most of Orange County. Groundwater in Orange County has been severely overdrawn. In the mid-1950s seawater intruded four miles inland because of overdraft (Toups, 1974). Several million acre-feet of Colorado River water have been imported to recharge groundwater basins so as to prevent seawater intrusion. Additional salt, however, enters the system from these waters. The Colorado River waters contain salts from industrial discharge and agricultural and domestic sources. Many programs have been initiated to alleviate this problem. The salt generated from manure and wash-water from the Chino-Corona-Ontario area, the highest dairy-concentrated area in the world, has affected the mineral balance of the groundwaters. By far, the most significant groundwater-quality problem in this area has been adverse salt-balance conditions (California State Water Resources Control Board, 1984).

Other contamination problems of major concern include the wastes deposited at the abandoned Stringfellow Class 1 Hazardous-Waste Site. This facility dumped approximately 32 million gallons of hazardous wastes between 1956 and 1972 into disposal ponds. Wastes have escaped on several occasions, and now the entire groundwater basin downstream is threatened by contamination (California State Water Resources Control Board, 1984). High concentrations of nitrates have been a problem in many areas, particularly in the Redlands, Upper Santa Ana River, and Chino basin areas (California State Water Resources Control Board, 1984).

Region Nine, San Diego

Along this 85 miles of fully urbanized coastline, the contamination problems of major concern include a serious overdraft that is responsible for the high salinity of groundwater as a result of saltwater encroachment, and degradation from agricultural and other non-point sources. The primary threat to the groundwaters of the San Diego region, however, is from sewage flow containing organics and nutrients that enters this region from Mexico. (California State Water Resources Control Board, 1981, 1984).

Connecticut

Connecticut uses very little groundwater to meet its freshwater needs—150 mgd, representing 11% of its total water use. Nearly

Table 6-7. Connecticut: Summary of Information from Documented Case Histories

Contaminants	Total
Pesticides	417 (36%)
Solvents	255 (22%)
Leachate	139 (12%)
Oil and Gas	163 (14%)
Road salt	61
Nitrates	95 (8%)
Others	25
	1155

Source: Personal Communication, Elsie Patton, Water Compliance Unit; Connecticut Department of Environmental Protection, 1986.

45% of the groundwater used, or 65 mgd, goes for public supply; the domestic rural supply comes almost entirely from groundwater, about 53.4 mgd; 27 mgd is used by industry. The remaining 1.6 mgd is used for irrigation (USGS, 1985).

Connecticut has two major types of aquifers—unconsolidated and bedrock—which produce water with a natural quality that is good to excellent. Calcium and bicarbonate are the principal ions in most samples; sodium, sulfate, chloride, and silica are the other major inorganic constituents. The water is generally soft to moderately hard. Iron and manganese are the most common problem constitutents. Groundwater quality varies among aquifers and within relatively short distances in the same aquifer.

During the 15-year period from 1971 through 1986, a total of 1,155 well-contamination incidents were recorded by the Connecticut Department of Environmental Protection Water Compliance Unit (Table 6-7). Pesticides were by far the most significant group of contaminants, accounting for 36% of all known cases. The second most common type of well contaminant was solvents, comprising 22% of the recorded incidents. Oil and gas (14%), leachate (12%), nitrates from agricultural practices (8%), and road salt (5%) were responsible for the majority of the remaining well-contamination problems. Most contaminated wells were private wells.

Approximately 1,200 public water-supply wells are currently being regularly monitored for potability (Connecticut Department of Environmental Protection, 1986). Personal communication with DEP personnel revealed that the primary initial action taken in the event of well contamination is supplying the persons affected with potable drinking water, usually in the form of bottled water.

Florida

Florida is a major user of groundwater, requiring more than 3,500 mgd in 1980. Ninety percent of Florida's population depends on groundwater for drinking water. More than one-half of the total fresh water used in Florida comes from groundwater. In 1980, 1,200 mgd of groundwater were used for public water supply, 290 mgd for rural use, 710 mgd for industrial uses, and 1,600 mgd for irrigation (USGS, 1985).

189

There are four major types of aquifers in Florida that naturally produce water of good quality—the sandy, chalky limestone Floridian Aquifer; the limestone, sandstone Biscayne Aquifer; the sand and gravel aquifer; and the shallow sand aquifer (Miller et al., 1977). The Floridian aquifer produces water that is both hard and low in iron and fluoride. The Biscayne aquifer contains very hard water, which is also more mineralized and more alkaline than that of the other aquifers. The sand and gravel aquifer has very soft water that is acidic, high in iron, and low in other mineral content. It is very vulnerable to nitrate contamination. The shallow aquifer contains alkaline, hard water, with a low iron content; it also is susceptible to nitrate contamination.

In 1983, only 92 cases of groundwater contamination were reported. At that time, the most commonly reported contaminant was chloride from saltwater intrusion and agricultural return-flow, accounting for 37% of the incidents. The second most significant source of contamination was industrial and manufacturing wastes. These wastes were responsible for 35% of the incidents reported.

Each year, Florida updates an inventory of its known cases of groundwater contamination, and 260 cases were reported in 1985. Of these, 43% affected or threatened water supplies (Table 6-8). At the time of this inventory the most significant source of groundwater contamination was petroleum products from sources such as leaking underground storage tanks, pipeline ruptures, and accidental spills. Petroleum problems accounted for 36% of the cases mentioned. The second largest contaminant was industrial and municipal wastes, which accounted for 32% of all incidents. Contamination due to chlorides from saltwater intrusion and agricultural return-flow is and has been a great problem; as of 1985, it had contaminated 33 water supply systems. Of all the

Table 6-8.   Florida: Summary of Information from Documented Case Histories

Contaminant	Total	Supplies		Public Health Effects
		Aff.	Thr.	
Industrial	84(32%)	10	8	
Landfill	18	6	1	
Petroleum Products	94(36%)	38	1	
Chlorides	34(13%)	33		
Organic	9	6		2
Pesticides/Fertilizers	9	1		
Radioactive	1			
Mine	2	1		
Other	9	7		7
	260	102(39%)	10(4%)	9(3%)

Means of Detection

Well cont.	72(28%)
Investigation	53
Surface water cont.	3
Spill	2
Leak	3
Other	8
None mentioned	66(25%)

Remedial Action

Direct		*Indirect*	
Groundwater pumped and treated	2	Extent determined	21
		Remedial action	18
		No action	3
Cont. soil removed	5	New water provided	15
Trench installed	1	Action being considered	84
Source elim.	18	Monitoring	38
Source repaired or replaced	18	Litigation	2
		Landfill closed	1
Other	22	None mentioned	77(30%)

Sources: Florida Department of Environmental Regulation 1980, 1981a, and 1981b, 1984, 1985e; Miller et al., 1977; U.S. EPA, 1981a; CDC, 1982, 1983, 1984, 1985.

cases listed, only 25% resulted in direct action aimed at eliminating the contaminant (Florida Department of Environmental Regulation, 1984, 1985; U.S. EPA, 1981a; CDC, 1982, 1983, 1984, 1985).

Idaho

In 1980, Idaho used 6,300 mgd of groundwater. This represents 35% of the total water used in the state. Sixty-five percent of this total was pumped for irrigation, and 33% was withdrawn for industrial purposes (USGS, 1985).

There are five major drainage basins in Idaho. The Snake River Basin, the largest, is the home of nearly two-thirds of the state's population. One of the world's most productive aquifer systems, the Snake Plain Aquifer, lies beneath the Snake Plain, a great structural depression covering 10,040 square miles. It is made up of a series of basaltic flows, alternating with thin beds of pyroclastic and sedimentary materials. In 1980, about 1,720 mgd of water was withdrawn from the Snake Plain Aquifer to irrigate almost 900,000 acres of farmland. Idaho's groundwater is generally of good quality, with localized incidences of fluoride, nitrate, chloride, dissolved solids, and coliform bacteria. The biggest natural groundwater problem is hardness (USGS, 1985).

There were relatively few documented cases of groundwater contamination in Idaho as of December 1985, but there were several potential sources of concern. The Idaho National Engineering Laboratory, a national reactor testing laboratory located in the eastern part of the Snake River Plain, deposited a major source of man-made radionuclides. A network of more than 100 wells monitors the migration of these wastes through the groundwater. As of September 1985, three major isotopes had been detected in the groundwater: tritium, strontium-90, and iodine-129. Fortunately, no contamination has yet been detected off the site (Idaho Department of Health and Welfare and Idaho Department of Water Resources, 1985). Mining is another potential source of contamination. There is large-scale mining for silver, lead, zinc, sand, gravel, and stone. Many abandoned mines are potential hazards. There are 11 abandoned coal mines, 1,749 abandoned metal mines, and 208 abandoned non-metal mines. The extensive agricultural industry in Idaho also presents several potential contamination hazards. Fertilizers and pesticides are used in large

Table 6-9. Idaho: Summary of Information from Documented Case Histories

Contaminant	Total	Supplies Aff.	Supplies Thr.	Public Health Effects
Industrial	12	7		1
Landfill	1			
Petroleum	58(57%)	2		
Organic	8 (8%)	4	1	4
Chlorides	1		1	
Pesticides/Fertilizers	13(13%)	2		
Radioactive	1			
Mine	3			
Other	4	4		4
	101	19(19%)	2(2%)	9(9%)

Means of Detection

Well Cont.	6
Investigation	1
Fumes in basement	1
None mentioned	93

Remedial Action

Direct		*Indirect*	
Source elim.	1	Extent determined	58
Cont. soil removed	3	Remedial action	46
Trench installed	1	No action	12(12%)
Sewage system modified	4	Monitoring	4
		Action being considered	11
Wells capped	2	Injection well abandoned	4
Lagoon sealed	2		
Winter spreading reduced/stopped	7	Mine closed	1
		Well deepened	1
		None mentioned	9 (9%)

Sources: Idaho Department of Health and Welfare, 1985; W. Graham IDWR & C. Brower IDHW, written correspondence; C. Brower, personal communication; U.S. EPA, 1981a; CDC, 1982, 1983, 1984, 1985.

quantities. Dieldrin, now banned but once used heavily for the control of Weiss worm in potatoes, still persists in the soil. Another potential contaminant is nitrate resulting from cattle ranches. In 1985, there were between 550 and 600 feedlots and dairies located over the Snake River Plain (Idaho Department of Health and Welfare and Idaho Department of Water Resources, 1985). In addition, there are at least 5,000 domestic and agricultural waste-disposal wells located in the Snake River Plain. A few cases of serious contamination have resulted due to the high permeability of the aquifer.

193

In 1983, only 29 incidents of contamination were reported. The major source of contamination was human and animal organic wastes, accounting for 48% of the incidents reported. Industrial and radioactive wastes together accounted for 24% of the cases. Of the incidents, 62% actually affected the water supplies, and 34% threatened them.

Of the 101 known contamination incidents recorded in 1985, 57% of the cases were from petroleum contamination resulting from leaking surface and underground storage tanks, pipeline ruptures, and surface spills (Table 6-9). Contamination from human and animal organic wastes accounted for 8% of the cases. Of the total incidents, 21% affected or threatened water supplies. In at least 79% of the cases some sort of remedial action was taken (W. Graham, IOWR and C. Brower, IDHW, written correspondence; U.S. EPA, 1981a; CDC, 1982, 1983, 1984, 1985).

Illinois

Groundwater is a major source of fresh water for nearly one-half of the population of Illinois. More than 980 mgd were withdrawn in 1980, 49% of which supplied public water systems. Twenty-two percent of groundwater used in 1980 served self-supplied industry, 19% was for rural use, and 10% was withdrawn for irrigation (USGS, 1985).

The principal aquifers in Illinois are of three major types— unconsolidated sand and gravel formations, partially cemented sandstone, and consolidated deposits of creviced limestone or dolomite (Walker, 1969). The shallow sand and gravel aquifers have a high potential for contamination, as do outcropped areas of creviced limestone found in western and southern Illinois and the shallow dolomite formations of northern Illinois.

Fifty-eight cases of contaminated groundwater had been documented in Illinois as of 1983. The summary of this information is shown in Table 6-10. Of the known incidents of contamination, 44 of them, or 76%, affected or threatened the water supplies. The most prevalent source of contamination was animal and human wastes, from which there were 20 cases, or 34%, followed by industrial waste and landfill leachate at 21% and 28%, respectively. Groundwater from shallow wells often recorded large concentrations of nitrate. Eighty-one percent of the dug wells that were less than 50 feet deep contained more than 10 mg/l nitrate-nitrogen, as opposed to 34% of the deeper, drilled wells in Washington County (NAS, 1977). Every incident of contamination from animal or human wastes affected or threatened the water supplies. The majority of incidents—45 cases, or 78%—were detected by well contamination, investigation, and outbreaks of illness. Illinois has a good record for applying remedial actions; 66% of the incidents received some sort of action. Unfortunately, similar information on updated cases of groundwater contamination, was not made available for this edition. So, again, it should be noted that this information is outdated and may not accurately represent the present groundwater situation in Illinois.

194

Massachusetts

One-third of the population of Massachusetts obtains water from groundwater sources. In 1980, 320 mgd was drawn for public, rural, industrial, and irrigational supplies. Fifty-nine percent of groundwater drawn went to public supply alone (USGS, 1985).

The most productive aquifers in Massachusetts are found in the Coastal Plain Province. This area is underlain by a continuous blanket of unconsolidated sediment 100 feet thick. These aquifers were deposited during the Wisconsin glaciation and provide virtually all water for groundwater use in the state (USGS, 1985).

Between 1978 and 1981, 25 public-supply wells, with a total capacity of 23 mgd, were taken out of service because of contamination. Sixty-eight cases of contamination are listed in Table 6-11. The greatest contaminant was industrial wastes (59%), followed by organic pollution (10%). Contamination was discovered by routine testing and investigation in the majority of the cases, and in all but 25% of the cases some sort of remedial action was taken, mentioned, or under way (U.S. EPA, 1981a; Massachusetts Department of Environmental Quality Engineering, 1984; CDC, 1982, 1983, 1984, and 1985).

Table 6-10. Illinois: Summary of Information from Documented Case Histories

| Contaminant | Total | Water Supplies | | Public Health Effects |
		Aff.	Thr.	
Industrial	12(21%)	7	5	
Landfill	16(28%)	2	6	
Petroleum	3	1		
Chlorides	6	1	1	
Organic	20(34%)	18	2	15
Mine	1	1		
Totals	58	30(52%)	14(24%)	15

Means of Detection

Well Cont.	10	None mentioned	5
Investigation	20	Fumes in basement	1
Spill on ground	3	Animal deaths	2
Leak discovered	2	Illness	15

Remedial Action

Direct		*Indirect*	
Groundwater pumped and treated	1	New water supply provided	7
		Action being considered	4
Trench installed	2	Monitoring begun	4
Source elim.	6	Damages awarded	2
Surface water collected and treated	1	Landfill closed	6
		None mentioned	16
Site regraded	1		
Impoundment upgraded	1		

Sources: Piskin et al., 1980; Walker, 1969; U.S. EPA, 1981a; Lidorff and Cartwright, 1977; U.S. Congressional Research Service, 1980.

195

Table 6-11.   Massachusetts: Summary of Information from Documented Case Histories

Contaminant	Total	Supplies Aff.	Thr.	Public Health Effects
Industrial	40(59%)	40		
Petroleum	4	4		
Landfill	5	5	1	
Organic	7(10%)	7		1
Chlorides	6	6		
Unknown	6	6		6
Totals	68	68(100%)	1	7(10%)

Means of Detection

Well cont.	10(15%)
Investigation	41(60%)
Soil cont.	2
Spill	1
Illness	1
None mentioned	13

Remedial Actions

Direct		*Indirect*	
Source elim.	2	Extent determined	8
Groundwater pumped and treated	9	No action	8(12%)
		New water provided	28
		Landfill closed	3
		Monitoring	7
		Water blended	3
		None mentioned	9(13%)

Sources: Massachusetts Department of Environmental Quality Engineering, 1984; U.S. EPA 1981a; CDC 1982, 1983, 1984, 1985.

Montana

Even though supplies are abundant throughout the state, groundwater provides Montana with less than 2% of its freshwater needs. Of the 200 mgd withdrawn in 1980, 27% went to public supply, 14% to industry, 11% to rural supply, and 48% to irrigation (USGS, 1985).

The aquifers of Montana consist of unconsolidated alluvial,

glacial, and basin-fill deposits and consolidated sedimentary rocks. In western and south-central Montana most of the groundwater used is derived from aquifers consisting of alluvial, glacial, and basin-fill deposits of unconsolidated to semi-consolidated gravel, sand, silt, and clay. Groundwater in eastern and north-central Montana is available from alluvial and glacial deposits and from deeper aquifers. Other aquifers are the Fort Union Aquifer, made up of consolidated and interbedded shale, siltstone, sandstone, and coal; the Fox Hills–Lower Hell Creek **197** Aquifer, made up primarily of sandstone with intertonging siltstone and shale; The Judith River, Eagle; Kootenai, and Ellis Aquifers, consisting mainly of sandstone separated by shale confining layers; and the Madison Aquifer, the lowermost widespread aquifer in eastern and central Montana, which consists mainly of limestone, dolomite, anhydrite, and halite (USGS, 1985).

Eighty-four cases of confirmed groundwater contamination were reported as of 1984, 56% of which were due to petroleum contamination (Table 6-12). Leaks beneath railroad centers, gasoline storage tanks, and delivery lines at service stations were the major sources of this petroleum contamination. The second most prominent source of contamination was landfill leachate, accounting for 19% of all incidents reported. In 65% of the cases no remedial action was mentioned (Montana Department of Health and Environmental Sciences, 1984; U.S. EPA, 1981a; CDC, 1982, 1983, 1984, and 1985).

Nebraska

This predominantly agricultural state used approximately 7,100 mgd of groundwater in 1980, accounting for 59% of its total water supply (USGS, 1985). In 1980, 6 percent of total groundwater withdrawn went to rural, public, and industrial uses. The remaining 94% withdrawn was used to irrigate 6.2 million acres of farmland (USGS, 1985).

Most of this is supplied by the extensive Ogallala Aquifer, which also underlies parts of Texas, Kansas, Colorado, Wyoming, Oklahoma, and New Mexico. Other aquifers of importance are the Holocene or Pleistocene aquifers that occur in the principal river valleys, and the Pleistocene sands and gravels (Engberg and Spalding, 1978). The natural quality of groundwater is generally good. The mean total dissolved solids count from 1,518 wells was

Table 6-12. Montana: Summary of Information from Documented Case Histories

Contaminant	Total	Supplies Aff.	Public Health Effects
Industrial	9	3	1
Landfill	16(19%)	1	
Petroleum	47(56%)		
Organic	3	3	3
Pest/Fert	1	1	
Mine	7	1	
Other	1	1	1
Totals	84	10(12%)	5

Means of Detection

Well cont.	8
Investigation	11(13%)
Surface water cont.	2
Other	1
None mentioned	62(74%)

Remedial Actions

Direct		*Indirect*	
Cont. removed	2	New water provided	2
Landfill capped	2	Action being considered	17
Regraded area	1	Extent determined	4
		Remedial action	4
		Monitoring	4
		Landfill closed	4
		None mentioned	55(65%)

Sources: Montana Department of Health and Environmental Sciences, 1984; U.S. EPA 1981a; CDC 1982, 1983, 1984, 1985.

474 mg/l. (The maximum contamination level for total dissolved solids as established by the National Secondary Drinking Water Regulations is 500 mg/l.) Most crops can tolerate this level.

The cases of groundwater contamination due, for the most part, to point sources are summarized in Table 6-13. Of the 140 incidents reported in 1984, 54% threatened or affected water supplies. Incidents involving pesticides and fertilizers accounted for 60% of the cases in 1984 and 43% in 1983; these compounds were also involved in non-point source contamination (Table 6-14). The second most significant source of contamination was

Table 6-13.   Nebraska: Summary of Information from Documented Case Histories

Contaminant	Total	Supplies		Public Health Effects
		Aff.	Thr.	
Industrial	19(14%)	9	1	
Landfill	3			
Petroleum	25(18%)	1		
Chlorides	5	1	1	1
Organic	4	1	1	
Pest/Fert	84(60%)	61(73%)		
	140	73(52%)	3(2%)	1

Means of Detection

Well cont.	4
Investigation	4
Other	3
None mentioned	130(93%)

Remedial Action

Direct		*Indirect*	
Groundwater pumped and treated	1	Extent determined	2
		No action	2
		New water provided	17
Other	21	Action being considered	1
		Monitoring	2
		None mentioned	96(69%)

Source: Krueger, 1984; U.S. EPA 1981a; CDC 1982, 1983, 1984, 1985.

199

petroleum and its constituents, accounting for 18% of the cases in 1984. These incidents were mainly due to spills and leaking underground storage tanks. In 1983, however, human and animal wastes were the second most frequently reported pollutant. In 69% of all of the cases reported in 1984, no remedial action was reported (Krueger, 1984; U.S. EPA, 1981a; CDC, 1982, 1983, 1984, 1985).

Various surveys in Nebraska sampled 4,350 wells and found that 700, or 16% of all those sampled, contained NO_3-N_2 contamination in excess of the standard of 10 mg/l (Table 6-14). Studies showed that of the contaminated wells, 575, or 82%, were contaminated by

Table 6-14. Nebraska: Summary of Groundwater Pollution Due to Nitrate-Nitrogen Pollution

Problem area identification started in December 1981 under the state ground-water protection program.
Nitrate contamination. Samples taken from wells from 1976–80.

Number of Wells Sampled	Number of Wells with Nitrate-Nitrogen Contamination		Non-Point	Point
34	20		19	1
537	118		118	—
557	61		3	58
581	17		—	17
558	45		5	40
566	102		102	—
575	29		29	—
615	55		55	—
53	53		53	—
18	17		17	—
256	183		174	9
Total 4350	700		575	125
% 100	16	as % of NO_3-nitrogen contaminated wells	82	18

Sources: Nebraska Department of Environmental Control, 1980a, b and 1981; Engberg and Spalding, 1978; U.S. EPA, 1981a; Spalding and Exner, 1980; Junk et al., 1980; Spalding et al., 1978a, b, 1979; Gormley and Spalding, 1979; Exner and Spalding, 1979; University of Nebraska, 1980.

non-point sources such as the nitrogen fertilizer contained in irrigation return-flow. Septic tanks, barnyards, and feedlots are point sources that accounted for the remaining 125, or 18%, of the cases reported.

New Jersey

New Jersey, one of the most densely populated regions in the United States, has more than 16,000 potable wells (Tucker, 1981). In 1983, nearly 3.5 million people—45% of New Jersey's population—depended on groundwater. In the southern half of the state, more than 90% of the population received its drinking-water supply from groundwater. In 1980, it was estimated that 450 mgd of groundwater went to public supply, 77 mgd to rural supply, 160 mgd to industry, and 40 mgd to irrigation—a total of 727 mgd, or 25% of the total freshwater supply used in New Jersey (USGS, 1985).

The aquifers of the state produce water that is generally of good quality. In the northwestern half of the state, there are several types of consolidated aquifer formations. The most extensive formation is one composed of shale, slate, and sandstone; it produces poor well yields. The Kittatinny Limestone Aquifer is a major one with high well yields, many in excess of 100 gallons per minute (gpm). Sandstone and quartzite formations occur here and are poor aquifers. Gneiss, schist, and granite formations produce well yields in the range of 5–10 gpm. In the southeastern half of the state, unconsolidated aquifers of sand, gravel, clay, silt, and marl are found that produce very high well yields, of 500–700 gpm. Water from the northwest is hard to very hard, and water from the southeast is soft to moderately hard. In some areas there are high concentrations of iron and chloride. Some formations on the coastal plain contain saline water below a depth of 1,000 feet.

201

In 1981 New Jersey completed an extensive sampling of its groundwater aquifers for toxic chemical contamination. The aquifer samples for chemical analysis were collected from a random sampling of wells throughout the state, a sampling that was designed to cover all the different areas. Where problems were found, additional investigations for contamination were conducted (New Jersey Department of Environmental Protection, 1981a).

A comprehensive inventory of groundwater contamination cases in New Jersey is updated annually by the NJ Geologic Survey. The inventory, however, is by no means a complete listing of all known groundwater contamination incidents in New Jersey; rather, it consists largely of cases that have had moderate to severe impact on groundwater quality. Tables 6-15 to 6-19 summarize these incidents of contamination as of April 1984 (New Jersey Department of Environmental Protection, 1984).

Of the 304 cases of contamination due to legal industrial, domestic, and municipal dumping (Table 6-15), 78% were due to dumping of industrial wastes, and 33% of these threatened or affected the water supplies. Contamination due to the disposal of animal and human wastes accounted for 6% of the incidents; 59% of these cases affected the water supply. Of the cases listed, 89% have led to some form of remedial action or monitoring program (New Jersey Department of Environmental Protection, 1984; U.S. EPA, 1981a; CDC, 1982, 1983, 1984, 1985).

Table 6-15. New Jersey: Summary of Information from Documented Case Histories—Industrial and Municipal Dumping

Contaminant	Total	Supplies Aff.	Thr.	Health
Industrial	237(78%)	56(24%)	18(8%)	
Landfill	1			
Petroleum	29(10%)	1	1	
Chlorides	6	6		
Organic	17 (6%)	10(59%)		4
Pest/Fert	4	1		
Radioactive	3	1		
Mine	2	1		
Other	5	5		4
	304	81(27%)	19(6%)	8(3%)

Means of Detection

Well cont.	58(19%)
Investigation	172(57%)
Surface water cont.	21
Spill	5
Leak	4
Soil cont.	7
Fumes in air	2
Dead vegetation	1
Seepage	
Basement	1
Sewer	3
None mentioned	30

Remedial Action

Direct		*Indirect*	
Groundwater pumped and treated	33(11%)	Extent determined	47
		Remedial action	33
Soil removed	18	No action	14 (5%)
Source elim.	35(12%)	New water provided	20
Trench installed	10	Action being considered	65
Surface water treated	3	Monitoring	149(49%)
Cont. area capped	7	Litigation	3
Treatment facility installed	14	Lagoon closed	8
		Plant closed	8
Hookup to sewer	8	None mentioned	19 (6%)
Area regraded	2		

Sources: New Jersey Department of Environmental Protection, 1984; U.S. EPA 1981a; CDC 1982–85.

Table 6-16 shows incidents of contamination attributable to accidental spills. Eighty-three percent involved petroleum products, and 29% of these threatened or affected water supplies. Of the 222 total incidents, 37% affected or threatened water supplies. In all but 20% of the incidents some form of remedial action or monitoring program had been initiated (New Jersey Department of Environmental Protection, 1984; U.S. EPA, 1981a; CDC, 1982, 1983, 1984, and 1985).

Table 6-17 shows 46 cases of groundwater contamination resulting from sanitary landfills. Of these incidents 37% threatened or affected water supplies. Only 4 cases had no remedial action taken or mentioned or monitoring program started by April 1984 (New Jersey Department of Environmental Protection, 1984; U.S. EPA, 1981a; CDC, 1982, 1983, 1984, 1985).

Table 6-18 summarizes information on cases of illegal dumping. Thirteen percent of these affected or threatened water supplies, and one posed a threat of fire. In 77% of the cases remedial action had been taken or was under consideration, or monitoring programs had been started (New Jersey Department of Environmental Protection, 1984; U.S. EPA, 1981a; CDC, 1982, 1983, 1984, 1985).

Table 6-19 is a summary of the information discussed above. The most numerous of incidents were those involving industrial wastes and petroleum products, accounting for 50% and 35% of the cases, respectively. The 1983 study reported that these same sources were the major contaminants at that time as well (New Jersey Department of Environmental Protection, 1984; U.S. EPA, 1981a; CDC, 1982, 1983, 1984, and 1985).

New Mexico

Groundwater provided 47% of New Mexico's water in 1984, and 86% of this was used for irrigation; 1,800 mgd of groundwater was pumped to serve 89% of New Mexico's population with water supplies (USGS, 1985).

The general natural condition of groundwater in New Mexico is fair to good. Mineralization is the most widespread and common problem of quality. Nearly all the groundwater is derived from infiltration of precipitation and seepage from streams and is at least slightly mineralized because of its contact with soil and rock. Over 60% of New Mexico's groundwater is saline, having a total content of dissolved solids greater than 1,000 mg/l. Hardness is

Table 6-16. New Jersey: Summary of Information from Documented Case Histories—Accidental Spills

Contaminant	Total	Supplies Aff.	Supplies Thr.	Fire/ Explosion
Industrial	30(14%)	16(53%)	2	
Petroleum	184(83%)	53(29%)	4	5
Chlorides	1	1		
Organic	3	3		
Pesticides	1	1		
Radioactive	2	1		
Other	1			
	222	75(34%)	6(3%)	5

Means of Detection

Well cont.	58(26%)
Investigation	17
Surface water cont.	15
Spill	30(14%)
Leak	62(28%)
Soil surface cont.	3
Fumes	
basement	5
ground	2
sewer line	4
building	7
Explosion	5
Seeps	
basement	3
sewer	5
None mentioned	5

Remedial Action

Direct		*Indirect*	
Groundwater pumped and treated	59(27%)	Extent determined	40
		Remedial action	6
Soil removed	12	No action	34(15%)
Trench installed	13	New water provided	28
Source elim.	33	Action being considered	25
Surface water collected and treated	6	Monitoring	96(43%)
		Litigation	1
Hookup to sewer	2	Carbon filters	1
Cont. area capped	1	None mentioned	10 (5%)
Area flooded	1		

Source: New Jersey Department of Environmental Protection, 1984; U.S. EPA 1981a; CDC 1982–85.

Table 6-17. New Jersey: Summary of Information from Documented Case Histories—Sanitary Landfills

Contaminant	Total	Supplies Aff.	Thr.	Fire
Landfill leachate	46	12(26%)	5(11%)	1

Means of Detection

Well cont.	8
Investigation	24(52%)
Fumes in basement	1
Surface water cont.	8
Fire	1
None mentioned	4

Remedial Action

Direct		*Indirect*	
Treatment facility installed	1	Extent determined	10
Trench installed	5	Remedial action	8
Source elim.	4	No action	2 (4%)
Gas evacuation	1	New water provided	1
Stream relocation	1	Action being considered	14(30%)
Landfill capped	6	Monitoring	15(33%)
Area regraded	2	Landfill closed	7
		None mentioned	2(4%)

Source: New Jersey Department of Environmental Protection, 1984; U.S. EPA 1981a; CDC 1982–85.

also a problem in many areas, as are fluorides, nitrates, and occasionally arsenic. The best quality water in New Mexico comes from the High Plains and the Ogallala Aquifer. The Rio Grande Valley has the largest supply of fresh water in the state, but it is of only fair quality.

Groundwater quality is threatened by overpumping in the eastern part of the state; by mining for uranium, copper, molybdenum, and potash in various areas throughout the state; and by oil production in the northeastern and southwestern parts of the state.

There were 131 reported incidents of groundwater contamination in New Mexico as of February 1984 (Table 6-20). The most significant source of contamination was chloride from oil-field brines, representing 31% of the total. Ninety percent of these cases affected water supplies. Animal and human wastes accounted for

Table 6-18. New Jersey: Summary of Information from Documented Case Histories—Illegal Dumping

Contaminant	Total	Supplies Aff.	Thr.	Explosion
Industrial	39	5(13%)	1(3%)	1
Petroleum	1			
Landfill	4			
Total	44	5(11%)	1(2%)	1

Means of Detection

Well cont.	4
Investigation	21
Surface water cont.	3
Spill	8
Explosion	1
None mentioned	7

Remedial Action

Direct		*Indirect*	
Soil removed	5	Extent determined	5
Source elim.	5	Remedial action	5
Cont. area capped	1	Action being considered	16(36%)
		Monitoring	17(39%)
		None mentioned	10(23%)

Source: New Jersey Department of Environmental Protection, 1984; U.S. EPA 1981a; CDC 1982–85.

Table 6-19. New Jersey: Summary of Information from Documented Case Histories—Totals

Contaminant	Totals	Supplies Aff.	Thr.	Fire/ Explosion	Public Health Effects
Industrial	306(50%)	77(25%)	21(7%)	1	
Landfill	51	12	5	1	
Petroleum	214(35%)	54(25%)	5	5	
Chlorides	7	7			
Organic	20	13			4
Pest/Fert	5	2			
Radioactive	5	2			
Mine	2	1			
Other	6	5			4
Totals	616	173(28%)	31(5%)	7	8

Table 6-19.   New Jersey: Summary of Information from Documented Case Histories—Totals (*continued*)

Contaminant	Totals	Supplies Aff.	Thr.	Fire/ Explosion	Public Health Effects
Means of Detection					
Well cont.	128				
Investigation	234				
Surface water cont.	47				
Spill	43				
Leak	66				
Soil cont.	10				
Fumes					
basement	6				
ground	2				
air	2				
building	7				
sewer	4				
Explosion/fire	7				
Seeps					
basement	4				
sewer	8				
Dead vegetation	1				
None mentioned	46				

Remedial Action

Direct		*Indirect*	
Groundwater pumped and treated	92(15%)	Extent determined	102
		Remedial action	52
Soil removed	35	No action	50 (8%)
Trench installed	28	New water provided	49
Source elim.	77(13%)	Action being considered	120(19%)
Surface water collected and treated	9	Monitoring begun	277(45%)
		Landfill/lagoon closed	15
Treatment facility	15	Facility closed	3
Hookup to sewer	10	None mentioned	41 (7%)
Cont. area capped	9		
Landfill capped	6		
Area regraded	4		
Area flooded	1		
Gas evacuation	1		
Stream relocated	1		

Source: New Jersey Department of Environmental Protection, 1984; U.S. EPA 1981a; CDC 1982, 1983, 1984, 1985.

Table 6-20. New Mexico: Summary of Information from Documented Case Histories

Contaminant	Total	Supplies Aff.	Public Health Effects
Industrial	10	4	2
Landfill	1	1	
Petroleum	30(23%)	17(57%)	
Chlorides	40(31%)	36(90%)	
Organic	35(27%)	31(89%)	4
Mine	14	6	
Other	1	1	1
Totals	131	97(74%)	7

Means of Detection

Well cont.	91(69%)
Spill	5
Leak	6
Animal deaths	1
Illness	5
None mentioned	15

Remedial Action

Direct		*Indirect*	
Groundwater pumped and treated	1	Extent determined	7
		Remedial action	5
Soil removed	1	No action	2
Source elim.	6	New water provided	6
Surface water collected and treated	4	Action being considered	14
		Monitoring	2
Sewage system modified	1	None mentioned	83(63%)
Evaporation ponds lined	2		

Source: New Mexico Environmental Improvement Division, 1980; U.S. EPA, 1981a; Goad, 1982; Jercinovic, 1982; McQuillan, 1984; Souder, 1983; CDC 1982, 1983, 1984, 1985.

35 cases, or 27% of the total; 89% of these incidents resulted in contaminated drinking-water supplies. Petroleum products also caused a significant number of incidents, 57% of which affected water supplies. The majority of incidents were discovered by contamination of wells, and in 63% of the cases of groundwater contamination there was no mention of remedial action that had taken place. The results of the 1983 survey of contamination sources showed similar results. Chlorides from petroleum

production caused 38% of the contamination reported, and 30% of
the cases were due to animal and human wastes (New Mexico
Environmental Improvement Division, 1980; Goad, 1982; Jercinovic,
1982; Mcquillan, 1984; Souder, 1983; U.S. EPA, 1981a; CDC, 1982,
1983, 1984, 1985).

North Dakota

Sixty-two percent of the state's population relies on groundwater for
domestic supply. For rural domestic use alone, 100 percent of all
fresh water used is from groundwater sources. Irrigation utilized
46% of all groundwater withdrawn in 1982; industry, only 2%
(USGS, 1985).

Unconsolidated (glaciofluvial and glaciolacustrine) deposits and
sedimentary bedrock are the two principal types of aquifers found
in North Dakota. Aquifers made up of unconsolidated deposits are
the most productive in the state. These aquifers are often linear in
shape and resemble surface-water drainage systems. The
Englevale, Oakes, Page, Spiritwood, Sundre, and West Fargo are
some of the most productive aquifers of this type. The Fort Union
sedimentary aquifer is unconfined and underlies the western half of
the state. It tends not to be very productive, and in many areas the
groundwater is highly mineralized. The Fox Hills Aquifer is
confined under the Fort Union Aquifer and provides an abundance
of good quality water to much of the central and northern regions of
the state (USGS, 1985).

Nitrates from sources such as septic tanks, agricultural
practices, and animal feedlots tend to be North Dakota's greatest
groundwater problem, according to data collected by the North
Dakota State Department of Health. Nitrates caused more than 40%
of the groundwater contamination incidents reported. Petroleum
contamination—mainly from leaking underground storage tanks—
was also a major groundwater problem in the state. Remedial
action was taken in more than 50% of the cases reported (Glatt et
al., 1986; U.S. EPA, 1981a; CDC, 1982, 1983, 1984, 1985). See
Table 6-21.

Oregon

Sixty percent of Oregon's population was dependent on
groundwater for water supplies in 1980, and this percentage is
predicted to increase. Of the 1.1 bgd of groundwater withdrawn in
1980, 75% was used for irrigation, 12% for rural domestic and

Table 6-21. North Dakota: Summary of Information from Documented Case Histories

Contaminant	Total	Supplies Aff.
Industrial	2	1
Petroleum	16(28%)	11(69%)
Pest/Fert	12(21%)	
Nitrates (unknown sources)	23(40%)	22(96%)
Organic	2	2
Chlorides	1	
Natural (arsenic)	1	
Totals	57	36(63%)

Means of Detection

Well cont.	8
Investigation	16(28%)
Surface water cont.	1
Leak	5
Seepage into basement	1
Seepage out of hill	1
Flowing injection well	1
None mentioned	24(42%)

Remedial Action

Direct		*Indirect*	
Source elim.	6	Extent determined	4
Soil removed	2	Remedial action	1
Groundwater pumped and treated	4	No action	24(42%)
		New water provided	4
Trench installed	2	Action being considered	1
Sump holes	3	Monitoring begun	5
		None mentioned	1 (2%)

Sources: Glatt, 1986: U.S. EPA, 1981a; CDC 1982, 1983, 1984, 1985.

livestock, 7% by industry, and 6% for public water supply (USGS, 1985).

The principal aquifers in Oregon are of three types. First, there are the basin-fill and alluvial aquifers, found in all parts of the state, which consist of unconsolidated and consolidated basin-fill alluvium, coastal dune, and beach deposits. The water quality is generally good; however, where shallow water tables are present, contamination can easily occur. Second, there are the volcanic and sedimentary aquifers, made up of materials from volcanic eruptions or from the erosion of exposed volcanic vents and veins. These aquifers are generally found in mountainous areas where water tables are deep below the surface, and as would be expected, the water quality in these areas is of good quality. Third, there is the Columbia River Basalt Aquifer, which consists of numerous lava flows interlain with a few sedimentary beds. Development of these aquifers has been slight because of rough terrain; the water is generally of good quality (USGS, 1985).

Information from a draft inventory of groundwater quality problems put out by the Oregon Department of Environmental Quality, as well as from various other sources, is summarized in Table 6-22. Eighty-three cases of groundwater contamination were

211

Table 6-22. Oregon: Summary of Information from Documented Case Histories

Contaminant	Total	Supplies Aff.	Supplies Thr.	Public Health Effects
Industrial	27(33%)	7	7	
Petroleum	2	1	2	
Organic	22(27%)	19	9	11
Landfill	24(29%)	3	3	
Fert/Pest	1	1	1	
Natural	3	3		
Other	4	4		4
Totals	83	38(46%)	22(27%)	15(18%)

Means of Detection

Well cont.	7
Investigation	1
None mentioned	75

Remedial Action

Direct		*Indirect*	
Groundwater pumped and treated	4	Extent determined	20
		Remedial action	7
Source elim.	4	No action	13(16%)
Treatment facility	1	Monitoring begun	46
Site regraded	2	Landfill closed	3
Reduce nitrate loading	1	Action being considered	14
		New water provided	3
		None mentioned	19(23%)

Sources: Oregon Department of Environmental Quality, 1986b; U.S. EPA 1981a; CDC 1982, 1983, 1984, 1985.

reported as of 1986. Industrial wastes and landfill leachate show the greatest incidents of contamination in the state, at 33% and 29%, respectively. However, local groundwater degradation in four areas due to increasing nitrogen concentrations from on-site subsurface sewage-disposal systems is one of Oregon's greatest groundwater problems. These areas have underground conditions especially vulnerable to contamination: the Clatsop Plain is underlain by windblown sands and shallow soils; Central Maltnomah County is set on top of fluviolacustrine sediments; and both the River Road in Santa Clara and the La Pine area have soils that are of moderate to high permeability (Oregon Department of Environmental Quality, 1986b; U.S. EPA, 1981a; CDC, 1982, 1983, 1984, 1985).

212

Pennsylvania

Ninety percent of all fresh water used in Pennsylvania in 1980 was supplied by groundwater, which provided 45% of the state's population with drinking water. More than two-thirds of Pennsylvania's public and almost all of its private water supplies come from groundwater. Groundwater contributes 70% of stream baseflow under average stream-flow conditions, and up to 100% during low flow periods (USGS, 1985).

No major aquifer of large areal extent exists in Pennsylvania; however, among its complex hydrogeologic structure there are four principal types of aquifers. Unconsolidated sand and gravel aquifers consisting of sediments and glaciofluvial and alluvial deposits are found in the northwestern, northeastern, and extreme southeastern parts of the state and also in major stream valleys. Sandstone and shale aquifers underlie most of Pennsylvania. Carbonate aquifers are found mostly in the Valley and Ridge Province. Lastly, crystalline bedrock aquifers are found in the southeastern corner of the state (USGS, 1985).

The groundwater quality in most of the state is believed to be acceptable for drinking, with little or no treatment, with the exception of areas in western Pennsylvania. Heavy mining and oil and gas production have contributed to significant amounts of contamination in these areas. A comprehensive outline of major groundwater problems is issued on a biennial basis by the Pennsylvania Department of Environmental Regulation in accordance with Section 208 of the Clean Water Act. A summary of this information and additional data sets are given in Table 6-23.

Table 6-23.   Pennsylvania: Summary of Information from Documented Case Histories

Contaminant	Total	Supplies Aff.	Thr.	Public Health Effects
Industrial	156(27%)	59(38%)	1	
Landfill	65	3		
Petroleum	228(40%)	70(31%)	1	
Chlorides	24	2		
Organic	44 (8%)	17(39%)		15
Pest/Fert	5	2		
Mine	27	16		
Oil/Gas prod.	2			
Other	25	25		25
	576	194(34%)	2	40 (7%)

Means of Detection

Well cont.	64(11%)
Investigation	3
Spills	7
Leaks	20
Surface water cont.	4
Illness	1
Fumes in basement	1
None mentioned	476

Remedial Action

Direct

Groundwater pumped and treated	49
Cont. soil removed	12
Trench installed	11
Source elim.	79
Sewage system repaired	4
Impoundment lined	3
Covered salt	3
Surface water treated	
Injection wells plugged	2
Microbiological degradation	1
Landfill capped	1
Impoundment capped	2

Indirect

Extent determined	115
Remedial action	62
No action	53(9%)
New water	26
Action being considered	126
Monitoring begun	55
Landfill closed	21
Lagoon abandoned	3
None mentioned	163(28%)

Sources: Pennsylvana Department of Environmental Resources, 1984; U.S. EPA 1981a; CDC 1982, 1983, 1984, 1985.

Groundwater Contamination

Of the 576 cases of contamination reported in 1984, 34% contaminated water supplies. Petroleum products and industrial wastes posed the greatest threats to groundwater supplies, involving 40% and 27% of the cases, respectively. In nearly two-thirds of the reported cases, action was being taken to deal with the contamination (Pennsylvania Department of Environmental Resources, 1984; U.S. EPA, 1981a; CDC, 1982, 1983, 1984, 1985).

214

Rhode Island

In 1980, groundwater supplied 22% of all fresh water used in Rhode Island. Twenty-four percent of the state's population

Table 6-24.   Rhode Island: Summary of Information from Documented Case Histories

Contaminant	Total	Supplies Aff.	Supplies Thr.	Public Health Effects
Industrial	33(49%)	14(42%)		
Landfill	21(31%)	4(19%)		
Petroleum	1	1		
Chlorides	4			
Organic	2	1		1
Other	2	2		1
Unknown	5			
	68	22(32%)		2

Means of Detection

Wells cont.	20(29%)
Surface water cont.	3
None mentioned	45(66%)

Remedial Action

Direct		Indirect	
Source elim.	2	Extent determined	1
Leachate collection system	1	Remedial action	1
Area regraded	1	Action being considered	27(40%)
		Landfill closed	1
		None mentioned	22(32%)

Sources: Rhode Island Department of Environmental Management, 1984; U.S. EPA, 1981a; CDC, 1982, 1983, 1984, 1986.

depended on groundwater for drinking water. More than half of the groundwater withdrawn was used for public supply, 35% was used by industry, 13.3% for rural supplies, and only 1.3% for irrigation. Withdrawals for public supplies nearly doubled from 1960 to 1980, rising from 10 mgd to 19 mgd (USGS, 1985).

Groundwater typically occurs in unconfined conditions throughout the state. In certain locations, however, groundwater can be drawn from thick layers of silt and clay, stratified drift, till, and fractured bedrock, all under confined conditions. Generally, the groundwater is of good quality, soft, slightly acidic, with total dissolved solids of less than 150 mg/l. However, because of the high permeability of soils and a generally shallow water table, Rhode Island's groundwater is very vulnerable to groundwater contamination (USGS, 1985).

215

As of 1984, Rhode Island has had very few documented cases of groundwater contamination—68 incidents in all (Table 6-24). Forty-nine percent of the contamination was caused by industrial wastes, 42% of which contaminated water supplies. Landfill leachate contaminated groundwaters in 21 cases; 19% of these affected water supplies. Twenty-nine percent of the cases were discovered through the contamination of wells, and some sort of remedial action was being considered or was implemented in 68%, of the 68 cases reported (Rhode Island Department of Environmental Management, 1984; U.S. EPA, 1981a; CDC, 1982, 1983, 1984, 1985).

South Carolina

In 1980, although groundwater accounted for only 4% of South Carolina's total water use, 42% of the total population relied upon groundwater for its drinking-water supply. Groundwater use in 1980 was 210 mgd, 40% of which went to public supply, 31% to rural supply, 22% to industry, and 7% to irrigational supply (USGS, 1985).

Several types of aquifers are found in South Carolina. In the Piedmont-Blue Ridge region, the aquifers are composed of granite, schist, gneiss, slate, and phyllite. The groundwater is stored mainly in open fractures, and most wells are low-yielding. Along the coastal plain in the southeastern half of the state, the aquifers consist of sand, gravel, and limestone. In the extreme south, the Ocala limestone aquifer is the principal artesian one; it is a continuation of the principal artesian aquifer of Georgia and the

Table 6-25. South Carolina: Summary of Information from Documented Case Histories

| Contaminant | Total | Supplies | | Fire | Public Health Effects |
		Aff.	Thr.		
Industrial	55(34%)	16(29%)	8(15%)		
Landfill	14		2		
Petroleum	73(45%)	62(85%)		1	
Organic	10	9(90%)			3
Pest/Fert	4	2			
Radioactive	1				
Other	6	5	1		2
	163	94(58%)	11 (7%)	1	5(3%)

Means of Detection

Well cont.	79(48%)
Investigation	37(23%)
Surface water cont.	9
Spill	4
Leak	5
Fumes	
ground	1
sewer	1
Animal deaths	1
Other	8
None mentioned	19(12%)

Remedial Action

Direct		*Indirect*	
Groundwater pumped and treated	4	Extent determined	5
		Remedial action	3
Source elim.	6	No action	2 (1%)
Surface water collected and treated	1	New water provided	5
		Action being considered	22
		Monitoring begun	23
French drain installed	2	Landfill closed	2
		None mentioned	96(59%)

Sources: South Carolina Department of Health and Environmental Control, 1980, 1983; South Carolina Department of Health and Environmental Control (Draft), 1981; CDC, 1982, 1983, 1984, 1986.

216

Floridian aquifer of Florida. The Tuscaloosa Formation Aquifer
underlies nearly the entire coastal plain. The Peedee and Black
formations comprise one aquifer in the central and eastern part of
the southern coastal plain. The Black Mingo Formation is a major
aquifer in the central coastal area, and the Santee Limestone
Formation is used from the central coastal area for a distance 60
miles further inland (Scalf et al., 1973). The naturally good quality
of the groundwater in South Carolina makes it suitable for most
purposes. High iron concentration, low pH, high fluoride, excessive   **217**
hardness, and high chloride are the main problems, but alternate
sources of better-quality water are locally available from other
aquifers (Scalf and Dunlap, 1977).

Table 6-25 summarizes the sources of contamination, their
means of detection, and any remedial action taken on the 163
known cases of groundwater contamination reported in 1983.
Petroleum products were involved in the majority of the incidents,
accounting for 73 cases. Of these cases 85% affected the water
supply. Fifty-five incidents of contamination due to industrial
wastes were reported, 34% of which threatened or affected the
water supply. Nearly half of the cases were detected by well
contamination, and for the majority of the incidents, no form of
remedial action was mentioned. These results are similar to the
results obtained in 1983; at that time, however, only 89 cases of
contamination were documented. Again, petroleum products and
industrial wastes caused the majority of the groundwater pollution
cases reported (South Carolina Department of Health and
Environmental Control, 1980, 1981, 1983; U.S. EPA, 1981a; CDC,
1982, 1983, 1984, 1985).

Texas

In 1980, groundwater supplied 61% of the total water use in Texas,
or approximately 3,536 billion gallons per day. Of this, 11.9% was
used for municipal purposes, 2.3% went to manufacturing, and
82.5% was used for irrigation. As of 1980, recoverable reserves of
usable-quality groundwater, containing less than 3,000 mg/l of
dissolved solids, was 140,000 bgd. Eighty-nine percent of this was
found in the High Plains (Ogallala) Aquifer (Texas Department of
Water Resources, 1984b).

More than 80% of Texas is underlain by 7 major and 16 minor
aquifers. The remaining 20% of the state is underlain by

subsurfaces with small, undependable supplies. Of the 7 major aquifers, the Ogallala is by far the most productive. The other 6 are the Alluvium and Bolson Deposits, which occur in many parts of Texas; the Edwards-Trinity (Plateau) Aquifer; the Edwards Aquifer, which supplies the city of San Antonio; the Trinity Group Aquifer, which extends over a large area of northern and central Texas; the Carrizo-Wilcox Aquifer, which is the state's most geographically extensive aquifer; and the Gulf Coast Aquifer, which underlies most of the Coastal Plain from the Rio Grande Valley northeastward into Louisiana (Texas Department of Water Resources, 1984b).

218

As of 1984, 134 cases of groundwater contamination, had been identified, 69% of which contaminated water supplies (Table 6-26). The greatest sources of contamination were petroleum products, chlorides from oil fields and saltwater intrusion, and organic wastes (both human and animal), accounting for 22%, 20%, and 28% of the reported cases, respectively (Texas Department of Water Resources, 1984a; U.S. EPA, 1981a; CDC 1982, 1983, 1984, 1985).

Vermont

Based on 1980 statistics, approximately 54% of the population in Vermont relies on groundwater for drinking water. Of the 45 mgd of groundwater withdrawn in 1980, 37% went to by public water systems, 38% went to domestic rural use, 13% was used for agriculture, and 12% was consumed by local industry. Generally, the quality of Vermont's groundwater is suitable for most purposes (USGS, 1985).

Vermont's hydrogeology is fairly complex because of the varied deposits laid down by continental glaciations. There are four principal types of aquifers found in Vermont. The unconfined stratified drift aquifers, found scattered over most of Vermont, produce the highest yield of groundwater. Confined carbonate bedrock aquifers, found in the western lowlands of Vermont, produce the second greatest yield. The remaining bedrock consists of crystalline bedrock; water can be drawn from this type of confined aquifer from the fractured surface of the bedrock. The last type of aquifer found in Vermont is the unconfined glacial till (till is made up of heterogeneous sand, silt, clay, gravel, and boulders). These tills are generally poor aquifers, though they can sustain small rural populations (USGS, 1985).

In 1982 the largest single cause of groundwater contamination reported was the use and storage of road salt, representing 35% of

Table 6-26. Texas: Summary of Information from Documented Case Histories

Contaminants	Total	Supplies Aff.	Supplies Thr.	Public Health Effects
Industrial	24	13		
Landfill	3	1	1	
Petroleum	30(22%)	22(73%)		
Chlorides	27(20%)	21(78%)		
Organics	37(28%)	24(65%)	1	4
Pest/Fert	2	2		1
Radioactive	1	1		
Natural	6	5		
Other	4	3		1
Totals	134	92(69%)	2(1%)	6

Means of Detection

Wells cont.	81(60%)
Investigation	23
Fumes in water	1
Dead animals	1
None mentioned	28

Remedial Action

Direct		*Indirect*	
Source elim.	19	Extent determined	19
Groundwater pumped and treated	7	Remedial action	19
		Action being considered	65
		New water provided	5
		Monitoring begun	2
		None mentioned	24(18%)

Sources: Texas Department of Water Resources, 1984a; U.S. EPA, 1981a; CDC, 1982, 1983, 1984, 1985.

the incidents (Table 6-27). Twenty-six of these 27 cases contaminated water supplies. Leaks from underground gas and oil tanks represented the second largest cause of groundwater contamination—26% of all cases reported. Sixty percent of these cases contaminated water supplies. Of the 77 cases, 53 were detected by well contamination. A new source of water was supplied in 35% of the incidents and in 44% of the cases no form of remedial action was mentioned to have taken place (Marchfield

Table 6-27.   Vermont: Summary of Information from Documented Case Histories

		Supplies		Public Health Effects
Contaminant	Total	Aff.	Thr.	
Industrial	6	6		
Landfill	2	2		
Petroleum	20(26%)	12(60%)		
Chlorides	27(35%)	26(96%)		
Fert/Pest	5	5		
Organic	13(17%)	11(85%)		1
Unknown	4	4		3
Totals	77	66(86%)		4

Means of Detection

Wells cont.	53
Spring cont.	6
Fumes	
cellar	1
ground	2
Seeps in cellar	2
Other	3
None mentioned	10

Remedial Action

Direct		*Indirect*	
Groundwater pumped and treated	1	Extent determined	1
		No action	1 (1%)
		Action being considered	14
		New water	27(35%)
		None mentioned	34(44%)

Sources: Marchfield Engineering Services, 1982; U.S. EPA, 1981a; CDC, 1982, 1983, 1984, 1985.

Engineering Services, 1982; U.S. EPA, 1981a; CDC, 1982, 1983, 1984, 1985).

Washington

Seventy-one percent of the state is served by water-supply systems that rely on groundwater. Only 9% of total fresh water use in Washington is provided by groundwater, but in some areas groundwater is the only source of water available. Of all the fresh water withdrawn, groundwater provides 37% of the public supply, 78% of the rural domestic supply, 67% of the supply used for livestock, 5% of the industrial supply, and 4% of the supply used for irrigation (USGS, 1985).

Glacial drift and volcanic rocks make up the principal aquifers

Table 6-28. Washington: Summary of Information from Documented Case Histories

Contaminant	Total	Supplies Aff.	Thr.	Public Health Effects
Industrial	23(31%)	10(43%)	2	
Landfill	9	2		
Petroleum	10(14%)	4		
Fertilizer	5	1		
Organic	19(26%)	15(79%)	1	11
Chlorides	4	3(75%)		
Radioactive	1			
Mining	2			
Other	1			
Totals	74	35(47%)	3(4%)	11

Means of Detection

Wells cont.	20(27%)
Investigation	6
Surface water cont.	1
Spill	3
Leak	6
Fumes in basement	2
None mentioned	36

Remedial Action

Direct		Indirect	
Source elim.	5	Extent determined	7
Soil removed	1	Remedial action	4
Groundwater pumped and treated	6	No action	3 (4%)
		New water provided	10
Trench installed	2	Action being considered	19
Wastes capped	2	Monitoring begun	11
		Landfill closed	1
		None mentioned	19(26%)

Sources: U.S. EPA, 1984d, 1981a; CDC, 1982, 1983, 1984, 1985.

in Washington. Glacial-drift aquifers provide Puget Sound and Spokane Valley with most of their water. Terrace and valley-fill aquifers provide areas near Vancouver with water, and the Columbia River Basalt aquifer supplies the Columbia Plateau region with much of its water (USGS, 1985).

A list of contamination incidents of Region 10 in 1984 put together by the EPA indicates that the greatest groundwater concern in Washington is contamination from industrial wastes. The second greatest groundwater problem is caused by human and animal wastes, and the third by petroleum products from leaking underground storage tanks. In most of the incidents reported, (70%), some sort of remedial action was being considered or had been administered (U.S. EPA, 1984d; 1981a; CDC, 1982, 1983, 1984, 1985). See Table 6-28.

Conclusions

Bearing in mind that these summaries apply only to known incidents of contamination, and that if undertaken, a comprehensive national survey might well uncover other important sources of contamination or different frequencies of the same sources. The summaries show that the problems encountered to date vary from one region of the country to another. Table 6-29 summarizes the results of the survey of the states outlined above. From the inventories collected between 1984 and 1986, industrial wastes were among the three most frequently reported sources of contamination for all of the states except New Mexico, North Dakota, Texas, and Vermont. However, in our 1983 study, human and animal wastes proved to be the most frequent cause of groundwater contamination in the United States. California and Florida have reported, both in 1983 and 1986, that saltwater intrusion in coastal areas has been an important problem. California, Florida, Idaho, Nebraska, North Dakota, and Texas all have reported problems arising from agricultural practices. The industrial Northeast predictably has had problems with industrial wastes, petroleum products, and landfill leachate. New Mexico, Texas, and California have reported problems from oil exploration and the disposal of oil-field brines, as one would expect.

The significant sources of contamination identified from the state summaries (Table 6-29) differ in the order of importance from those identified by the EPA summaries (Tables 6-1 and 6-2) and the

Table 6-29. Most Frequently Reported Sources of Groundwater Contamination in the Nineteen States Covered in This Chapter

State	1980 GW Use (mgd)	Natural GW Quality	Sources of Contamination 1986	Sources of Contamination 1983	Known Contamination Incidents 1986	Known Contamination Incidents 1983	% Affecting or Threatening Water Supply 1986	% Remedial Actions Undertaken 1986
AZ	4,200	Generally good; some mineralization problems	*	1. Industrial and municipal wastes 2. Landfill leachate 3. Human and animal wastes	*	23	*	*
CA	14,600	Generally good	1. Industrial and municipal wastes 2. Nitrates from agricultural practices 3. Saltwater intrusion	1. Saltwater intrusion 2. Nitrates from agricultural practices 3. Brines & industrial wastes	*		*	*
CT	150	Good to excellent	1. Pesticides and fertilizers 2. Industrial and municipal products 3. Petroleum products	1. Industrial and municipal wastes 2. Nitrates from agricultural practices 3. Human and animal wastes	1155	64	*	*
FL	3,800	Generally good	1. Petroleum products 2. Industrial and municipal wastes 3. Saltwater intrusion	1. Chlorides from saltwater intrusion and agricultural return-flow 2. Industrial and municipal wastes 3. Human and animal wastes	260	92	43	69
ID	6,300	Good	1. Petroleum products 2. Pesticides and fertilizers 3. Industrial and municipal wastes	1. Human and animal wastes 2. Industrial and municipal wastes 3. Radioactive wastes	101	29	21	79

223

Table 6-29. Most Frequently Reported Sources of Groundwater Contamination in the Nineteen States Covered in This Chapter (continued)

State	1980 GW Use (mgd)	Natural GW Quality	Sources of Contamination 1986	Sources of Contamination 1983	Known Contamination Incidents 1986	Known Contamination Incidents 1983	% Affecting or Threatening Water Supply 1986	% Remedial Actions Undertaken 1986
IL	980	Good	*	1. Human and animal wastes 2. Landfill leachate 3. Industrial and municipal wastes	*	58	*	*
MA	320	Fair to good	1. Industrial and municipal wastes 2. Human and animal wastes 3. Landfill leachate	*	68	*	100	75
MT	200	Fair to good	1. Petroleum products 2. Landfill leachate 3. Industrial and municipal wastes	*	84	*	12	35
NB	7,100	Generally good	1. Irrigational and agricultural practices 2. Petroleum products 3. Industrial and municipal wastes	1. Irrigational and agricultural practices 2. Human and animal wastes 3. Industrial and municipal wastes	140	35	54	30
NJ	730	Generally good	1. Industrial and municipal wastes 2. Petroleum products 3. Landfill leachate	1. Industrial and municipal wastes 2. Petroleum products 3. Human and animal wastes	616	374	33	85
NM	1,800	Fair to good	1. Oil-field brines 2. Human and animal wastes 3. Petroleum spills during oil extraction	1. Oil-field brines 2. Human and animal wastes 3. Mine wastes	131	105	74	35

ND	110	Fair to good	1. Nitrates	*	57	*	63	56
			2. Petroleum products					
			3. Pesticides and fertilizers					
OR	1,100	Generally good; some brackish water	1. Industrial and municipal wastes	*	83	*	73	61
			2. Landfill leachate					
			3. Human and animal wastes					
PA	1,000	Fair to good	1. Petroleum products	*	576	*	34	63
			2. Industrial and municipal wastes					
			3. Landfill leachate					
RI	37	Good	1. Industrial and municipal wastes	*	68	*	32	68
			2. Landfill leachate					
			3. Road salts					
SC	210	Good to excellent	1. Petroleum products	*	163	89	65	40
			2. Industrial and municipal wastes					
			3. Landfill leachate					
TX	9,700	Fair to good: connate water	1. Human and animal wastes	*	134	*	70	82
			2. Petroleum spills during oil extraction					
			3. Oil-field brines					
VT	ɔ	Generally good	1. Road salts	*	77	*	86	55
			2. Petroleum products					
			3. Human and animal wastes					
WA	750	Very good	1. Industrial and municipal wastes	*	74	*	51	70
			2. Human and animal wastes					
			3. Petroleum products					

* Asterisk indicates data or updated data not available.

Source: Compiled from data provided by the respective states.

225

Groundwater Contamination

Water Quality Report Summary (Table 6-3). None of the surveys can be considered complete, as none are the result of a well-designed, comprehensive national survey. The information contained in these summaries is a best-case scenario; the situation can only get worse as new cases of groundwater contamination are discovered. The present assessment of groundwater contamination would be more useful if a method for estimating severity could be agreed upon.

Figure 6-1. California Groundwater Basins

Source: Adapted from figure used in California Department of
Water Resources, 1975

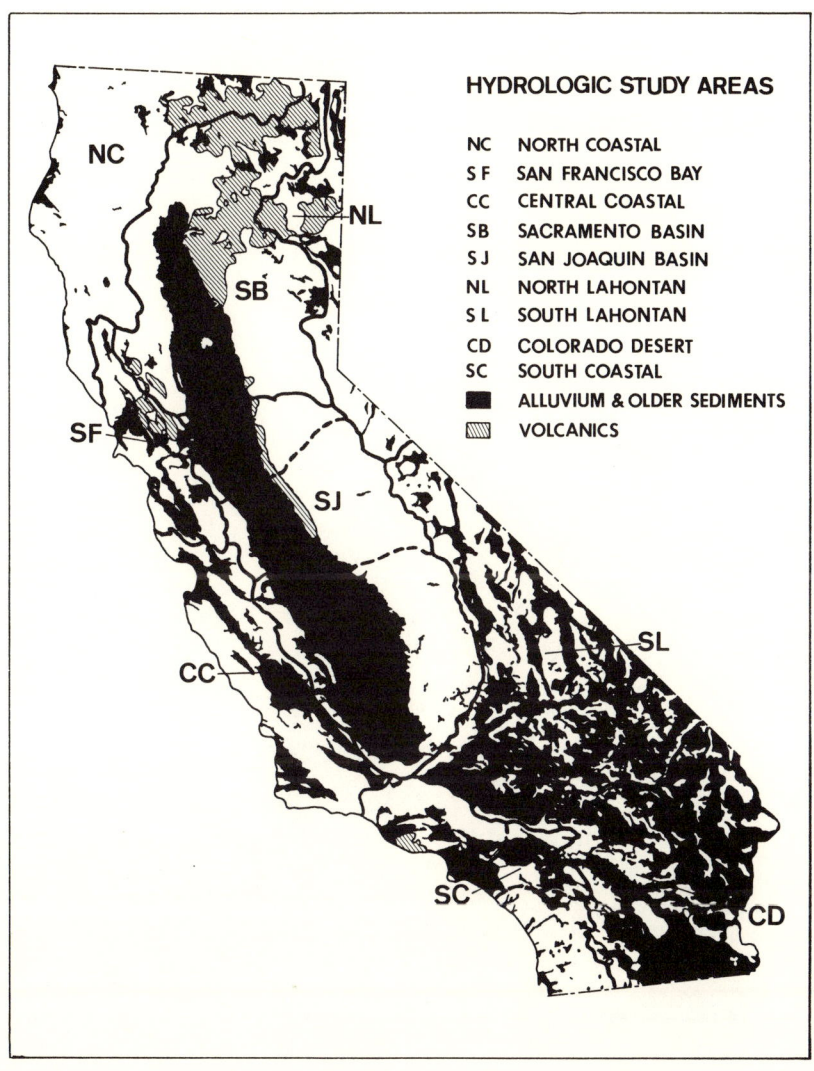

Figure 6-2. Major Aqueducts in California
Source: Adapted from figure used in California Department of
Water Resources, 1975

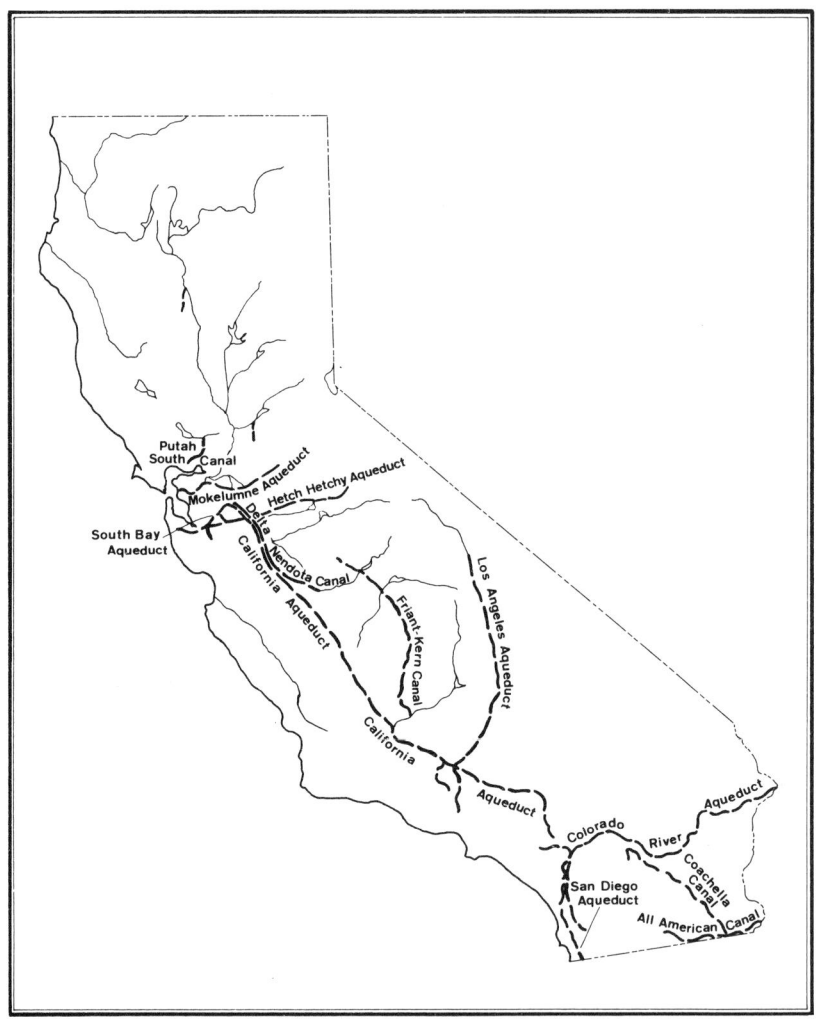

Seven

Monitoring the Quality of Groundwater

- Monitoring the quality of groundwater may help prevent, reduce, and eliminate groundwater pollution.
- The design of monitoring programs varies according to climate, population, hydrology, and pollution sources.
- Costs of monitoring programs may vary considerably from case to case. A thorough monitoring program is expensive, but necessary to appraise the problem of contamination.
- The usefulness of monitoring is highly dependent on the quality of the program.

Monitoring the quality of groundwater may help prevent, reduce, and eliminate groundwater pollution. Such monitoring can provide basic information on the extent and severity of groundwater contamination and can allow for measurement of groundwater quality, quantity, and flow. This information is important in establishing and enforcing a management program. A book by Everett (1980) reviews monitoring methodology, the costs of various methods, data management, and monitoring practices for disposal wells and concludes with illustrative examples. An earlier review, also put out by the General Electric Company (Tinlin, 1976), gives site-specific illustrative examples of monitoring practices for incidents involving oil-field brine disposal, plating-waste contamination, landfill-leachate contamination, oxidation ponds, and agricultural return-flow. The firm of Geraghty and Miller runs periodic seminars on the fundamentals of groundwater quality protection, in which monitoring practices are dealt with in detail (Miller, 1981b and c). The Office of Technology Assessment report (OTA, 1984) covers federal and state monitoring programs and reviews the general approach in conducting a hydrologic investigation. Gillham et al. (1983) prepared a report for the American Petroleum Institute that covers groundwater monitoring procedures and points out sampling biases that may occur during monitoring.

230

EPA Ground-Water Monitoring Strategy

In December of 1985 EPA published the Ground-Water Monitoring Strategy. The Office of Drinking Water, Office of Solid Waste, Office of Emergency and Remedial Response, Hazardous Waste Ground-Water Task Force, Office of Pesticide Programs, and Office of Management Systems and Evaluation prepared the strategy cooperatively under the leadership of the Office of Ground Water Protection. Together they analyzed and summarized the need for and use of monitoring data. They determined that in order to develop useful groundwater protection programs, information on a broad range of environmental problems must be acquired. Information must be gathered by federal, state, and local officials on the causes of these problems, their public health implications, and their potential for control (U.S. EPA, 1985a).

The monitoring strategy focuses on such fundamental activities of groundwater monitoring as improving the quality and management of groundwater data collected and analyzed by or on behalf of EPA programs, and upgrading existing agency monitoring activities and filling the gaps between them (U.S. EPA, 1985a).

The strategy is designed to: 1) be broad in scope and not address specific details of any specific monitoring program; 2) focus on a five-year horizon for implementation, with both short-term and long-term action items; 3) recognize that the job is complex and that EPA and others are interested in considering what constitutes an appropriate overall groundwater monitoring program; 4) provide a balance in the strategy-implementation plan between general action items and a detailed work plan; and 5) focus on solutions to the groundwater monitoring problems identified in the Office of Technology Assessment (OTA) report

Protecting the Nation's Ground Water from Contamination, and not merely restate their analyses.

The Ground-Water Monitoring Strategy recognizes and responds to a number of needs identified in the 1984 EPA Ground-Water Protection Strategy (U.S. EPA, 1984a). These are:

1. The need to strengthen state programs by assisting state decision makers in developing and gaining access to needed groundwater monitoring data.
2. The need to deal with contamination sources of national concern. The strategy addresses ways to acquire information on the extent and consequences of the many sources of groundwater contamination of national concern, so that decision makers may protect the public and the environment.
3. The need to create a policy framework. In order to implement consistency in a groundwater protection policy centered on differential protection, site specific and regional groundwater data are needed.
4. The need to strengthen EPA's internal groundwater organization. The strategy will help EPA by reviewing (a) its approach to monitoring, (b) the need for and use of groundwater monitoring data, and (c) how that data can be made accessible to other environmental decision makers.

The monitoring strategy addresses seven monitoring objectives:

1. To characterize the nation's groundwater resources—which entails collecting background data on groundwater resources and their sellings.
2. To identify new contamination problems—which will provide information on previously unidentified threats to groundwater quality.
3. To assess known problems to support regulatory development and standard setting and respond to site-specific problems— which involves monitoring sites with previously identified contamination in order to support regulatory and standard-setting requirements and to respond to site-specific problems.
4. To ensure compliance with regulations—which entails the collection of monitoring data so that program managers, permit writers, inspectors, and enforcement officials responsible for controlling or abating the consequences of contamination can ensure compliance.
5. To evaluate program effectiveness—which entails the collection of groundwater monitoring data to evaluate regulatory effectiveness in protecting groundwater quality.

231

6. To improve data quality—which involves improving quality assurance and quality-control procedures.
7. To develop a groundwater data management system—which includes activities such as determining what data is to be collected, developing protocols on data entry, and making the data accessible to users (U.S. EPA, 1985a).

232 ## Federal Statutes Requiring Groundwater Monitoring

There are 10 Federal regulations that require monitoring of groundwater (OTA, 1984):

1. Atomic Energy Act—requires low level radioactive facilities to conduct pre-operational, operational, and post-operational monitoring related to the development of geologic repositories; and as part of remedial action programs at storage and disposal facilities for radioactive substances.
2. Clean Water Act—requires treatment facilities to monitor on a case-by-case basis sites of land application of wastewater and sludge from sewage treatment plants.
3. Comprehensive Environmental Response, Compensation and Liability Act—monitoring may be conducted by a state or by EPA in response to the release of any hazardous substance, contaminant, or pollutant (as defined by CERCLA).
4. Federal Insecticide, Fungicide and Rodenticide Act—monitoring may be conducted by EPA in instances when certain pesticides are contaminating groundwater.
5. Federal Land Policy and Management Act—requires facilities to monitor geothermal recovery operations on federal lands for at least one year prior to production.
6. Resource Conservation and Recovery Act—requires facilities to monitor all hazardous waste land disposal facilities. Background quality, detection, compliance, and corrective-action monitoring are all required.
7. Safe Drinking Water Act—monitoring may be required for injection wells used for in-situ or solution mining of minerals (Class III wells) where injection is into a formation containing less than 10,000 mg/l TDS; monitoring may be required for wells injecting beneath the deepest underground source of drinking water (Class I).
8. Surface Mining Control and Reclamation Act—requires facilities to monitor surface and underground coal-mining

operations to determine the impacts on the hydrologic balance of the mining and adjacent areas.

9. Toxic Substances Control Act—requires facilities to monitor prior to the disposal of PCBs.

10. Uranium Mill Tailings Radiation Control Act—requires facilities to monitor active mill tailing sites. Background quality, detection, compliance, and corrective-action monitoring are all required; monitoring may be required for inactive sites if contamination has occurred. **233**

Interstate compacts such as those for the Delaware River Basin and the Susquehanna River Basin have the authority to monitor groundwater quality and to formulate and carry out plans to protect groundwater quality (Everett, 1980). State laws and regulations affecting groundwater vary and are summarized later in this report.

Federal, State, and Private Organizations That Carry Out Monitoring Activities

Federal, state, and private organizations use several different monitoring approaches to determine groundwater contamination. These include:

- Ambient-trend monitoring measures quality in relation to standards and reflects temporal and spatial trends in a groundwater area.
- Source monitoring measures effluent quality and quantity from pollution sources that could affect groundwater. Present methodology is usually geared toward this type of monitoring (Everett, 1980) and concentrates on the most important sources.
- Case-preparation monitoring is used to accumulate data for enforcement actions.
- Research monitoring is used for studies of groundwater quality and pollution occurrence and movement.

Federal

The Water Research Division of the USGS is currently the principal federal division conducting groundwater research (EPA, 1985a). The largest program, the Federal-State Cooperative Program, conducts groundwater investigations that involve the collection of hydrologic data and the investigation of water resources relevant to state and local needs. Within this program, costs are shared equally by the USGS and state and local agencies (U.S. EPA, 1985a). This program is active in all 50 states (OTA, 1984).

The USGS has also proposed the National Water Quality Assessment Program (NAWQAP), which will attempt to describe water quality on a national scale. Other USGS programs involving groundwater monitoring include the Regional Aquifer System Analysis Program (RASA); the National Water Resources Conditions Report, which is published monthly; the Toxic Waste Ground-Water Contamination Program; the Radioactive Waste Program; and the Coal Hydrology and Oil Shale Hydrology Programs (OTA, 1984; EPA, 1985a).

234

The EPA requires resource assessment information to support most of its activities. For example, monitoring is conducted on groundwater at Superfund sites and RCRA sites. The EPA also is involved with the identification and assessment of new groundwater problems. RCRA and Superfund programs conduct investigations on new or additional problems concerning groundwater contamination. Studies conducted under the Clean Water Act and the Safe Drinking Water Act have also been used to identify undiscovered contamination problems. Recently, the National Pesticide Monitoring Plan has been working on the identification of problems associated with pesticide use and exposure-related illnesses. To address known goundwater problems, several surveys and studies are now being conducted by the Office of Pesticides Programs (OPP) and the Office of Drinking Water (ODW) on pesticide contamination, by the ODW on inorganics and radionuclides, by the OTS on leaking underground storage tanks, by RCRA on non-hazardous landfills, and by the Office of Water on publicly owned treatment works/lagoons (EPA, 1985a.)

Other federal agencies involved with groundwater monitoring are the Bureau of Reclamation, the Army Corps of Engineers, the Soil Conservation Service, the Nuclear Regulatory Commission, the Office of Surface Mining, the Department of Defense, the Department of Agriculture, and the Department of Energy (U.S. EPA, 1985a).

State

In 1984, OTA reported that 47 states have conducted inventories for potential sources of contamination, most for waste related facilities.

Forty-nine states monitored sources for potential contamination, with efforts also focused on waste-related facilities and discharge areas. Monitoring priorities, however, vary from state to state. In addition, 10 states monitor private wells for suspected contamination. Ambient water quality is monitored in approximately 38 states by a number of different programs. These programs include USGS monitoring programs, monitoring required for permitted activities, statewide monitoring systems, and monitoring of special sites or areas of concern (OTA, 1984). Virtually all states **235** have programs that monitor activity to ensure compliance with groundwater protection regulations. States with primacy under the Clean Water Act, RCRA, and Safe Drinking Water Act are involved in ensuring compliance. In a recent study on RCRA sites, however, 85% of the states reported a considerable amount of non-compliance (OTA, 1984; EPA, 1986a).

Basic Procedures in Setting Up a Groundwater Monitoring Program

As factors such as climate, population, land use, geohydrology, and pollution sources vary from site to site, the design of an appropriate monitoring program should vary accordingly (Everett, 1980). A single set of guidelines for all monitoring programs would not suffice unless it were very general and flexible. Each program should reflect those specific variables identified at each site.

While designing monitoring systems, five general questions should be kept in mind (OTA, 1984):
- What information is required?
- What techniques are applicable?
- What should be the numbers and locations of measuring points?
- How frequently should samples be collected?
- How much money can be spent?

During all monitoring procedures, steps must be taken to limit any actions that may alter the physical and chemical environment of the samples. These changes may occur from (Gillham et al., 1983):
- physical disturbance during drilling, such as vertical mixing between contaminated and uncontaminated zones

- contamination caused by foreign materials used during drilling operations
- contamination by foreign materials such as dust or soil in the sampling installation or equipment
- contamination caused by materials of the sampling installation, sampling equipment, and sample containers
- adsorption onto materials in contact with the sample
- cross contamination between sampling points

236
- chemical alterations caused by degassing and volatilization
- direct or indirect chemical changes occurring as a result of atmospheric contamination of the sample

The following is a very brief and general outline of the steps that should be taken in implementing a groundwater monitoring program.

Initial Site Assessment

Vital hydrologic and groundwater quality information can be obtained through visual investigation, research of hydrologic and land use information, and in-situ tests of the site in question. Simply through visual examination of the site, hydrologic as well as contaminant information can be assessed. The location of hills and streams can help in determining hydrologic flow, just as areas of distressed vegetation, soil stains, and refuse can help in estimating groundwater contamination movement (Werth, 1985). Hydrologic and land use reports, topographic maps, and aerial photographs can also be of major importance in determining groundwater flow and potential sources of pollution (Werth, 1985). More popular methods that provide for a more detailed, site-specific data set are geophysical testing, temperature studies, tracer studies, soil boring and well drilling programs, and trench and excavation work (Werth, 1985). Geophysical testing methods include (Werth, 1985):

- electrical earth resistivity and electomagnetic terrain conductivity techniques to evaluate depth to water and areal extent of contaminants involving dissolved ions;
- borehole logging to determine vertical limits of aquifers and ionized contaminants;
- seismic surveys to identify subsurface geological conditions;
- proton magnetometers to locate buried metallic objects; and
- soil vapor (soil sniffing) surveys to identify areas of contamination with volatile and semi-volatile organic compounds.

Selecting Parameters

All known or suspected contaminants should be identified and considered for analysis. Table 7-1 shows the constituents in industrial and municipal wastewater that may affect groundwater. The substances chosen should be limited to those for which approved analytical methods exist (Werth, 1985). In the absence of known or suspected contaminants, substances from the area's mineralogy, substances listed in the drinking water standards list, or EPA's priority pollutants may be tested (Werth, 1985). Where cost is a consideration, which is in most cases, testing should be limited to the most mobile contaminants (Werth, 1985). Costs of chemical determinations are shown in Table 7-2 and 7-3.

237

Types of Wells

Groundwater sampling wells can be divided into two general types: single level installations, which are the most common type, and multilevel installations (Gillham et al., 1983). Single level installations consist of one single inlet interval for each borehole (Figure 7-1). This type of well provides no information on the vertical distribution of contaminants. Multilevel installations are used for more detailed mapping of contaminated areas. They consist of either a bundle of small diameter tubes buried together in a single borehole, or a single well with a long screened interval designed for the collection of samples at various depths (Gillham et al., 1983) (Figure 7-2). The advantages and disadvantages of both types of sampling installations are summarized in Table 7-4.

Table 7-1. Constituents in Industrial and Municipal Wastewater Having Significant Potential for Groundwater Contamination

Mining

Metal and Coal Mining Industry

pH	Zinc	Magnesium
Sulfate	Tin	Silver
Nitrate	Vanadium	Manganese
Chloride	Radium	Calcium
Total dissolved solids	Phenol	Potassium
Phosphate	Selenium	Sodium
Copper	Iron	Aluminum
Nickel	Chromium	Gold
Lead	Cadmium	Fluoride
	Uranium	Cyanide

Paper and Allied Products

Pulp and Paper Industry

COD/BOD	Phenols	Nitrogen
TOC	Sulfite	Phosphorus
pH	Color	Total dissolved solids
Ammonia	Heavy metals	Biocides

Chemicals and Allied Products

Organic Chemicals Industry

COD/BOD	Alkalinity	Phenols
pH	TOC	Cyanide
Total dissolved solids	Total phosphorus	Total nitrogen
	Heavy metals	

Inorganic Chemicals, Alkalies, and Chlorine Industry

Acidity/alkalinity	Chlorinated	Chromium
Total dissolved solids	benzenoids and	Lead
Chloride	polynuclear	Titanium
Sulfate	aromatics	Iron
COD/BOD	Phenols	Aluminum
TOC	Fluoride	Boron
	Total phosphorus	Arsenic
	Cyanide	
	Mercury	

Table 7-1. Constituents in Industrial and Municipal Wastewater Having Significant Potential for Groundwater Contamination (*continued*)

Plastic Materials and Synthetics Industry

COD/BOD	Phosphorus	Ammonia
pH	Nitrate	Cyanide
Phenols	Organic nitrogen	Zinc
Total dissolved solids	Chlorinated	Mercaptans
Sulfate	benzenoids and	
	polynuclear	
	aromatics	

Nitrogen Fertilizer Industry

Ammonia	Sulfate	COD
Chloride	Organic nitrogen	Iron, total
Chromium	compounds	pH
Total dissolved solids	Zinc	Phosphate
Nitrate	Calcium	Sodium

Phosphate Fertilizer Industry

Calcium	Acidity	Mercury
Dissolved solids	Aluminum	Nitrogen
Fluoride	Arsenic	Sulfate
pH	Iron	Uranium
Phosphorus	Cadmium	Vanadium
		Radium

Petroleum and Coal Products

Petroleum Refining Industry

Ammonia	Chloride	Nitrogen
Chromium	Color	Odor
COD/BOD	Copper	Total phosphorus
pH	Cyanide	Sulfate
Phenols	Iron	TOC
Sulfide	Lead	Turbidity
Total dissolved solids	Mercaptans	Zinc

Table 7-1. Constituents in Industrial and Municipal Wastewater Having Significant Potential for Groundwater Contamination (*continued*)

Primary Metals

Steel Industry

pH	Cyanide	Tin
Chloride	Phenols	Chromium
Sulfate	Iron	Zinc
Ammonia	Nickel	

Electric, Gas, and Sanitary Services

Power Generation Industry

COD/BOD	Copper	Phosphorus
pH	Iron	Free chlorine
Polychlorinated biphenols	Zinc	Organic biocides
	Chromium	Sulfur dioxide
Total dissolved solids	Other corrosion inhibitors	Heat
Oil and grease		

Municipal Sewage Treatment

pH	Nitrate	Sulfate
COD/BOD	Ammonia	Copper
TOC	Phosphate	Lead
Alkalinity	Chloride	Tin
Detergents	Sodium	Zinc
Total dissolved solids	Potassium	Various organics

Source: U.S. EPA, 1978b.

Table 7-2. Cost of Chemical Determinations of Water Samples, 1986

Parameter	Method of Analysis		Price per Sample ($)
Metals and Minerals			
Aluminum	Atomic Absorption—Flame		13
Antimony	"	Furnace	13
Arsenic	"	Furnace	13
Barium	"	Flame	13
Beryllium	"	Furnace	13
Boron	"	Flame	13
Cadmium	"	Furnace	13
Calcium	"	Flame	13
Chromium	"	Furnace	20
Cobalt	"	Furnace	13
Copper	"	Furnace	13
Gold	"	Flame	13
Iron	"	Flame	13
Lead	"	Furnace	13
Magnesium	"	Flame	13
Manganese	"	Flame	13
Mercury	"	Cold vapor	20
Molybdenum	"	Furnace	13
Nickel	"	Furnace	13
Palladium	"	Flame	13
Phosphorus	"	Furnace	13
Platinum	"	Flame	13
Potassium	"	Flame	13
Selenium	"	Furnace	13

Table 7-2. Cost of Chemical Determinations of Water Samples, 1986 (*continued*)

Parameter	Method of Analysis	Price per Sample ($)
Silicon	" Flame	13
Silver	" Furnace	13
Sodium	" Flame	13
Thallium	" Furnace	13
Tin	" Furnace	13
Titanium	" Furnace	13
Uranium	" Flame	20
Vanadium	" Furnace	13
Zinc	" Flame	13
Drinking Water and Waste Water		
Acidity	Electrometric titration	12
Alkalinity—total	Auto analyser colorimetric (methyl orange)	15
—phenolphthalein	Titration	12
—bicarbonate	"	12
—carbonate	"	12
—hydroxide	"	12
Bromide	Solvent extraction with colorimetric development	25
Carbon dioxide—free	Titrimetric	12
—total	"	12
Carbon—total	Combustion	25
—inorganic	"	25
—organic	"	25
Chemical oxygen demand	Low valve COD ampule method—water	25
	—sediment	35

Parameter	Method	Charge
Chloride	Auto analyzer colorimetric ferricyanide	15
Chlorine—total residual	Amperometric titrimetric	10
—free available	"	10
Chlorophyll A,B,C, Phaeopigments	Spectrophotometer, acetone extraction	40
Hexavalent chromium	APDC/MIBK extraction with graphite furnace atomic adsorption quantitation	20
Color	Visual comparison—platinum cobalt	9
Specific conductance	Conductance cell, Wheatstone bridge	8
Cyanide	Auto analyzer distillation—water	25
	—sediment	35
Fluoride	Potentiometric, ion selective electrode	12
Specific gravity	Method #210	12
Total hardness	Auto analyzer (as $CaCO_3$)	15
Iodide	Titrimetric	12
MBAs (surfactants)	Extraction and colorimetric spectrophotometer	30
Odor	Threshold measurement	12
Oil and grease	Gravimetric separatory funnel extraction—water	25
	—sediment	35
		Field time charge
Dissolved oxygen	Electrode	Variable
Particle size distribution	Sieve analysis test	5
pH	Electrometric	25
Phenolics	Auto analyzer distillation—water	35
	—sediment	
Total residue	Gravimetric	12
Dissolved residue	Gravimetric	12
Total non-filterable residue	Gravimetric	12
Volatile residue	Gravimetric	12

243

Table 7-2. Cost of Chemical Determinations of Water Samples, 1986 (continued)

Parameter	Method of Analysis	Price per Sample ($)
Reactive silica	Colorimetric	12
Sulfate	Auto analyzer barium chloride	12
Sulfide	Titrimetric (iodine)	12
Temperature °C	Calibrated glass thermometer	Field time charge
Turbidity	Nephelometric	12
Nutrient Parameters		
Ammonia nitrogen	Auto analyzer colorimetric phenate	15
Total Kiehldahl nitrogen	" " " "	25
Nitrate nitrogen	" " " "	15
Nitrite nitrogen	" " " "	15
Organic nitrogen	" " " "	30
Ortho phosphorus	Auto analyzer colorimetric molybdate	15
Dissolved ortho phosphorus	Color development w/ spectrophotometer	15
Total soluble phosphorus	" " "	12
Total phosphorus	Auto analyzer colorimetric molybdate	12
Sanitary Parameters		
Fecal coliform bacteria	Membrane filter	15
Total coliform bacteria	Membrane filter	15
Fecal streptococci	Membrane filter	15
Biochemical oxygen demand	5-day incubation	30
Organics		
Trihalomethanes	Gas chromatography/mass spectrometry (GC/MS)—water	150
Chloroform		
Dibromochloromethane		

Parameter	Method	Cost
Bromodichloromethane		
Bromoform		
Polychlorinated biphenyls (PCBs)	Gas chromatography—water	80
	—sediments	85
Polybrominated biphenyls (PBBs)	Gas chromatography—water	75
	—sediments	85
Herbicides (2,4-D/2,4,5-TP, Silvex)	Gas chromatography—water	185
	—sediments	220
Pesticides—organochlorine (Endrin, Lindane [BHC], Methoxychlor, Toxaphene)	GC/MS or gas chromatography—water	75
	—sediment	125
Trichloroethylene (includes solvents and related compounds)	Gas chromatography/Electron capture Detection (GC/ECD) or gas chromatography—water	75
	—sediments	100
Hydrocarbons and "finger printing"	Gas chromatography	125
Priority Pollutants		
Acid extractables	GC/MS—water	225
	—sediments	250
Base/neutral extractables	GC/MS—water	275
	—sediments	350
Purgeables	GC/MS—water	200
	—sediments	250
Pesticides	GC/MS and electron capture detection—water	200
	—sediments	250
Identification of unknown peaks NBS library search		10 per peak

Source: Information assembled from personal communications

Table 7-3. Cost of Priority Pollutant Analysis (in $), October 1981

Samples per Batch	Volatile	Acid Extractables	Base/Neutral Extractables	Pesticides and PCBs	Total
1–10	235	155	225	155	770
>10	215	140	205	140	700

Samples per Batch	13 Metals and Total Cyanides	PCE/TCE
1–10	120	48
>10	110	42

Source: Information assembled from personal communications

Table 7-4. Advantages and Disadvantages of the Main Types of Sampling Installations

Advantages	Disadvantages
Single-Level Installations	
• Suitable in any type of formation. • Simple design. • The installation itself, and where applicable, the packing and sealing materials, can be implanted more easily than in the multilevel installations. • No problems of vertical communication between sampling points due to leaky seals. • Maximum permissible diameter is limited by size of borehole only. Common sizes vary from 1.2 to 15 cm (½–6 in.) for monitoring installations. • Most common sizes of installation 5–10 cm (2–4 in.) do not restrict the choice of sample-collection methods.	• No information can be obtained regarding vertical distributions of groundwater constituents from a single installation (piezometer nests can be used for this purpose but at a much higher cost than multilevel installations). • High cost per sampling point (especially at great depths) as compared to multilevel installations. • Contaminated water may bypass installations with short screened intervals. The problem is also present, but to a lesser extent, in multilevel installations. • In most situations, long-screened installations provide concentration and hydraulic-head values that are spatially averaged over the length of the screen. Therefore they may not give accurate measurements of maximum concentrations. • Because of possible dilution in the well, long-screened installations can be used to confirm the presence but not the absence of a contaminant.

247

Table 7-4. Advantages and Disadvantages of the Main Types of Sampling Installations (*continued*)

Advantages	Disadvantages
Single-Level Installations	
	• Long-screened installations can contribute to aquifer contamination by providing a passage from contaminated zones to uncontaminated zones.
Multi-Level Installations (*General*)	
• Provide information on vertical distribution of groundwater constituents. • Lower cost per sampling point than single-level installations, especially for large numbers of sampling points of great depths. • In general, only small volumes of water need to be removed to purge the installation because of the usually small diameters.	• The installation itself, and where applicable, the packing and sealing materials, are more difficult to place than for single-level installations. • The short screened intervals commonly used in many multilevel installations can miss small zones of contaminated water in heterogeneous materials. The likelihood of this problem can be reduced considerably by using reconnaissance methods (e.g., destructive sampling). • Possibility of vertical communication between sampling points due to leaky seals. Problems of leaky seals can occur in single-level installations, but the consequences are usually less severe than for multilevel installations.

Source: R. W., Gillham, M. J. L. Robin, J. F. Barker and J. A. Cherry. *Groundwater Monitoring and Sampling Bias*. Prepared for the American Petroleum Institute. Washington, D.C., 1983. Reprinted by permission of the American Petroleum Institute.

248

Monitoring Network

The validity of a monitoring effort depends upon the extent to which point measurements are representative of the spacial distribution of contamination (Gillham et al., 1983). The location and number of wells must be compatible with the objectives of the program, the physical characteristics of the hydrogeologic environment, and the physical and chemical characteristics of the contaminant species (Gillham et al., 1983). At most monitoring sites a single monitoring well cannot adequately determine the extent of contamination; it is necessary to have several wells at different depths (Figures 7-3 and 7-4) (Gillham et al., 1983). Groups or clusters of wells, each in separate boreholes, are referred to as piezometer nests. Several nests at a site can provide a three-dimensional representation of groundwater flow and chemical composition of the groundwater (Gillham et al., 1983). Werth (1985) discourages the use of linear alignment of wells unless, for example, the flow along identified bedrock fractures is to be investigated. It is also suggested that one well be installed off site to provide background water-quality information for the area examined.

249

Well Drilling Techniques

Drilling methods should reflect the well construction details, the soil sampling requirements, and the proposed water-quality sampling program (Werth, 1985). Werth (1985) and Gillham et al. (1983) have listed several well drilling methods and point out the positive and negative aspects of each:

1. Shallow monitor wells are usually constructed with an *auger.* This technique is simple and cheap, requires no drilling fluids, and provides good soil samples from core materials. However, both the well depth and diameter are limited, contamination is possible from the metal components of the drill, and the method can only be used in unconsolidated geologic material.
2. When using a *mud rotary,* drillers are not limited to any specific depth or geologic material, and undisturbed core materials can be obtained for sampling. However, the method is slow going, contamination from the metal parts is possible, and fluids are used in the process.
3. *Air rotary* rigs are quite fast. This method is not limited to depth or geologic material, and drilling fluids are not

necessary. However, the method is expensive, the air used in the drilling process may contaminate the samples, and it is not a suitable method in areas contaminated by toxics.

4. A *cable-tool* rig can be used to hammer down a well shaft. It provides good geologic samples, requires little or no drilling fluid, and is not limited to certain geologic materials. However, it is slow.

5. The *barber* rig is a very good method for drilling wells. It is rapid, provides good soil and water samples, and uses very little or no fluids in the process. However, in 1985, there were only three of these rigs in the United States.

6. *Diamond drilling* is a good technique for drilling into crystalline rock. It provides good geologic samples, and there is no depth limit. However, it is used only in consolidated rock or compacted tills; foreign water is used during the process, which may contaminate the formation.

250

All drilling processes have the potential to physically and chemically contaminate samples through both disturbing the hydrogeologic conditions of the subsurface, and through the use of drilling equipment and fluids. There are no general guidelines for the prevention of contamination because of the wide variety of methods and tools used. However, preventive measures can be taken according to the specifics of each case (Gillham et al., 1983).

Casing Materials

The types of well casings and screens used for monitoring depend upon which contaminants are being monitored and how the wells are to be set in place. Most important, the materials chosen should demonstrate resistance to corrosion from the suspected contaminants. Werth (1985) lists six materials from which the screens and casings are made and the pros and cons of using each:

1. PVC—is inexpensive and easy to work with and reasonably durable; however, it is not totally inert.

2. Steel—is also inexpensive and strong; however, it sorbs organics and corrodes readily.

3. Galvanized steel—is inexpensive, reasonably resistant, and strong, but it leaches some metals.

4. Stainless steel—is very strong and resists corrosion; however, it is expensive to use and leaches some metals.

5. Teflon—is inert and corrosion proof, but it is expensive and not a very strong material.
6. Fiberglass—is corrosion resistant, but is not very strong.

Glues and solvents, applied to join PVC pipes together or cloth filters to well casings, should not be used with monitoring wells, for they may contaminate samples.

Annular seals should be placed around all monitoring wells. They are used to guard against infiltration of surface waters to the well screens and to protect the integrity of the confining unit penetrated by the well (Werth, 1985).

Sampling Procedures

There are a large number of sampling devices available for monitoring groundwater. Their principal operating methods include: down-hole collection (or boring), suction lift, positive displacement, gas-drive, gas-lift, jet-pumps, and destructive sampling. For further discussion of these methods, look to *Groundwater Monitoring and Sample Bias,* by Gillham et al.

Careful measures should be taken throughout sampling procedures to prevent bias in sampling results. Potential bias can occur from sample contact with foreign materials used with the sampling device, cross contamination between installations, chemical alterations as a result of degassing or atmospheric contamination, and sampling stagnant water (Gillham et al., 1983). Pressure changes affect CO_2 and pH concentrations. Oxygen and temperature changes alter the solubilities of solids and gases, and also alter biological activity (Gillham et al., 1983; Werth, 1985). These changes can be minimized if samples are pumped directly into units containing monitoring devices.

Evaluation of Data

Reviewing and interpreting monitoring results may enable quality trends, new problems, improvements, and effectiveness of remedial actions to be determined. Methods of groundwater and contaminant modeling may be implemented to interpret these results. Assessment of substandard water occurrences and prediction of future alteration in quality due to population and land use changes should also be considered. The final step in a well-organized monitoring program is the dissemination of the results in a readily understandable and usable form.

Variabilities involved in the groundwater monitoring process must be taken into account when evaluating groundwater data (Saar, 1985). Saar (1985) lists five factors in monitoring that can lead to inaccurate data results: actual variability in the environment; variability resulting from monitoring system setups; variability introduced during sampling; variability caused by laboratory procedures; and lastly, variability introduced during calculations and reporting of data.

252

Monitoring Costs

Depending on several interrelated factors, the costs of a monitoring program may vary considerably from case to case. The two key factors are the type and quality of survey used to locate and identify the contamination and the type and number of monitoring wells. These, in turn, are influenced by site size, geophysical aspects such as depth and type of formations, and often by whether a public agency or the site owner is requesting the program. Surveys of the contaminated site may vary from a preliminary assessment describing whether contamination is present and the general direction of groundwater flow, to a detailed assessment describing the distribution and quantity of each contaminant and the quantity, rate, and direction of the groundwater flow. The monitoring wells vary in configuration and construction, and both are described earlier in this chapter. It has been estimated that 1% of the total capital costs would provide adequate maintenance for the wells. Where a typical monitoring well for a waste-disposal site may cost several thousand dollars, wells monitoring the underground injection of contaminants may cost ten times that much. Monitoring programs for underground injection wells will vary considerably according to formation type, depth, and the location of any hydrologically connected aquifers that also must be monitored. For any monitoring program, costs can only be judged on a case-by-case basis.

Data Management and Information Retrieval

There are several computerized information storage systems suitable for groundwater. The STORET system, developed by the U.S.

Public Health Service and now operated by EPA, has a water-quality file and a general point-source file. It was developed to help ensure compliance with the Federal Water Pollution Control Act Amendments (PL 92-500). This data base contains water-quality data for both surface waters and groundwaters. It includes chemical, physical, biological, and radiochemical water analyses, as well as many other types of information. STORET receives data from several federal agencies, from more than 40 state agencies, from USGS–state cooperative investigations, and, on a monthly basis, from USGS WATSTORE. EPA's Office of Pesticides is currently investigating the use of STORET for the National Pesticides Monitoring Program. RCRA and the Superfund program are also investigating the use of STORET (EPA, 1985a).

253

Drawbacks to the STORET system result from a lack of quality control over the data entered and an absence of hydrologic and well information associated with water quality data. In other words, much of the data received reflects the needs of the investigation from which the data has come, which results in a wide variety of information (EPA, 1985a).

WATSTORE, the National Water Data Storage and Retrieval System, established in 1971, is operated by the USGS. Groundwater data is found in the Ground-Water Site Inventory file (GWSI) of the WATSTORE system. GWSI contains an inventory of wells, springs, and other sources of groundwater, and includes data on site locations, geohydrologic characteristics, well construction history, and various field measurements. This data is collected from USGS–state cooperative investigations, and other USGS, state and local groundwater investigations (EPA, 1985a).

WATSTORE information is limited in that the data is derived, for the most part, from investigations of point sources, and most of the information is related to inorganic contamination (EPA, 1985a).

NAWDEX, The National Water Data Exchange, operated by the Water Resources Division of the USGS, is an index of water data locations used by the USGS and is designed to assist users of water data with the identification, location, and aquisition of needed data (EPA, 1985a).

The USGS Water Resources Scientific Information Center (WRSIC) has computerized abstracts and bibliographic information on water related scientific literature and maintains a data base on water research in progress. In addition, the USGS is now developing a National Water Information System (NWIS), incorporating WATSTORE, WRSIC, NAWDEX, and other computerized bibliographic information into a master file (EPA, 1985a).

The National Water Well Association in Worthington, Ohio, has a detailed computerized library for all groundwater literature. An additional bibliographic data base for computers is being put together by the National Groundwater Information Center, located at the University of Oklahoma. Many states have developed or are in the process of developing their own data management systems.

254

Possibilities for Improving Monitoring Systems

Both states and regulated facilities have encountered problems while implementing or upgrading monitoring systems. The greatest difficulty states encounter seems to be a lack of funds for covering the cost of well construction, sampling, and analysis of data (OTA, 1984). Other problems include the lack of skilled personnel, the lack of laboratory facilities, the lack of accessible data for interpretation, the questionable credibility of some existing data, and the lack of consistency in monitoring techniques and selected parameters, which results in a wide diversity in available data (OTA, 1984; EPA, 1985a).

One obstacle to successful monitoring is choosing the parameters that will reveal vital information on the extent of contamination. The parameters that have been monitored historically may not be the best indicators for current groundwater contamination problems. Many of the items measured, such as total suspended solids (TSS), were chosen because they are easy to monitor. TSS represent an index of a contaminant and are not, in themselves, specific contaminants. The important index parameters that are routinely measured in the field are specific electrical conductance, pH, redox potential, and dissolved oxygen. Specific electrical conductance gives a general indication of total dissolved solids (TDS) (Freeze and Cherry, 1979). Recent groundwater monitoring has shown that toxic organic chemicals are a growing problem (CEQ, 1981a), and to date only about 10% of the organic chemicals contaminating drinking water in the United States have been identified (Page, 1982). As of 1975 EPA reported finding more than 250 different organic chemicals in drinking waters in the United States (Greenberg et al., 1981). Although it would be possible to establish a water-quality baseline for standard water-quality parameters, there are no guidelines or sample sets of carcinogens (Greenberg et al., 1981). It has recently become possible to measure many substances in concentrations of parts per billion (ppb) by means of gas and liquid chromatography and mass spectrometry. It has been suggested that the sophistication of the analytical methods has outstripped our sophistication in interpreting exactly what such low-level contaminant concentrations imply for human and environmental health (A. Wolman, Johns Hopkins University, personal communication; Greenberg et al., 1981).

255

Research Needs

In response to Section 2(a) of the Environmental Research, Development and Demonstration Authorization Act of 1981 (PL 96-569), EPA prepared the *Ground Water Research Plan* (U.S. EPA, 1981b). In this plan, EPA recognizes a need to develop non-contaminating drilling techniques, pumps that accurately sample trace organic compounds from deep groundwater, methods for detecting biological activity that may indicate pollution, and methods for locating possible contamination sources. Geophysical techniques capable of detecting changes in groundwater quality are at present not an adequate replacement for monitoring wells. Efforts to find indicators representative of classes of contaminants should be increased, as should research into non-contaminating tracers. In addition, EPA includes in the plan refinement of advanced detection techniques such as glass fiber and laser optics to reduce monitoring costs and the development of monitoring systems for unsaturated zones for the early detection of pollution.

In addition to research needs for monitoring methods, the plan also includes research programs to (U.S. EPA, 1981b):

- Understand the movement, transformation, and fate of contaminants in the subsurface.
- Determine changes in behavior, movement, and characteristics of pollutants as they move through the subsurface.
- Identify the type and extent of groundwater pollution resulting from specific sources of contamination.
- Assess aquifer rehabilitation strategies and cost.
- Enhance the availability and accessibility of technical information related to groundwater pollution.

Figure 7-1. Typical Monitoring Well Screened Over a Single Vertical Interval
Source: After R. W. Gillham, M. J. L. Robin, J. F. Barker and J. A. Cherry. *Groundwater Monitoring and Sampling Bias*. Prepared for the American Petroleum Institute. Washington, D.C., 1983. Reprinted by permission of the American Petroleum Institute.

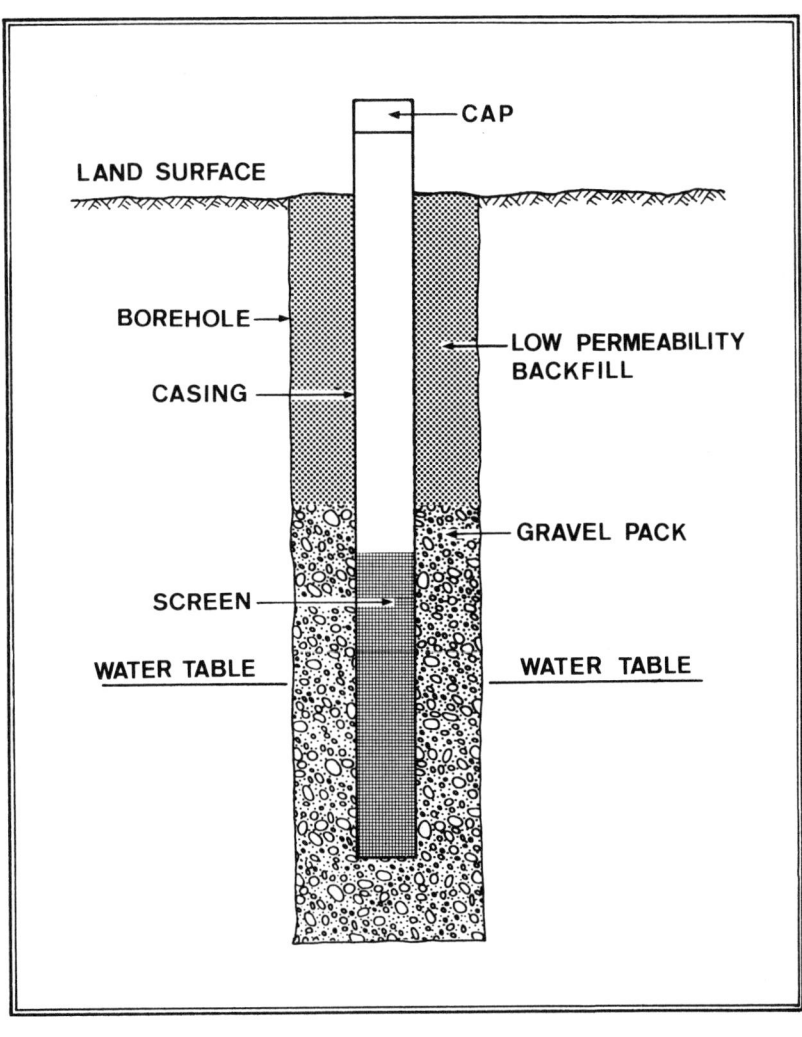

Figure 7-2. Effect of Layered Heterogeneity on the Distribution of Contaminants
Source: After R. W. Gillham, M. J. L. Robin, J. F. Barker and J. A. Cherry. *Groundwater Monitoring and Sampling Bias.* Prepared for the American Petroleum Institute. Washington, D.C., 1983. Reprinted by permission of the American Petroleum Institute.

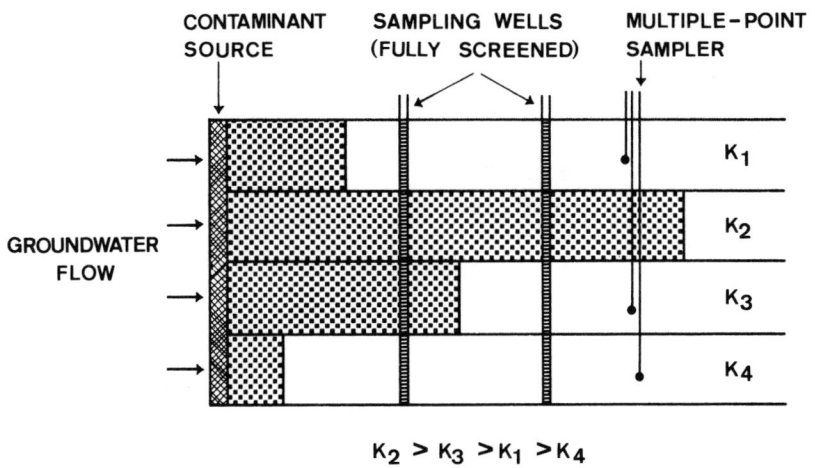

Figure 7-3. Typical Plezometer Nest or Cluster
Source: After R. W. Gillham, M. J. L. Robin, J. F. Barker and J. A.
Cherry. *Groundwater Monitoring and Sampling Bias.* Prepared for
the American Petroleum Institute. Washington, D.C., 1985.
Reprinted by permission of the American Petroleum Institute.

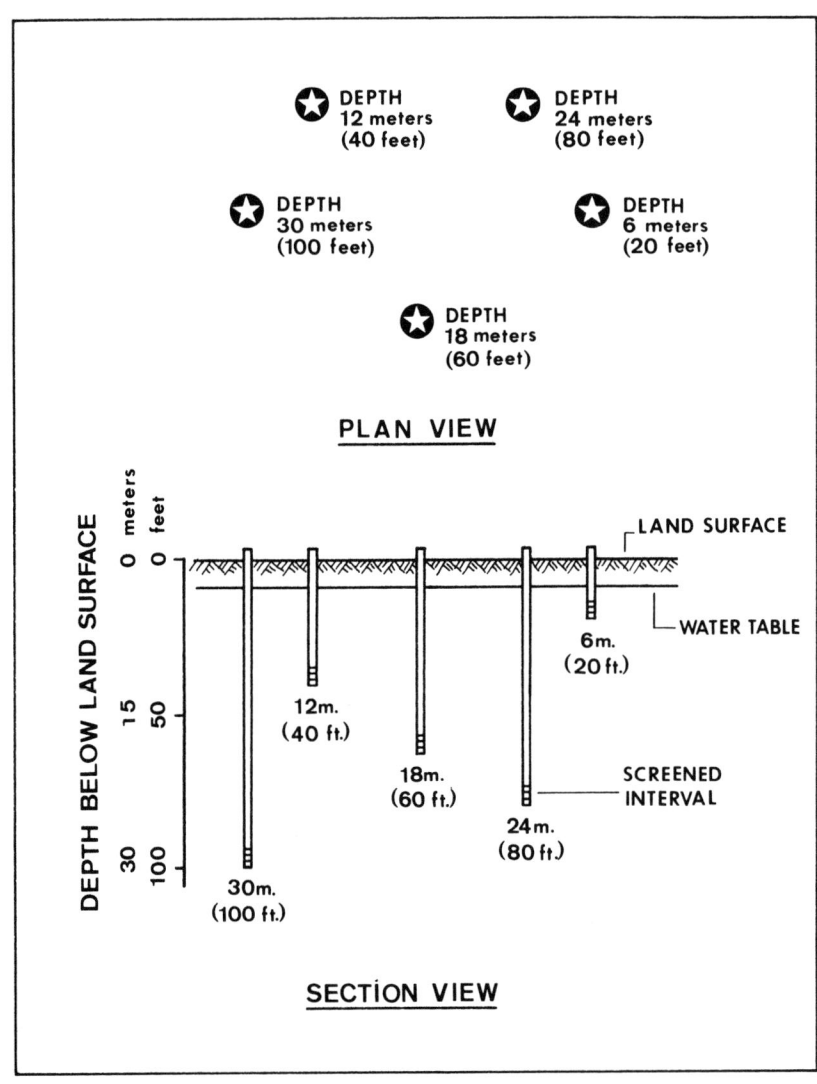

Figure 7-4. Selection of the Screened Interval for a Monitoring Well
Source: After E. P. Werth, Setting Up Monitoring Programs. In *The Fundamentals of Ground-Water Contamination.* Geraghty and Miller, Inc., American Ecology Services, Inc. Syosset, N.Y., 1985.

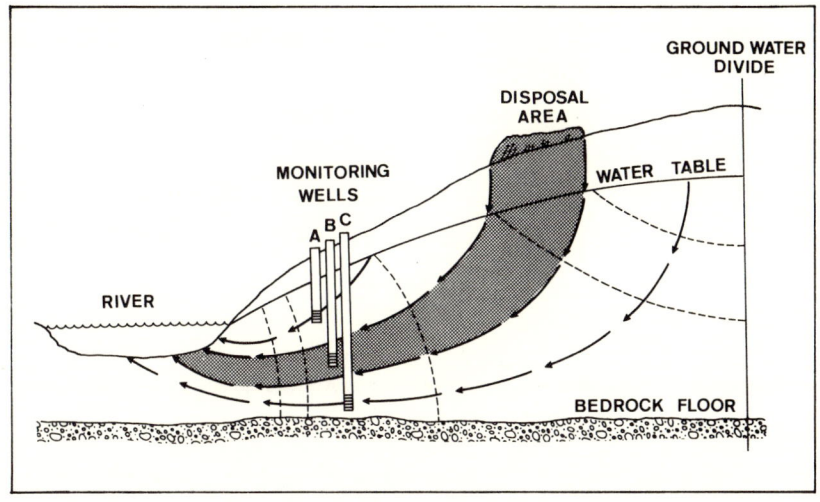

Eight

General Practices for Preventing and Controlling Groundwater Contamination

The number and diversity of potential sources of groundwater contamination make it clear that groundwater contamination will never be totally eliminated everywhere. We can, however, minimize or reduce the impact of many potential sources of contamination to groundwater. Many strategies for the management and protection of groundwater involve policy issues that are generic to the field of resource management in the United States, and in particular to the field of water resources.

The protection of groundwater is not a simple matter, because the potential sources of groundwater contamination are numerous and highly diversified, and they vary greatly from region to region or locality to locality. Variable environmental factors, such as soil types, geological strata, and climatic conditions, only complicate the picture further. This complexity means that there is no uniform pattern to groundwater problems. Each situation must be analyzed in the context of its particular circumstances. Techniques for the protection of groundwater must be tailored to specific conditions.

Water use and allocation in many parts of the country have been determined by historical practice and by pricing policies, taxes, and land use. Policies aimed at modifying the use of groundwater or surface water may include incentives, such as the alteration of the price of water, tax policies to encourage or discourage water connections and use, and institutional structures to manage or allocate the resource. These policies should recognize that many individuals or groups may currently tap, or may in the future tap, the same aquifer for varying amounts of water for different uses.

Regulations of land use, either at the local level or as part of a more comprehensive regional planning endeavor, can also contribute to groundwater protection and management strategies. Land use controls, which allocate or zone certain kinds of activities through positive choice or exclusion, are able to achieve two goals. **261** Land use restrictions may prohibit the siting of potentially polluting activities in the recharge areas of aquifers that contain water of high quality, or may encourage the siting of potentially contaminating point sources, such as waste-disposal activities, in hydrogeologically suitable or less vulnerable areas. Changing the demand for water in different locations may alter surface runoff and residual flow, which may ultimately affect the quality of groundwater.

The management of waste or residuals resulting from human activities must include an assessment of the appropriate method of disposal, whether it be by land, air, or water. For example, a decision to incinerate a waste might well reduce the waste volume remaining, but incineration may increase the concentrations of unwanted constituents in the residual and may contribute to air pollution. Federal regulation over the past decade has restricted air and surface water contamination, thus restricting the use of these two media for the disposal of wastes. Consequently the land disposal of waste has become prevalent. This, in turn, has intensified the problems of waste management and the dangers of infiltration of waste constituents deposited on the land surface that may adversely affect the quality of groundwater. The policies should recognize that contaminated aquifers are not easily rehabilitated and that the slow rate of groundwater movement does not result in the rapid dilution of contaminants in the aquifer. The regulations concerning landfills and impoundments under the Resource Conservation and Recovery Act of 1976 (RCRA) have undoubtedly changed many contaminating practices within the last decade.

Many products of our society are difficult to dispose of without harming the environment. Some sources of groundwater contamination can be ascribed to the existence and use of such products. It is possible, in some instances, to alter the composition of the product in order to reduce or eliminate its undesirable effects on the environment. An example of this approach is the reduction or removal of phosphorus from detergents, which resulted in a reduction of the eutrophication of some surface waters. Water conservation is another strategy that may help to reduce the disposal of wastewater and thereby enhance the protection of groundwater.

Alternative policies for the prevention or mitigation of groundwater pollution might involve a broad range of incentives and disincentives. Among the choices for public policy are different types of regulation at the federal, state, and local levels: market mechanisms; taxes; subsidies; land-use controls; and use modification due to different behavioral patterns and philosophical goals. Many of these potential remedies and tools have long been discussed in the context of water-resource management, in some cases for more than a century. Other approaches are of more recent origin.

The listing of these options is intended to demonstrate the diversity of possible methods or strategies to protect groundwater, and to examine the variety of the most common types of groundwater contamination and the steps that might be followed to reduce these problems. No attempt is made in this report to evaluate either the political feasibility or the social desirability of any of the possibilities described below that may result in the protection of groundwater from contamination. It is hoped, however, that it will illustrate the range of choices open to society in dealing with this problem, and stimulate further investigation into which approaches will most effectively protect groundwater in different situations.

This report does not set forth a rigid system of priorities but does recognize that without some attention to priorities little progress can be made toward the adoption of intelligent policies for groundwater management. Priorities must take into account the natural occurrence of groundwater, its regional variations both in the volume available and in quality, present and historical demands upon the resource, and prospective demands dictated by the interests and geographic mobility of American society. It may be necessary to determine which aquifers are the most important resource for both present and future uses. An obvious example is an aquifer that has a good yield of high-quality water situated in an area where it provides the major source of available fresh water for a large population. It is necessary to judge which threats to the quality of groundwater are the most serious. Such a judgment would involve assessing the nature of the threat to the environment and to public health, the expected time period of the threat, the number of persons involved, the hydrogeology of the site, and the feasibility of mitigating the danger.

In the following discussion we will consider some of the most frequently encountered types of groundwater contamination along with some of the more effective options for dealing with them. It is not intended that these examples should give a complete overview of all the problems that might be encountered, but rather that they serve as selected illustrations of a range of difficulties.

263

Disposal of Hazardous Wastes

Hazardous wastes are generated by industrial, domestic, and governmental activities. They may be part of material disposed of or stored in a landfill, dump, impoundment, or other facility, or they may be generated by chemical reactions or biological activity within the disposal site itself. Under the Comprehensive Environmental Response, Compensation and Liability Act of 1980 (CERCLA), the U.S. Environmental Protection Agency (EPA) requires owners and operators of inactive sites—and under RCRA requires owners and operators of privately owned active industrial sites—to report the site locations and the amount and type of wastes stored. As there may be numerous disused sites whose previous operators are no longer in business, the estimate of the quantity of hazardous wastes generated may not be complete.

Groundwater contamination resulting from the disposal of hazardous wastes has been reported in several parts of the United States, particularly along the eastern seaboard. Contamination occurs when leachate formed in a landfill or dump percolates down to an aquifer or when a surface impoundment develops a leak that allows its contents to percolate through the soil into an aquifer. The quantity of leachate produced is further increased by rainfall and snowmelt as it percolates through the disposed wastes. Many impoundments and landfills are lined with impermeable plastic or clay seals to separate their contents from the soils below, but these seals may develop leaks over time. As leachate passes through the soils, some is attenuated, or taken up, by the soils. The extent to which this occurs is dependent on the type of soil, the hydrogeologic conditions, and the type of contaminant. Thus, deeper aquifers, both confined and unconfined, are less vulnerable to serious contamination by hazardous wastes than are shallow aquifers. Shallow unconfined aquifers in the vicinity of hazardous-waste-disposal sites are extremely difficult to protect. Deep-well injection of hazardous wastes may also present a threat to deep, confined, potable aquifers if this means of disposing of wastes is improperly used.

The control of the contamination of groundwater by hazardous wastes can be approached in several ways. The most obvious solution—but usually one that is unrealistic—is the complete removal of the problem by banning the production, use, and disposal of hazardous material. Even a reduction in the amount of hazardous wastes generated is, however, very important. A reduction can be achieved by recycling, detoxifying, dewatering, and by the substitution of non-hazardous materials. A second method of control is to change the means of disposal, either by incinerating the wastes or by discharging them into bodies of surface water, but to do so would only alter the focus of the pollution problem from groundwater to the air or surface water. A third possibility is to use the best management practices known in the siting, design, and operation of facilities for the disposal of hazardous waste. Sites could be chosen that are away from shallow unconfined aquifers, in areas of low population density, and on relatively impermeable geologic formations. A system of monitoring wells should be installed to detect leaks, as is required under RCRA, since liners and other sealants are not necessarily leakproof. These and numerous other controls are called for by recent federal regulations, although the application of such requirements to existing facilities is at an early stage. Monitoring systems should be allied with the capacity for an effective response to both actual and threatened contamination of groundwaters. One way to reduce the risk of contamination is to retire old facilities that cannot meet the new safety requirements.

The reuse of water is also becoming increasingly attractive. Recycle/reuse technology is considered by industry to be the ultimate solution to the problem of disposing of contaminants, as it can reduce those problems by converting wastes into usable products or into energy.

The disposal of hazardous wastes by injection wells raises specialized problems. In 1983 the U.S. Library of Congress estimated that there were more than 221,000 injection wells operating in the United States, disposing of approximately 11% of the nation's wastes (U.S. Library of Congress, 1984). Injection wells are commonly used for the disposal of hazardous wastes, agricultural and urban runoff, sewage, and wastes produced during oil and gas recovery. In addition to disposal practices, injection wells are often used to recharge aquifers artificially and to extract minerals in solution mining. The injected fluid is usually of a hazardous nature; often it is radioactive or a toxic chemical, petrochemical, or pharmaceutical waste that is difficult to treat. Care must be taken to site deep disposal wells where they can neither inject waste into potable aquifers nor pose the threat of contamination by the upward migration of contaminants through unplugged dry boreholes. The proper construction and maintenance of such wells is also essential.

265

Perhaps the best method for preventing the contamination of groundwater by hazardous wastes is to consider carefully adopting ways to reduce the use and production of hazardous wastes. In addition, different types of disposal for different types of hazardous wastes should be considered. Landfills, dumps, or impoundments may be more suitable for certain wastes than for others. An effort could be made to compact or condense extremely harmful wastes and to place them in inert media such as glass or ceramics before disposing of them.

Septic Tanks

Septic tanks are a widely used method of disposing of residential or other domestic wastewater and sewage, and are sometimes used to dispose of commercial and industrial wastes. When operating correctly, they efficiently renovate wastewater, as the sewage is decomposed by bacterial action and other natural processes. The use of septic tanks is widespread in the United States. A 1980 census estimated that in the United States there were aproximately 22 million operating septic systems serving almost 30% of the population and discharging an approximate total of 1 trillion gallons of waste directly into the ground per year (U. S. EPA, 1986c). In the eastern United States, where the population density is high and there are many shallow unconfined aquifers, the potential for groundwater contamination is greater than in any other part of the country.

At present septic tanks are regulated by all states. Owners of new septic tanks are usually required to have a permit for their construction and to report on the soil, percolation rate, and details of the design and maintenance of the equipment. There is often a minimum-lot-size regulation that prevents septic tanks from exceeding a certain density. In some instances the lot size is set large enough to discourage any excessive development over particularly susceptible aquifers. Some states also require that septic-tank installers and pumpers be registered, and some have maintenance and sludge-disposal rules. Thus, a measure of protection against groundwater contamination is provided by regulations currently in place, although we do not have an accurate idea of how well these regulations are enforced or of the efficiency of the systems that were installed prior to the existence of the regulations. Even the best designed and maintained septic tank may cause groundwater contamination.

266

There are several options that can further reduce the potential of septic tanks to contaminate groundwaters. Of these, the major solution to the problem is the installation of sewers. This is a costly alternative to septic-tank usage, especially to new users in low-density areas. In addition, it obviously creates a need for the construction and operation of sewage-treatment plants, which entails great expense and yet still requires the discharge of certain waste matter, such as nutrients, heavy metals, and complex organic chemicals, into bodies of surface water. Furthermore, these sewage pipes often leak. The EPA estimated a 5% leakage rate from sewage pipelines in the United States (U.S. EPA, 1977). Using this percentage it is estimated that 280 billion gallons of sewage may be leaching into groundwater each year (OTA, 1984).

Another choice that might be feasible in some situations is the use of waterless toilets, which operate by acid incineration, gas incineration, or composting. This choice would eliminate the organic solids and liquids produced by septic tanks, but would not deal with the majority of domestic waste water that results from bathing, laundry, food preparation, etc. In addition, using these toilets results in the formation of ash or compost, the disposal of which might still present a problem.

If home sites are on soil that is not suited to the use of traditional septic tanks, alternative designs can be used. These designs include a dosing system, which distributes effluent evenly over a drainage area, and a mound system, which may be used in sandy or clay soils. The latter method uses a built-up mound of permeable soil for a drain field.

Agricultural Practices

Modern agricultural practices have resulted in the increased production of food and feed grain, and in an increase in domestic animal husbandry. The large increase in the use of agricultural chemicals and the prevalence of irrigation in the last twenty-five years has made agriculture a very significant non-point source of groundwater contamination, particularly by pesticides and nitrate in fertilizers.

Non-point sources of contamination are more difficult to control than point sources, and often in these cases the goal would be to minimize the effect of such contamination rather than to seek its total prevention. Contamination from the nitrate in fertilizers could be minimized by avoiding excessive applications of fertilizers, by supplying only the amount of nutrients that the plants require, by applying fertilizer only in the growing season, by using slow-release fertilizers, and by employing crop rotation systems utilizing nitrogen fixing plants. Pesticides have been found in groundwater in Arizona, California, New York, and elsewhere. Again the threat of contamination from pesticides could be reduced by closely controlling their application rates and timing, by using biodegradable pesticides, and by using pesticides only as part of an integrated approach to pest management in order to reduce the applications of such chemicals. Biological control of pests is becoming the preferred method in many instances.

Confined animal feedlots and their disposal facilities, a relatively recent innovation, are a potential source of groundwater contamination through runoff and infiltration from the waste products they generate. When a great many animals are confined to a small area, the natural assimilative capacity of its soil and vegetation is severely stressed. Various animals produce different ratios of ammonia and phosphate, while nearly all animal waste contains large amounts of nitrate. Nitrate is not attenuated in the soil but may percolate down into the groundwater. A certain amount of attenuation of other heterogeneous compounds does take place naturally in the underground environment. The management and control of concentrated animal feedlots, including stock rate and density, play an important role in protecting groundwater against contamination. Siting restrictions and a reduction in the density of livestock are both possible ways to alleviate the problem. Other safeguards against contamination include mandatory specifications for water diversion and containment, for operation of the facility, and for the storage and disposal of wastes.

The total of irrigated land in the United States is about 58 million acres, and the irrigation of agriculture requires nearly 70% of the groundwater used annually. Irrigation in arid areas may cause an increased mineralization of groundwater. Bouwer (1981) has discussed some options for dealing with these problems. The ways to avoid mineralization include a reduction of the volume of water used and the growing of a crop with a lower evapotranspiration rate. Water use could be reduced by maximizing crop yields per unit of irrigation water, by growing cool season crops rather than warm season ones, by growing crops with low water requirements, and by developing and using antitranspirants. Certain vegetables might not be available at certain times of the year if these practices

268 were followed. The use of subirrigation systems (which are not susceptible to evaporation) in place of furrow and sprinkler systems is one method for reducing solute concentrations in irrigation return flow. When the groundwater does not lie far below the surface of the land, the amount of salts that reach the aquifer can be reduced by using underground drains. The water thus removed can be recycled or disposed of elsewhere. Where the aquifer is deep, the water percolating down can be removed by pumping wells that penetrate only the upper portion of the aquifer, ensuring that the bulk of the groundwater remains uncontaminated. This method is costly and requires a high-energy input.

The overpumping of groundwater in agricultural areas has caused saltwater intrusion in coastal areas, and it has also reduced the quality of groundwater in inland areas where water of poor quality has migrated into a usable aquifer. Land subsidence has been documented in regions where the withdrawals of groundwater exceed the recharge of its aquifer. This has occurred, for example, in some areas of California, Florida, and Texas. Overpumping may be reduced by conservation methods such as lining irrigation canals or using pipes to transfer water to the crop. Another approach is to make the pumping of large quantities of groundwater uneconomical, which could be achieved, in some instances, by eliminating subsidies, imposing high water charges, or limiting the amount of water that can be used.

Accidental Leaks and Spills

Leaks and spills may occur at airports, industrial sites, highways and railroads, gas stations and refineries. There is always an inherent risk of leaks and spills in the storage and transportation of liquids. Organic chemicals are the most frequently reported groundwater contaminant resulting from leaks and spills. A number of states have reported that leaking underground petroleum storage tanks are a leading cause of groundwater contamination (U.S.

General Accounting Office [U.S. GAO], 1984). OTA (1984) estimates that there are 2.5 million underground storage tanks across the nation containing liquid petroleum and non-hazardous chemical substances. At least 2,000 more storage tanks containing hazardous substances were regulated under RCRA in 1981. A state survey, conducted by EPA in 1984, revealed 12,444 incidents of releases from underground storage tanks nationwide. Sixty-five percent of these were petroleum-related incidents from retail gasoline stations, with three percent from chemicals other than petroleum fuels.

269

Pipelines are widely used to transport a variety of fluids, the most common being petroleum. In 1978 there were 700,000 miles of sewage pipelines in the United States transporting 5.6 trillion gallons of sewage (OTA, 1984), and 175,000 miles of non-waste pipeline transporting approximately 10 billion barrels of petroleum products in 1976. It is not known how many of these incidents resulted in contaminated groundwater.

Small hydrocarbon spills may be absorbed by the unsaturated soil zone, but larger ones may reach the aquifer. Very low concentrations of hydrocarbons in water render it unpalatable and unfit to drink. Immediate response and proper cleanup procedures are therefore essential to prevent groundwater contamination. If spills are soaked up with absorbent substances that are then properly disposed of, spilled materials could be prevented from entering the soil and subsequently contaminating the aquifer.

Methods for detecting leaks in underground storage tanks, such as the careful inventorying of stock to bring to light unexplained losses that could be due to leaks, reduce the threat of groundwater contamination. Siting storage facilities on natural or constructed impermeable material could also reduce the problem. The use of non-corrodible materials or liners for pipelines can assist in preventing or minimizing leaks. Many pipelines now have pressure gauges at regular intervals that facilitate the speedy detection of leaks. Pipelines are often located close to the surface of the land so that evidence of leaks can be spotted by aerial surveillance carried out at frequent intervals. Positioning pipelines deeper would not only decrease the possibility of detecting leaks, but would also place them closer to the groundwater. It is important to protect pipelines from rupture by heavy equipment; this could be achieved by educating and licensing the operators of heavy equipment. Careful maintenance and inspection of storage and pipeline facilities could detect system failures before they become a major problem.

Land Spreading and Spraying of Sludges

Sludges are the semi-solid residues that remain after the disposal of wastewater from domestic, commercial, and industrial operations. Groundwater may be contaminated by sludge when it is disposed of by land spreading because precipitation may leak chemical constituents, bacteria, and viruses into the soil.

Sludges can be disposed of by spreading or spraying over land. The use of municipal sludge on agricultural land is more readily accepted in Europe than in the United States. Sludges from manufacturing processes may contain agricultural fertilizers, but they also may contain toxic chemicals and heavy metals that vary from degradable to very refractory. The major hazardous constituents of sludges are heavy metals, organic chemicals, and biological components such as bacteria or viruses.

Groundwater contamination caused by the land spreading or spraying of sludges could best be controlled by a careful consideration of the site and its physical capacity to handle the wastes. The rate of application should be dependent on the depth and richness of the soil, its attenuation capacity, and the use of the land. Sludge disposal may best be sited on land where there is a mild gradient and a good soil cover that allows slow percolation and a high attenuation capacity, where there is a low water table, and where the vegetation is not to be used for food. The rate of application and loading of sludge needs to be carefully regulated so that the soils' capacity for renovation is not exceeded.

Mining

Domestic mining for minerals and coal are multibillion-dollar industries in the United States. Mining operations change the natural environment, either permanently or temporarily, and may disturb groundwater. Waste products from mining can also cause a variety of problems in the quality of groundwater. Because the contamination of surface water by mining activities is of such a serious nature, relatively little attention has been given to the contamination of groundwater, although it too may be a serious problem.

Mining can be hazardous to the quality of groundwater in several ways. Uncontrolled mining may pollute streams and rivers in the aquifer-recharge area, or it may contaminate an aquifer by intersecting it, thereby introducing contaminants. The artificial lowering of the water table, especially in coal-mining operations, may expose sulfur-bearing minerals to oxidation, leading to the formation of an acid solution that may enter the groundwater. Mining for metals may introduce toxic or radioactive contaminants into groundwater. The tailings ponds used for the disposal of wastes from a mine can contribute significantly to the contamination of groundwater. Quarrying may increase the total amount of suspended solids in groundwater.

There are many ways in which the threat of contamination from mining operations could be reduced or, in some cases, eliminated. Since mining and its products are important to society, it would be unrealistic to restrict mining activities nationwide. Nevertheless, it may be beneficial to restrict these activities in particularly sensitive areas. Unlike many contaminating activities, mining cannot usually be shifted to a site that is less vulnerable hydrogeologically. Mining has to take place where the minerals occur. Instead, the best management techniques known and the best technology applicable to mining could be used to minimize groundwater pollution. This can be, and often is, achieved in part by requiring permits for mining operations, well drilling, and the disposal of mine wastes. The Surface Mining Control and Reclamation Act requires that all permitted surface coal-mining operations be in compliance with environmental-protection performance standards, including the special measures designed to minimize disturbances to the hydrologic balance and to prevent toxic drainage from entering the groundwater. The act also gives the secretary of the interior the authority to carry out whatever actions are necessary to abate the adverse effects of past coal-mining practices upon water resources. This act does not cover mining operations other than coal.

Other methods of mitigating the threat to the quality of groundwater from mining include the neutralization of acid mine drainage where practicable and the sealing of aquifers that have been cut through wherever it is technically feasible. The use of impermeable liners and berms in tailings ponds is important, as it is for hazardous-waste impoundments. When the ponds are backfilled, the surface should be covered by soil and revegetated. Long-term monitoring and maintenance is required to keep the heavy metals in the ponds from migrating into the groundwater. The use of quarry sites and abandoned mines as disposal sites should be discontinued unless it can be proven that the disposal technology employed at these sites poses no serious threat to the quality of groundwater.

Highway Deicing Salts

The use of salts (sodium chloride and calcium chloride) to melt ice on highways is a common practice in many of the northern states. In the winter of 1978–79, over 12 million tons of salt were used on roads in the United States. Groundwater is particularly vulnerable to salt contamination because salts are highly soluble and relatively mobile in groundwater.

Salts may be washed off paved surfaces into storm drains or onto adjacent ground, where they may percolate down into an aquifer. Salts for use on highways are often stored in uncovered piles along roadsides and can cause further contamination if the salt is dissolved by rain and snowmelt. Since salt is cheap, the loss of deicing salt by this means was largely ignored until recently.

It is unlikely that the use of highway deicing salts will ever be eliminated. At present, it is the cheapest and most effective method of clearing ice from highways. There are several ways to minimize their use, however, and thereby to minimize groundwater contamination. One available option involves installing heat elements under the pavement. Efforts are being made to make spreading equipment more precise and accurate in the application of deicing salts so that smaller quantities of salt can be used. More and more states are storing salt in enclosed structures in order to minimize its exposure to precipitation. Planning to develop priority systems for salting so that primary and secondary roads are treated according to their use can be encouraged. Heavily used roads require better deicing than roads receiving less use.

Infiltration of Surface Water

Aquifers are often in hydraulic connection with bodies of surface water. Under certain conditions, the polluted surface water of a lake or stream can percolate into a water-table aquifer beneath it, thereby degrading the quality of the groundwater. The development of groundwater near a body of surface water may also draw contaminated water into the aquifer.

A major national effort is now under way to reduce the pollution of surface waters under the mandate of the Clean Water Act. That effort includes the use of pollution-control equipment to reduce discharges of pollutants into public water bodies with a particular emphasis on cleansing wastewater from municipal sewage treatment plants and from industrial facilities. As improvements are made in the quality of surface waters, additional protection will also be afforded to groundwater.

Brine Disposal Associated With the Petroleum Industry

Oil production is accompanied by the production of brine wastes. Groundwater contamination resulting from the disposal of oil-field brines has been documented in 21 states (U.S. EPA, 1985c). In the past it was common practice to dispose of brines in evaporation pits, which often leaked their contents into the soil and raised the possibility of groundwater contamination. This procedure is now normally prohibited.

Current practices in most oil fields have made progress toward eliminating present and future problems associated with brine disposal. The practices include the injection of the brine into deep formations that are generally deemed unsuitable for other uses and the reinjection of brines into oil-producing formations to enhance oil recovery. Both of these practices may pose a potential threat to groundwater quality if they are not properly regulated or if the regulations are not enforced. Reinjection of brines is considered to be potentially a less contaminating method of disposal than the use of evaporation pits.

Development of Groundwater

The simple action of pumping water from wells may bring lesser-quality water into their zone of influence. In inland areas pumping can cause saline water to migrate from saline aquifers through leaky aquitards into potable water. In coastal regions the problem is often one of saltwater intrusion into potable water.

With a clear understanding of the hydrogeologic conditions of the area where groundwater is pumped, these problems can be anticipated and pumping accordingly restricted to a safe level. The restriction of pumping can itself limit the growth of communities, and land zoning may also be used to prohibit excessive development in areas where saltwater intrusion may occur. In coastal areas the injection of freshwater barriers can reverse the hydraulic gradient so that the flow is toward the sea rather than toward the pumping wells, thus effectively reducing or eliminating saltwater intrusion. The injection wells required to create a freshwater barrier are an expensive undertaking, however, and their efficient operation requires a source of water of reasonable quality which is sufficient for continuous injection. In some situations physical barriers between the aquifer and the seawater can be built in the ground to protect groundwater from saltwater intrusion.

274

Contamination From Radioactive Sources

Radionuclides occur both naturally and in waste products from hospitals, academic institutions, utilities and other commercial reactors, defense operations, industries, uranium milling operations, and fuel-recycling facilities. In addition, they may occur from atmospheric fallout, and from nuclear accidents. These radionuclides may pose a threat to the integrity of groundwater quality.

The primary way to protect groundwater from naturally occurring radioactive contaminants is to avoid the development of groundwater in those areas where the minerals occur. Contamination from radioactive wastes can often be mitigated by proper storage. Many sites and various methods of storage are being investigated. For low level radioactive waste disposal, it is generally agreed that the technology exists for siting and safe packaging, handling, transport, and isolation of the wastes (Hileman, 1982). Currently, low level radioactive wastes are concentrated at three commercially operated sites and federally maintained disposal sites across the country. By 1986, under the National Low-Level Radiation Waste Policy Act of 1980, each state will be responsible for the disposal of all low level radioactive wastes generated by commercial operations within its borders (D. Siefken, Nuclear Regulatory Commission, personal communication). However, this goal has not been achieved.

A great deal of research has sought to determine the best method for isolating concentrations of highly radioactive wastes from the biosphere. The wastes may be solidified in glass or ceramics and may be buried in geologically stable formations that can withstand high heat for at least 10,000 years (Hileman, 1982). The ultimate disposal for high level radioactive wastes is actively being sought. Three deep-mined burial sites (1,980 to 2,970 feet deep) are being considered for the first repository scheduled to be completed by 1998 (Environmental Reporter, 1986). These are located in Yucca Mountain, Nevada, Deaf Smith County, Texas, and Hanford, Washington. These wastes are presently in temporary storage facilities at utility and defense sites. A search for a second repository is centered on granite formations found in selected eastern and midwestern states (League of Women Voters, 1985). Once a designated amount of waste has been placed in the first repository construction of the second will begin (League of Women Voters, 1985).

275

Conclusion

This review of options for protecting groundwater from various types of contamination illustrates the complexity of establishing effective overall programs that will protect the quality of groundwater. The dominant characteristic both of the potential sources of contamination and of the alternative approaches for the protection of groundwater is their diversity.

It must be realized that several of these procedures are still in experimental stages, and their viability as options must still be determined. Each situation must be evaluated clearly in its own particular context. Any comprehensive range of programs developed to deal with the problem of groundwater contamination must be prepared to deal with the wide range of the situations described.

Nine

Remedial Action and the Rehabilitation of Aquifers

- Aquifer rehabilitation is difficult, expensive, and time-consuming, with no guarantee of complete success.
- Given the present state of technology, remedial actions may be impossible in many cases.
- Where contamination has occurred there are many possibilities for dealing with the problem, ranging from leaving the contamination for degradation by natural processes, to implementing detoxification techniques directly at the site, to removal of the contaminants.
- Many methods of aquifer rehabilitation are in the experimental stages only, and relatively few are presently cost effective.
- The choice of remedial action depends on the physical characteristics of the site, the nature of the contamination, the importance of the aquifer, and the resources available.
- Costs of remedial action must be estimated on a case by case basis and may vary from several thousand to several billion dollars.

Upon the discovery that an aquifer is contaminated, a decision must be made as to what kind of remedial action to take, if any. Several recent reports have reviewed and discussed various methods of remedial action that can be applied to contaminated aquifers (Lindorff and Cartwright, 1977; U.S. EPA, 1980; Hajali and

Canter, 1980; Geraghty, 1981; OTA, 1984; Canter et al., 1985; Nyer, 1985). All of these reports agree that remedial action is complicated, time-consuming, and expensive, and that the best solution to groundwater contamination is prevention. When contamination is extensive and has occurred over a long period of time, the cost of remedial action can be very high. When the source is unknown, a policy of no direct action to remove or halt the contamination may be the most suitable course of action. In both of these cases, a cost-effective solution may be to locate a new source of water (Lindorff and Cartwright, 1977). For rehabilitation of contaminated groundwater where remedial action is deemed necessary, as in the case of a sole-source aquifer, a wide range of treatments including detoxification and stabilization alternatives exist. These remedial management alternatives may involve in situ remedial techniques, withdrawal treatment techniques, and final disposal options.

277

In Situ Remedial Techniques

Several techniques of in situ treatment, either through detoxification, stabilization, or immobilization, are currently being used, while others are still under development. In general these techniques are difficult to apply. The suitability of the various treatment options depends on the chemistry of the contaminants that are present either in a heterogeneous leachate, in the affected groundwater, or bound to the shallow unsaturated soil matrix. Other factors influencing the suitability of these in situ treatment alternatives include the subsurface mixing regime and reaction kinetics, competing or interfering reactions with the soil matrix or other injected materials, formation of toxic by-products, and environmental conditions such as pH, temperature, and solubility (Sills et al., 1980). In experimental stages are some promising options that introduce biological cultures, chemical reactants, or sealants into the affected soil or groundwater.

In Situ Detoxification

Methods for in situ detoxification include biological degradation, chemical detoxification, in situ physical treatment, and remediation via natural processes, many of which are in experimental stages of development.

Biological Degradation

Biological degradation occurs when microbial organisms present in or introduced into the subsurface environment utilize contaminating substances and produce non-toxic degradation products. Recent research indicates that microorganisms can be found throughout subsurface profiles (Ward and Thomas, 1987). Many strains are capable of degrading both simple and complex organics through either aerobic or anaerobic degradation (Suflita, 1987). Nutrient supplements or adjustments of temperature and dissolved oxygen may be required for biologically efficient environments (Nyer, 1985). The pH, redox potential, pore space, hydrostatic pressure, availability of water, salinity, and concentration of the contaminant itself also influence the control of biodegradation (Canter et al., 1985; Hall, 1984; Ward and Thomas, 1987). A method presently being considered by the EPA as an approved land treatment procedure involves blending preselected biodegradable semi-solids (e.g., some petroleum products) into the top layer of appropriate soil by disc harrowing, and then letting it degrade naturally. Nitrogen and phosphorus are often added to the soil before blending to enhance the process.

Problems encountered with biological degradation result from attempting to achieve uniform distribution and mixing of introduced microorganisms in the contaminated matrix and to achieve and sustain ideal conditions for optimal degradation, and also from the cost involved (Miller and Sgambat 1985). Present research is directed towards the characterization and adaptation of both indigenous and introduced subsurface microbial populations in aerobic and anaeorbic environments, the manipulation of degradation, and the determination of their degradation by-products (Hall, 1984; Ward and Thomas, 1987).

Chemical Detoxification

In situ chemical detoxification involves the injection or application of specific chemicals that react or otherwise alter contaminants (OTA, 1984). In situ chemical detoxification may be accomplished by the following methods:

- Injection of neutralizing agents into the contaminated plume to neutralize acid or caustic leachates.
- Addition of oxidizing agents such as ozone, chlorine, or hydrogen peroxide, for destruction of organic compounds (Nyer, 1985). The addition of hydrogen peroxide into the soil subsurface also stimulates the aerobic activities of organisms.
- Leaching techniques, which utilize super-critically hot fluids

to leach organics and metals from the soil matrix. These techniques are being tested for their feasibility in the field as well as in treatment facilities (Robinson, 1987).

- Land spreading of chemical agents onto an unsaturated contaminated area accompanied by soil shredding. This technique enhances chemical detoxification, aerobic activity, and volatilization (Robinson, 1987; Nyer, 1985).

The greatest problem encountered with in situ chemical treatment is in the creation of good contact through mixing of the contaminant and the detoxifying agent (Nyer, 1985).

279

Physical Detoxification

One physical treatment technique involves water table adjustments (achieved by pumping water into or out of the saturated zone). Lowering the water table and thus expanding the unsaturated zone can increase natural degradation processes. Likewise, artificially raising the water table can induce cleansing through flushing actions (OTA, 1984).

Another physical in situ method involves the recovery of volatile contaminants from the unsaturated zone by using vacuum extraction. Volatile contaminants, such as chlorinated solvents, may be extracted from the unsaturated layer (the vadose zone) by constructing wells into the contaminated zone and extracting the volatiles by vacuum. This method of extraction has been successful in contaminated clayey soils to depths ranging from 75–180 feet, and also in the vadose zone of fractured bedrock to a depth of 300 feet (Malot, 1985). Monitoring wells are usually installed at the same time that the extraction wells are put in place. If the monitoring wells indicate that the groundwater (aquifer) is contaminated, down-gradient wells should be pumped until they are free of the contaminants. This procedure is usually carried out after vacuum extraction is in place, so that all contamination is removed and/or prevented from reaching other down-gradient wells (Agrelot et al., 1985).

The in situ venting of volatiles to the atmosphere (air stripping) may be accomplished in several ways. One way forces volatile organics out of the saturated zone into the unsaturated zone by injecting air below the water table. The volatiles are then released into the atmosphere (Dr. John Corey, personal communication). Another method involves driving wells into the contaminated zone and allowing the volatiles to diffuse from the unsaturated zone into the wells and then into the atmosphere (Dr. Glenn Paulson, personal communication).

Other physical processes, most of which are in the experimental stages of development, involve thermal techniques using various radio frequencies to drive off volatiles directly from the soil (Robinson, 1987).

Natural Detoxification
Passive remediation, leaving the contaminated area alone to allow the contaminants to biodegrade naturally, may be the best solution in many instances where public and environmental health are not directly affected (Robinson, 1987). Often passive remediation is accompanied by containment measures.

280

In Situ Stabilization

Stabilization alternatives prevent the plume of contamination from extending. This is usually accomplished by converting contaminants to insoluble forms or by encapsulating them into a highly insoluble matrix. The chemical conversion of contaminants to an insoluble form is limited to inorganic species with insoluble salts. For example, immobilization of heavy metals and metalloids by metal hydroxide precipitation results from pH modification within the subsurface environment. Immobilization of heavy metals also results from the modification of reduction/oxidation conditions in order to convert heavy metals into their low-soluble reduced form (Mutch, 1987). This means of treatment can be effective in the removal of the last few percent of contaminants present within the plume (Mutch, 1987).

Encapsulation in an insoluble matrix can be achieved by using several commercially available products, such as silicate, sulfur, or organic polymer-based cementaceous compounds. Vitrification of contaminants and the soil matrix through heat fusion has been used in the field, but is not commonly practiced nor practical economically (Robinson, 1987). However encapsulation of hazardous and radioactive waste into glass, ceramic, or other inert solid material is potentially viable and is now being used in some cases (Canter et al., 1985).

In Situ Immobilization

In situ remedial immobilization involves the partial or total containment of the contaminated groundwater either as a single remedial stabilization technique or to enhance the efficiency of other collection, withdrawal, treatment, or detoxification options (Sills et al., 1980).

A low permeable cap can be used to prevent precipitation from leaching contaminants into groundwater and partially contain contamination, but it will do nothing to reduce contamination that existed prior to the capping (Geraghty, 1981). Another method used to partially contain contaminants involves surface water control through the construction of drainage ditches or diversion berms. **281** This method may also reduce the risk of contamination from infiltration (Canter et al., 1985). Hydrodynamic control may also be used to limit horizontal migration of contaminants through selective pumping by creating artificial pressure troughs and sumps (OTA, 1984).

Methods of total containment involve the impoundment of contaminants behind physical barriers (cut-off walls) or the use of injection wells to drain contaminated fluid into a non-potable zone below the contaminated aquifer (Figure 9-1). With the use of this type of injection well, there is no need to handle, treat, or dispose of the contaminated fluids at the surface (Geraghty, 1981). Impoundment barriers include slurry walls, grout curtains, sheet steel, and floor seals. Slurry walls, composed of concrete slurry or gels, are emplaced by digging a trench (Figure 9-2) and are, therefore, difficult to use in bedrock or for contamination more than 80–90 feet deep. Grout curtains are formed by injecting a grout of cement, bentonite, epoxy resins, rubber, lime, fly ash, or bitumen under pressure through numerous wells (Figure 9-3). The uncertainty as to whether the injected grout forms an impermeable barrier and the depth limit of 50 to 60 feet are the major drawbacks to the use of grout curtains. Sheet steel can be driven into the soil to form a wall to a depth of about 100 feet, but the formation of impermeable joints between sheets is difficult (Figure 9-3). Slurry walls, grout curtains, and sheet steel are usually placed below the water table and surround the site in order to limit the horizontal migration of contaminants in the saturated zone (OTA, 1984). Bottom sealing, another in situ containment technique, involves the injection of a layer of grout under the contaminated area to prevent contamination from spreading downward (Mutch, 1987). This process may be necessary when the aquifer is deep and the plume of contamination extends only part way through the thickness of the aquifer (Figure 9-3). This procedure is costly, however, and there is little guarantee that a tight seal will result. Another containment procedure in experimental stages involves soil freezing. This technique temporarily prevents contaminants from moving and enhances the ease to which they may be excavated (Robinson, 1987).

Withdrawal Treatment Techniques

282

Methods of contaminant withdrawal include removing or facilitating the removal of contaminated groundwater or soils from the subsurface (OTA, 1984).

Methods of Withdrawal for Treatment

The feasibility of collection and withdrawal techniques is usually governed by the size and geohydrology of the site. Important characteristics of the geohydrology to be considered include the site surface and groundwater topography, soil hydraulic characteristics, depth to watertable, and depth to impervious layer. Various collection and withdrawal options include the following:

Pumping involves the removal of contaminated groundwater by pumping from wells or drains. Pumping is an effective method for withdrawal which is not limited to plume depth and can also control the lateral and vertical movement of the contaminated plume (OTA, 1984). The major concern with this method, however, is the amount of time that fluid has to be pumped in order for concentrations of contaminants to reach acceptable levels (Figure 9-4).

Gravity drains can be used to intercept the plume and utilize the force of gravity to collect contaminants for their removal from the subsurface (OTA, 1984). This technique also has the ability to control the lateral and vertical movement of the contaminated plume (OTA, 1984). Unfortunately drains are not very effective in deep aquifers or areas of hard rock (Figure 9-5).

Withdrawal enhancement uses various methods to enhance the ability to withdraw either groundwater or contaminants from the subsurface. These methods usually involve increasing the solubility of contaminants in water with the injection of steam, solvents, surfactants, nutrients, bacteria, or heat (OTA, 1984).

Excavation involves the direct removal of contaminated soil and/ or groundwater (OTA, 1984).

Cut-off trenches may intercept the contaminated water, if the plume is not too deep, and drain the water for further treatment or for disposal in a suitable body of surface water (Figure 9-6).

Treatment Methods

Physical, chemical, and biological methods can be used to detoxify contaminants found in groundwater. Treatment takes place after contaminated soil or groundwater has been removed from the subsurface, and may be applied at the source, at on-site treatment units, at off-site wastewater treatment facilities, or at the point of use (OTA, 1984). The feasibility and performance standards for a treatment method depend on the susceptibility to treatment of the leachate, of the contaminated soil, or of the groundwater. Treatment methods are listed and summarized below; many of these are still experimental and not in general use (Sills et al., 1980; OTA, 1984). Table 9-1 summarizes the suitability of various methods.

283

Table 9-1. Suitability of Treatment Processes

	Volatile Organics	Non-Volatile Organics	Inorganics
Air stripping	Suitable for most cases	Not suitable	Not suitable
Steam stripping	Effective concentrated technique	Not suitable	Not suitable
Carbon adsorption	Inadequate removal	Effective removal technique	Not suitable
Biological	Effective removal technique	Effective removal technique	Not suitable— metals toxic
pH adjustment precipitation	Not applicable	Not applicable	Effective removal technology
Electrodialysis	Not applicable	Not applicable	Inefficient operation/ inadequate removal
Ion exchange	Not applicable	Not applicable	Inappropriate technology— difficult operation

Source: After L. W. Canter and R. C. Knox, eds. *Ground Water Pollution Control.* Chelsea, Mich: Lewis Publishers, Inc., 1985. Copyright © 1987 by Lewis Publishers, Inc.

Physical Treatment

Skimming involves the removal of floating contaminants in multi-layer solutions (e.g., oil, grease, and hydrocarbons).

Filtration involves the physical retention and subsequent removal of contaminants present as suspended solids.

Reverse osmosis (RO) converts the leachate to fresh water by using pressure in excess of the osmotic pressure of the leachate to force fresh water through a membrane that is permeable to the water molecules but not to the leachate molecules. This process is effective for the removal of most dissolved organics, inorganic salts, heavy metals, and emulsified oils.

Ultrafiltration (UF) also uses a pressure-driven method of separation through a membrane, but it operates at lower temperatures and is suitable only for leachates with larger molecules.

Air stripping. Contaminated water or soil that has been brought to the surface may be air stripped by air injection to facilitate the volatilization of contaminants and their removal to the atmosphere.

Steam stripping involves the fractional distillation of volatile organics or gases by heating.

Chemical and Biological Treatment

Precipitation, coagulation, and clarification processes precipitate and coagulate dissolved and colloidal particles but are ineffective in removing soluble organic and inorganic substances. Frequently the chemical processes are used in conjunction with other treatment options.

Ion exchange uses a resin bed to remove selected toxic or undesired ions and replace them with harmless ions. The system is mobile and can be loaded with various types of resins (exchange resins or absorptive resins) to demineralize water that is low in organics and to remove organic or inorganic substances. The resins may then be regenerated using acids, bases, or brine solutions. The suitability of various resins depends largely on their ability to be regenerated.

Activated carbon (AC) can remove a wide range of organic and inorganic material from liquid and aqueous-phased streams by sorption on activated carbon (AC) columns (granular) or in suspension in batch reactors (powdered).

Electrodialysis separates and removes positive or negative ions under the action of an electrical field.

Chemical transformation involves oxidation-reduction reactions for the chemical conversion of contaminants to less toxic substances (e.g., by ozone treatment, hydrogen peroxide treatment, ultraviolet photolysis, and chlorination).

Thermal detoxification involves raising of water, soils or contaminants to high temperatures in the presences of oxygen. At temperatures between 1500 and 2000 degrees Fahrenheit essentially all organic compounds are oxidized (Nyer, 1985). Many innovative thermal treatment techniques are currently in the experimental stages of development or recently applied in the field. They include electric reactors and infrared incinerators and use of molten glass and salts to thermally change waste or drive off volatiles (Robinson, 1987).

285

Biological treatment processes involve the transformation and removal by microorganisms of dissolved and colloidal biodegradable contaminants; they include both aerobic and anaerobic processes by introduced and indigenous microorganisms. Many experiments with anaerobic and aerobic biological degradation, are being researched for both in-situ remediation and groundwater treatment.

A treatment process may also include more than one of the above options. For example, biological treatment processing may include an activated carbon step in order to cleanse the contaminated leachate of all types of toxic substances.

Final Disposal

The feasibility of any treatment depends on the amount of end-product and the method of final disposal. The three primary final disposal options are discharge to a municipal sewage-treatment plant, discharge to a surface-water body, or land application. Any discharge to a sewage-treatment plant requires pretreatment to remove contaminants that could damage collections sewers, pumping stations, or the plant itself (Sills et al., 1980). In order to discharge to a body of surface water (Figure 9-6), certain planning criteria must be met, such as water-quality standards, federal or state discharge-permit requirements, and overlapping federal and state water-quality laws and regulations. Land-application processes include slow-rate irrigation, overland flow, and rapid infiltration. The feasibility of these techniques depends on data collected about the decomposition/application rates, toxic effects on crop or volunteer vegetation systems, and soil-renovation capabilities (Sills et al., 1980). Often with organics, the method of final disposal is to burn or rebury.

Determination of Suitable Remedial Management Alternatives

Sills et al. (1980) have suggested a two-phase evaluation procedure for determining the suitable remedial management alternatives. The first phase, a preliminary screening, determines the most viable management alternatives for a given site. The use of computer simulation to evaluate potential remedial action options may be considered. In the second phase, these viable alternatives are evaluated according to technical, economic, and environmental criteria. An iterative screening procedure can compare alternatives as to their effectiveness in treating the contaminants, the establishment of their performance criteria, their cost-effectiveness, and their potential environmental impacts.

In determining the cost-effectiveness of each alternative, unit costs, obtained from published guides, can be applied to each activity in the management alternative. These costs will include short-term costs of capital, labor, and installation and long-term costs of maintenance and monitoring (Paige et al., 1980). Total costs management alternatives can be compared by converting all capital, operational, and maintenance costs at the end of the life period to an equitable present-worth cost at the first day which, if invested at a given interest rate, would exactly yield the necessary monies to cover annual costs of the management practice chosen (Sills et al., 1980). Depending on the selected remedial management alternative and the characteristics of the site, the cost of remedial action may vary from several thousand to several billion dollars. An estimation of the average costs of various remedial methods are listed in Table 9-2.

The appropriate response to a groundwater-pollution problem will depend upon the physical characteristics of the site and the nature of the contamination. At times, no remedial action is necessary or, at least, cost-effective, and some indirect action, such as locating a new water source, is the best solution. In other cases, some remedial action is needed to halt the contamination and, possibly, to rehabilitate the aquifer. Two categories of management alternatives for remedial action exist: in-situ detoxification, stabilization, and immobilization and conventional withdrawal treatment and final disposal. Selection of viable management alternatives will depend on the nature and extent of contamination, site-specific feasibility, including cost-effectiveness, and potential environmental impacts. In any groundwater pollution problem, prevention would have been the best solution. Prevention

Table 9-2. Average Costs and Characteristics of Direct Remedial Methods, 1980

Method	Average Estimated Costs[a]	Characteristics
Surface Water Control		
Contour grading	510	Increases runoff, reduces infiltration.
Surface water diversion	55	Diverts surface water from fill.
Surface sealing (Clay, fly ash, concrete, PVC)	639–1,336	If locally available, native clay is an economical means of retarding infiltration.
Groundwater Flow Control		
Bentonite slurry trench (wall 1,700 ft. long × 60 ft. deep)	1,860	Simple construction methods; retards groundwater flow. Mainly used for shallow, unconsolidated aquifers.
Grout curtain (wall 1,700 ft. long × 60 ft. deep)	3,880	Very effective in permeable soils.
Sheet piling (wall 1,700 ft. long × 60 ft. deep)	2,218	Widely used for shoring. Mainly used for shallow unconsolidated aquifers.
Bottom sealing (4 ft. deep)	11,000	Leachate collection may be needed: difficult to accomplish results under unconsolidated aquifers.
Plume Management[b]		
Drains	64	Effective in lowering water table a few meters in unconsolidated materials; can be used to collect shallow leachate.
Well point dewatering	514	Suction lift limits depth to 20–30 ft.; inexpensive installation; can be used to collect shallow leachate.
Deep well dewatering	508	Used in lowering deep water tables; high maintenance costs.

Table 9-2. Average Costs and Characteristics of Direct Remedial Methods, 1980 (*continued*)

Method	Average Estimated Costs[a]	Characteristics
Injection/extraction barrier	552	Creates a hydraulic barrier to stop leachate movment; operation and maintenance costs are high.
Chemical Immobilization		
Chemical fixation of cover	403	Uses chemically fixed sludge to provide a top seal; provides means of disposal for sludge; helps stabilize landfill.
Chemical injection	239	Immobilizes a single pollutant; in most cases not feasible; results unpredictable.
Excavation and Reburial		
Excavation and reburial	12,686	Very expensive; difficult construction.

[a] For a 22-acre landfill (prices in thousands of 1980 dollars). Original costs updated using the Consumer Price Index.
[b] Costs include present worth of 20 years, operation, maintenance, and, where applicable, power for a 22-acre landfill.

Source: After M. M. Sharefkin, Shechter, and A. Kneese. Impacts, Costs, and Techniques for Mitigation of Contaminated Ground Water: A Review. *Water Resources Research* vol. 20, no. 12(1984), pp. 1771–1783. Copyright © by the American Geophysical Union.

through management, source control strategies, land zoning, effluent charges and credits, aquifer standards, and through guidelines for construction and operation of groundwater-threatening activities would prove to be the most cost-effective alternatives to the very expensive remedial action options (Knox et al., 1984).

Figure 9-1. Drainage by Gravity of Contaminated Groundwater Through a Deep Injection Well
Source: Geraghty & Miller, Inc., *Seminar Proceedings on the Fundamentals of Groundwater Quality Protection* (1981), p. 17. Reprinted by permission of James J. Geraghty.

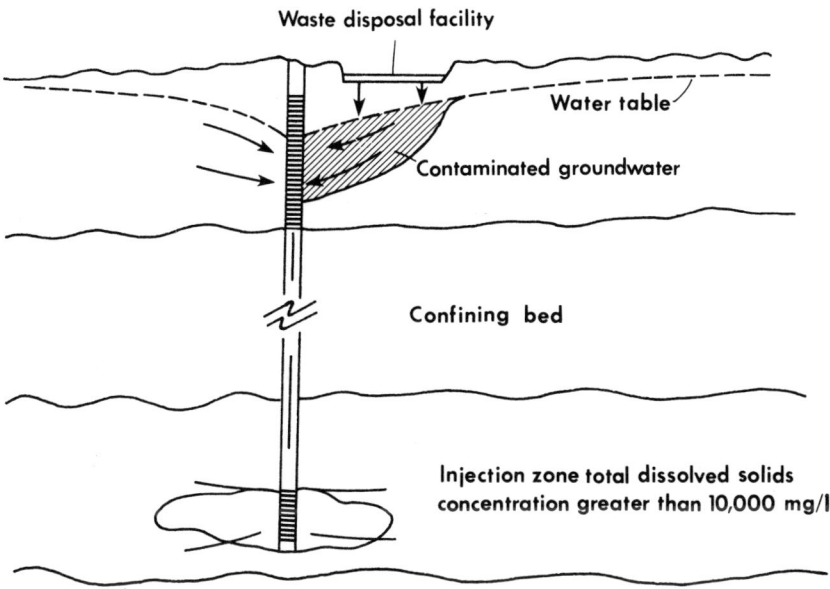

Figure 9-2. Cross Section of Cutoff Wall (Slurry Trench, Sheet Piling, Grout Curtain)
Source: Geraghty & Miller, Inc., *Seminar Proceedings of the Fundamentals of Groundwater Quality Protection* (1981), p. 10. Reprinted by permission of James J. Geraghty.

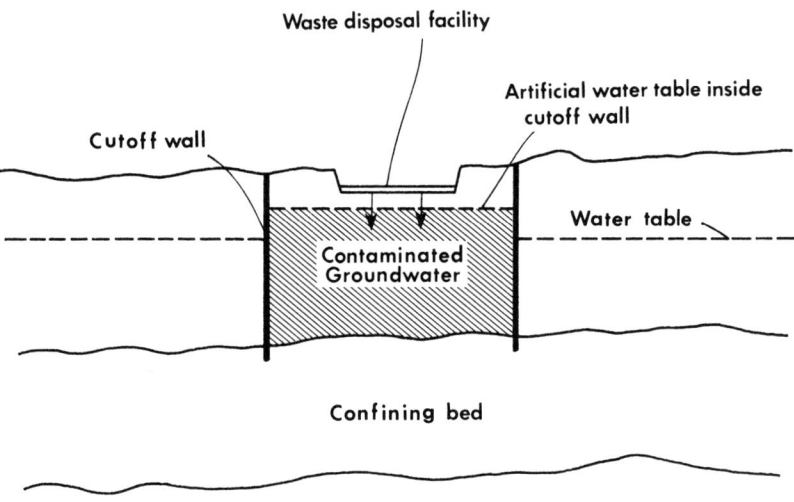

Figure 9-3. Cutoff Wall With Injected Bottom-Seal Grout
Source: Geraghty & Miller, Inc., *Seminar Proceedings of the Fundamentals of Groundwater Quality Protection* (1981), p. 11. Reprinted by permission of James J. Geraghty.

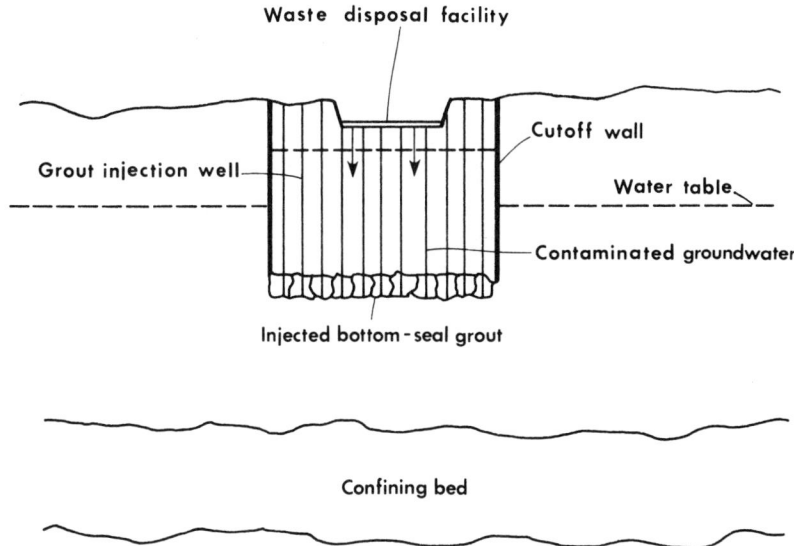

Figure 9-4. Control of Plume Movement by Pumping
Source: Geraghty & Miller, Inc., *Seminar Proceedings of the Fundamentals of Groundwater Quality Protection* (1981), p. 15. Reprinted by permission of James J. Geraghty.

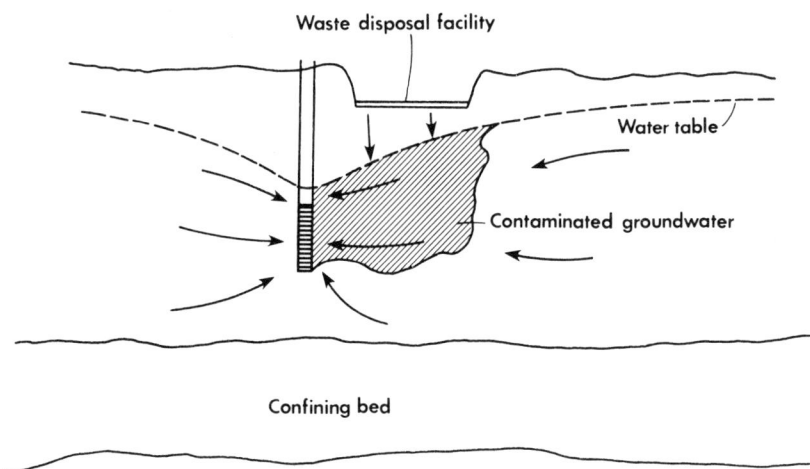

Figure 9-5. Control of Plume Movement Using Buried Drains
Source: Geraghty & Miller, Inc., *Seminar Proceedings of the Fundamentals of Groundwater Quality Protection* (1981), p. 14. Reprinted by permission of James J. Geraghty.

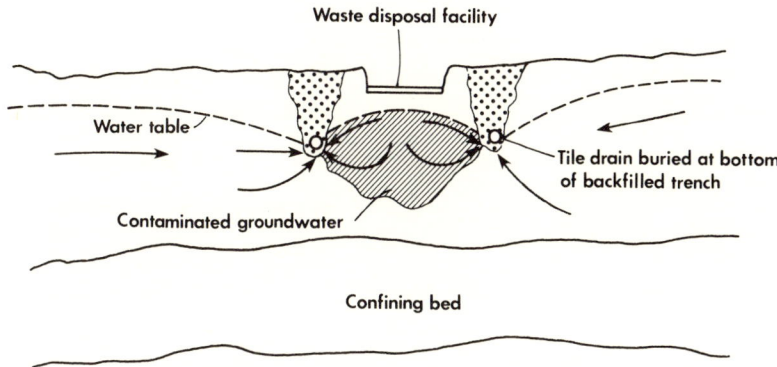

Figure 9-6. Blockage of Plume Movement by a Drainage Ditch, Brook, or Stream

Source: Geraghty & Miller, Inc., *Seminar Proceedings of the Fundamentals of Groundwater Quality Protection* (1981), p. 13. Reprinted by permission of James J. Geraghty.

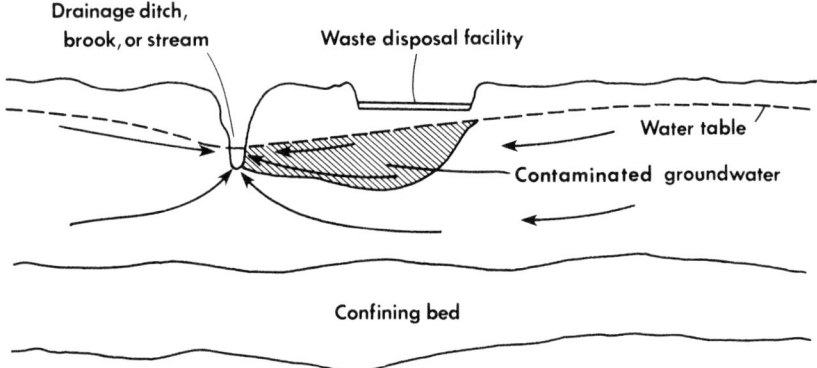

Ten

Proposed Strategies for the Protection of Groundwater

- Because of the difficulty and uncertainty of remedial action, protection of useful, potable aquifers from contamination is a better strategy than curative efforts.
- The EPA 1984 Ground-Water Protection Strategy is summarized.
- Recommendations for the management of groundwater developed by the National Ground Water Policy Forum, the Environmental and Energy Study Institute, the National Wildlife Federation, the Chemical Manufacturers Association, and the National Academy of Science's Committee on Ground Water Quality are reported.

EPA 1984 Ground-Water Protection Strategy

In August 1984 the EPA's Office of Ground-Water Protection published the Ground-Water Protection Strategy. The EPA strategy established a differential protection policy for groundwater based on the use, value, and vulnerability of ground water. The Strategy includes four major objectives that address critical needs for effective groundwater protection (EPA, 1984a):

- strengthen state groundwater programs;

- address groundwater contamination problems of national concern;
- create a policy framework for guiding EPA programs; and
- strengthen internal groundwater organization.

Strengthen State Groundwater Programs

296   EPA used this strategy to encourage states to broaden their institutional capacity to protect groundwater by developing groundwater protection programs and strategies. For fiscal years 1985 and 1986, $7 million in Clean Water Act grants were provided to states by Congress to develop groundwater protection strategies. These funds have been used by nearly all states and territories to establish or begin state strategies, to support necessary program development and planning, to create needed data systems, to assess legal and institutional impediments to comprehensive state management, and to develop state regulatory programs such as permitting and classification. It is anticipated that these funds will be renewed on an annual basis. EPA Regional Administrators have worked with governors in developing groundwater programs and strategies by providing state agencies with technical assistance in solving groundwater problems. EPA has also sponsored a number of other activities and published information designed to help states with groundwater issues, such as (Marian Mlay, personal correspondence):

- A study by the National Research Council was recently published, discussing 10 exemplary state groundwater protection programs (covered at the end of this chapter).
- Published information is available on all state groundwater programs, selected state classification systems, and state pesticides programs.
- A series of symposia on public policy issues bringing together both academia and state officials, are being conducted. A book of the symposia papers will be published as a result of this series.

Address Groundwater Contamination Problems of National Concern

EPA, through this strategy, has undertaken reviews of a number of groundwater contamination sources that have appeared to be of major national concern. For example, in 1984 EPA reviewed the effects of contamination from underground storage tanks. The study identified the nature, extent, and severity of leaking underground storage tanks. Further, the agency issued chemical advisories to alert owners and operators about the problem and worked with states and industry to develop voluntary steps to reduce contamination. In late 1984 Congress amended the Resource Conservation and Recovery Act, establishing a program for the regulation of underground storage tanks; it is now being implemented. EPA also recognized the potential environmental problems that could arise from land disposal facilities, including surface impoundments and landfills. Consequently, EPA initiated studies in cooperation with the states on impoundments and landfills and on the degree of danger they present. By statute, EPA and states are to set priorities for their control, review the regulatory options available, and determine if additional federal controls are needed.

297

EPA also noted in the Strategy the need to increase efforts to protect groundwater from pesticide and nitrate contamination. Over the past several years the agency has initiated a number of steps to assess the leaching potential and health impacts of individual pesticides and to develop and implement a program designed to mitigate these threats. In the area of pesticides/nitrates, EPA is now (Marian Mlay, personal communication):

- conducting a nationwide survey of more than 60 pesticides in well water in order to 1) determine the extent of contamination and its relationship to certain hydrogeologic characteristics, and 2) to determine how all EPA authority can be used to prevent and respond to this type of contamination; and
- developing a groundwater strategy for agchemicals.

EPA also noted in the Strategy the need for improved groundwater monitoring and data management. EPA has prepared a monitoring strategy that focuses its attention on improving the fundamental activities of groundwater monitoring. The strategy identifies seven monitoring objectives:

1. To characterize the nation's groundwater resources.
2. To identify new contamination problems.
3. To assess known problems to support regulatory development and standard setting and respond to site-specific problems.
4. To assure compliance with regulations.
5. To evaluate program effectiveness.
6. To improve data quality.
7. To develop groundwater data management systems.

Presently, EPA is working with USGS to expand the availability of aquifer characterization data. An interactive data system is being developed within EPA and with states in order to access existing groundwater data (via STORET). EPA is also in the process of identifying a series of environmental indicators, using existing data, which will eventually help determine the present status and overall trends in groundwater contamination. The monitoring strategy is covered at length in Chapter 8.

298

Create a Policy Framework for Guiding EPA Groundwater Programs

EPA recognized the great need for guidelines to ensure consistency in its groundwater protection programs. The guidelines are to reflect the policy's rationale that not all groundwater is of equal value, and that protection should consider the highest beneficial use to which groundwater can currently or potentially be put. These guidelines are intended to provide a framework for the decisions that EPA and the states will have to make in implementing EPA programs. The guidelines are to be used by EPA and the states to make decisions on levels of protection and cleanup under existing EPA regulations, to guide future regulations, and to establish enforcement priorities for the future. The guidelines define three classes of groundwater, based on their respective value and their vulnerability to contamination. The Final Draft of the Guidelines for Ground-Water Classification Under the EPA Ground-Water Protection Strategy (1986) defines the three classes and describes the procedures and informational needs for classification. The classes of groundwater are defined as follows (U.S. EPA, 1986a):

Class I: Special Groundwaters. These groundwaters are of unusually high value and in need of special protective measures. They are highly vulnerable to contamination because they have a relatively high potential for contaminants to enter and/or to be transported within the groundwater flow system. They are also characterized as either:

   a. an irreplaceable source of drinking water, in that no reasonable alternative source of drinking water is available for a substantial population; or

   b. ecologically vital, in that the aquifer provides the base flow for a particularly sensitive ecological system that, if polluted, would destroy a unique habitat.

It should be noted that sites designated as sole source aquifers under the Safe Drinking Water Act (SDWA) are not automatically categorized as Class I groundwaters. The criteria for sole source aquifers are less strict than those for Class I.

Class II: Current and Potential Sources of Drinking Water and Water Having Other Beneficial Uses. These groundwaters are all other groundwaters not in Class I that are currently used, or potentially available, for drinking water and other beneficial use. This class is divided into two subclasses:

Subclass IIA: Groundwaters currently used as a source of drinking water

Subclass IIB: Groundwaters that are a potential source of drinking water

Class III: Groundwater Not a Potential Source of Drinking Water and of Limited Beneficial Use. These groundwaters are saline or are otherwise contaminated beyond levels that allow use for drinking or other beneficial purposes. This class includes groundwaters with total dissolved solids (TDS) levels over 10,000 mg/l, or those that are so contaminated, naturally or by human activity, that cleanup using methods reasonably employed in public water system treatment is not possible. Class III groundwaters are subcategorized primarily on the basis of the degree of interconnection with other surface waters or adjacent groundwater units containing groundwater of a higher class:

Subclass IIIA: Groundwaters with a high to intermediate degree of interconnection. They have yields insufficient to meet the needs of an average sized family and therefore are not a potential source of drinking water.

Subclass IIIB: Groundwaters with a low degree of interconnection to adjacent surface waters or groundwaters of a higher class. They are naturally isolated from sources, so that there is little potential for producing additional adverse effects on human health and the environment.

The task of actually determining whether the groundwater in a particular location fits the criteria for Class I, II, or III will be site-specific. A first step to classify groundwaters must be a review of their characteristics. A segment of groundwater, the Classification Review Area (CRA), is first delineated based initially on a two-mile radius from the boundaries of the "facility" or "activity" in question. Within the CRA, a preliminary inventory is performed of public

299

water supply wells, populated areas not served by public supply, wetlands, and surface waters. Classification criteria are then applied to the area, and a classification determination is then made (U.S. EPA, 1986a).

To prevent contamination of Class I groundwaters, EPA will target its prevention efforts toward achieving Maximum Contaminant Levels (MCLs) within the groundwater or providing higher levels of protection. The regulatory requirements of programs are now being reviewed to ascertain how this can be achieved (Marian Mlay, personal communication).

300

Groundwaters that are current and potential sources of drinking water (Class II) will receive levels of protection consistent with those provided under EPA's existing regulations. While EPA's cleanup policy will ensure drinking water quality levels that protect human health, exemptions to requirements will be available under certain circumstances to allow less stringent levels when the protection of human health and the environment can be demonstrated (U.S. EPA, 1986a).

EPA's classification system is not designed for state use, except where the state is making a decision on behalf of an EPA program. EPA will develop a mechanism for states to use their own classification system, but they must demonstrate that it is "no less stringent" than EPA's in order to qualify for authorization for implementation (Marian Mlay, personal communication).

Strengthen Internal Groundwater Organization

EPA has established an Office of Ground-Water Protection to oversee the implementation of this strategy. EPA regional offices have also established regional groundwater units. These units coordinate regional groundwater policy and program development, and assist the states through grants and technical assistance to increase their institutional capabilities for the management of groundwater. In addition EPA has commissioned a Science Advisory Board Committee to review EPA's groundwater research; many of the committee's recommendations are now being carried out. Several coordinating committees have also been established to strengthen internal groundwater organization, including one composed of 23 other federal agencies (Marian Mlay, personal communication).

Recommendations of the National Groundwater Policy Forum for the Control of Groundwater Contamination

The National Groundwater Policy Forum was convened and sponsored by the Conservation Foundation in cooperation with the National Governors' Association in the fall of 1984. The policies and recommendations for the management of groundwater contamination are set forth below.

301

1. A new environmental partnership to pursue the goals of groundwater protection must be formed. It should consist of an environmental partnership of state and local governments, the federal government, private industry, and public interest groups. This group should dedicate itself to forming a comprehensive management plan for groundwater resources. Environmental organizations and other interested groups should educate the public about the nature of groundwater problems and ensure responsive actions by appropriate parties.

2. Role of the Federal Government: Congress should enact a law that would declare that the goal of the United States is to protect the physical, chemical, and biological integrity of the nation's groundwater resources and ensure that they are not degraded in any way that may be harmful to humans or the environment. The federal government should assure consistency and equity among the states by stimulating state and local action and providing technical support and oversight. The federal statute that establishes the National Groundwater Protection Goal should require the states to enact programs to achieve the goals. This legislation should allow for limited reviews of the state programs, provide technical and financial assistance to the states, and create inducements to the states to take effective actions.

3. The states must bear the primary responsibility for the management of groundwater with local governments bearing a major responsibility for program implementation. This is necessary because of the great variability of the resource in different states.

Every state that has not already done so should enact legislation to authorize and implement a comprehensive groundwater protection program. This program should include the following.

1. Comprehensive mapping of aquifer systems and their associated recharge and discharge areas.
2. Anticipatory classification of aquifers.
3. Ambient groundwater standards.

4. Authorities for imposing controls on all significant sources of potential contamination.
5. Programs for monitoring, data collecting, and data analyses.
6. Effective enforcement provisions.
7. Surface-use restrictions to protect groundwater quality.
8. Programs to control groundwater withdrawals so as to protect groundwater quality.
9. Coordination of groundwater and surface water management.

302

10. Coordination of groundwater programs with other relevant natural resource protection programs.

The Forum recommends that when the states have completed the development phase of their groundwater program, the states be able to apply to EPA for matching grant funds to undertake the implementation of the program. For such funding eligibility EPA would have to find the state program adequate for such a grant based upon whether it contains six of the ten components listed above.

The state should consider offering technical assistance as well as financial inducement to local governments to help them carry out their responsibilities under a comprehensive monitoring program. The state should also provide legal authority for local units of government to implement land use controls where these may be necessary to further protect the goals of the management program.

Congress should authorize EPA to provide technical and financial support to the states. Congress should also require EPA to recommend health based ambient groundwater guidelines for potential toxic compounds found in groundwater. These guidelines should be based upon toxicological reviews conducted by an independent scientific body such as the National Academy of Science. The law should also urge the United States Geological Survey to direct its cooperative program toward support of state groundwater management efforts. It is the duty of Congress to provide adequate funding for EPA and USGS to carry out these responsibilities and to provide research and technical assistance to the states. Statutes ratified by Congress should provide provisions for effective coordination of the federal groundwater program with other environmental and water resource management programs. It should include deadlines and incentives that are adequate to induce all parties to implement their program requirements promptly and effectively.

Recommendations of the Environmental and Energy Study Institute for Groundwater Protection

The Environmental and Energy Study Institute recommends the following agenda for groundwater protection. The Institute recommends strategy for management.

First Recommendation

The Institute recommends that Congress adopt a phased approach to groundwater protection, enacting limited legislation now to boost federal, state, and local protection levels. The first phase should take restrained but significant steps to meet clear, immediate protection needs and provide for evaluation and revision in a timely manner. The second phase would be based upon evaluation of the first phase.

The first phase would consist of a prevention-based national goal; redirection and coordination of federal actions to meet that goal; reliance on state governments with federal financial help to work toward a national goal through state and local protective sections; civilian education and participation initiatives, and information and technical assistance to decision makers; a special program to help local land use decision makers and to promote local technologies; and a special initiative to protect groundwater from agricultural sources of contamination.

Second Recommendation

The Institute recommends that Congress enact a national groundwater protection goal which would drive federal policies, programs, and projects, and state and local protection efforts. State and local governments, however, should be free to establish more protective goals.

Third Recommendation

The Institute recommends that Congress enact a prevention-based goal for groundwater protection. This goal would be to prevent contamination to the maximum extent possible, i.e., groundwater would not be contaminated by man's activities if it can be avoided. In defining the prevention goal, Congress should consider the feasibility of technologies and practices, costs, existing vs. new sources of contamination (i.e., differentiate between what new sources of pollution are required to do and what existing sources are required to do), and siting options (e.g., refrain from locating potential sources of contamination in critical areas).

304

Fourth Recommendation

The Institute recommends that Congress direct each federal agency to evaluate its policies, programs, regulations, and projects for their possible impact on groundwater quality and to make any changes needed to make them consistent with the national goal.

Fifth Recommendation

The Institute recommends that the federal government leave states great flexibility and authority to develop strategies and programs to achieve the national goal, and that Congress provide financial and technical assistance for their work. There is general agreement that the federal government should provide states with financial assistance since they are being asked to meet the national goal. To enable states to meet their obligations, Congress should authorize five-year development grants. Within six months EPA should develop guidlines for evaluating the effectiveness of state strategies and programs to meet the goals and criteria. In developing these guidelines EPA should also seek the participation of other relevant federal agencies, congressional representatives, state and local government representatives, industry representatives and knowledgeable members of the public.

Recommended Policies and Practices of the National Wildlife Federation for the Protection of Groundwater

In recognition of the severity of the problem of groundwater contamination in the United States, the National Wildlife Federation has set forth the following statement of policies and practices for the protection of groundwater:

- Emphasis should be placed on conservation of resources and the development of processes to detect and protect potential contamination.

- Before carrying out a comprehensive effort to detect and prevent groundwater contamination, the public should have an accurate and complete characterization of our groundwater resources.
- Efforts should be concentrated first on prevention and cleanup where they are most needed, e.g., focusing first, but not exclusively on prevention and cleanup of public drinking water supplies in order to insure protection of these key resources.

305

- Technologies should be developed and used to minimize the amount of hazardous waste being generated and/or requiring disposal. When disposal cannot be avoided, current federal and state rules for the tracking and disposing of hazardous and non-hazardous wastes from "cradle to grave" should be adhered to scrupulously in order to ensure minimum contact of these materials with groundwater supplies.
- Industry should examine its processes and the operation of its contractors for potential and actual impacts of wastes on groundwater. These include any potential spillage, leakage, or operational discharges of products or processed water in ways that could eventually contaminate groundwater.
- Programs should be put in place to detect and prevent leaks from underground storage tanks. New tanks should be designed to prevent structural or corrosive failure and provide mechanisms for detection of the problem.
- Evaluation of potential to contaminate groundwater should be part of the siting process of any new facility.
- The Council of the National Wildlife Federation encourages the establishment of a broad based task force and advisory committees to examine local and regional issues and develop a groundwater protection plan.
- The Council further encourages efforts now underway to clarify the roles of federal, state, and local governments to appropriate legislation.

Summary of the Chemical Manufacturers' Association Program for State Groundwater Management

In 1987 the Chemical Manufacturers' Association drafted a review that provides a framework for industry representatives who are responsible for evaluating and commenting on groundwater issues at the state level (CMA, 1987).

CMA believes that to ensure effective management each state should establish a formal state groundwater management program. The program should coordinate all state regulatory programs with the purpose of protecting human health and the environment. It should define program objectives, be flexible, and identify the responsibilities of state and local agencies. CMA lists and describes the five basic components that it believes should be included in every state groundwater management program. They include (CMA, 1987):

1. Groundwater management planning process
2. Objectives and standards
3. Protection measures
4. Groundwater remediation
5. Groundwater monitoring

306

Groundwater Management Planning Process

Effective planning is an essential foundation when developing a comprehensive state groundwater management program. The following steps should be taken when implementing the planning process (CMA, 1987):

1. Assess the quality and quantity of the state's groundwater.

2. Establish uniform procedures to be used in the planning process.

3. Place the state's groundwater into the following categories to facilitate necessary groundwater planning decisions:
   a) areas not needing formal management at present (uncategorized);
   b) unusable groundwater;
   c) special management areas (both sole source aquifer and limited use areas); and
   d) designated use aquifers, including existing and potential drinking water aquifers.

4. Designate aquifer management planning areas where aquifers have been or potentially could be impaired or where there is a need to protect a particularly sensitive aquifer.

5. Identify all sources that have a reasonable potential to affect the state's groundwater, rank their significance, and establish guidelines on how to manage them.

6. Provide a method for establishing interstate agreements wherever aquifers have interstate management needs.

Objectives and Standards

Within aquifer management planning areas, objectives describing the intended use of the aquifer, along with standards needed to maintain that use, must be established. The following concepts should be taken into account when setting and using objectives and standards (CMA, 1987):

307

1. Groundwater use objectives should be developed to establish descriptive goals.

2. Multiple-use standards should be established to determine the suitability of water for a particular use. These standards are generally numerical and are applied at the point of use.

3. For groundwater designated as drinking water, states should establish standards according to federal maximum contaminant levels (MCLs); where MCLs are not available, federal health advisories (HAs) should be set on an interim basis.

4. To obtain needed MCLs and HAs from the EPA, states should use the formal petitioning process provided under the Federal Administrative Procedure Act.

5. Standards established by states for uses other than drinking should be based on criteria developed by the EPA.

6. Requirements for remediation should be developed on a case by case basis, considering use, applicable standards for uses, potential exposures, cost-effective engineering and remediation approaches, and alternative sources of water.

7. Groundwater protection programs are established with the goal of preventing groundwater contamination, and should take into account applicable objectives and standards (CMA, 1987).

Protection Measures

Prevention is the most effective means of protecting groundwater. As part of the planning process, states should identify all sources that have a reasonable potential to affect the state's groundwater. Protection programs include several key elements (CMA, 1987):

1. Existing federal and state regulatory requirements should be incorporated in protection programs.

2. Those sources that are not effectively managed, such as septic tanks and sanitary landfills, should be included in the protection program.

3. Groundwater protection plans should be established for sources that have the potential to interfere with the beneficial uses of groundwater.

4. Aquifer management plans may include some form of land use control only when the resource cannot otherwise by protected.

308   5. States should decide whether groundwater withdrawal activities should be managed.

6. Groundwater protection plans should be maintained and periodically reviewed by facilities and non-facility activities and should be made available for periodic review by the state (CMA, 1987).

Groundwater Remediation

Remedial action must be based on site specific information and data, and should consider the risks imposed on health and the environment, the cost of remediation, technological limitations, and the need to maintain or recover beneficial groundwater uses. The state groundwater management program should establish a lead agency to be responsible for groundwater remediation requirements (CMA, 1987).

Groundwater Monitoring

Monitoring data must be scientifically valid and must accurately characterize the resource in order to be useful for management decisions. To accomplish this, states should establish performance standards, implement procedures for the different stages of monitoring, develop quality assurance and quality control procedures for each of these activities, and provide a uniform management system with criteria for data quality (CMA, 1987).

Additional Issues

Other programs and issues should be considered as components of state groundwater management programs, such as public water supply monitoring and the role of federal and state governments. States should maintain the lead role in managing their groundwater resource. They should integrate state water quality and water rights laws and determine the most appropriate beneficial uses for groundwater. States should incorporate input from local governments. The federal role should be to provide technical and financial assistance to the states for development of a comprehensive and systematic groundwater management plan (CMA, 1987).

309

Criteria and Recommendations for Effective Groundwater Quality Protection by the Committee on Ground Water Quality Protection

In November 1984, the Committee on Ground Water Quality Protection was established by the National Research Council at the request of the U.S. Environmental Protection Agency. The Committee was asked to identify, review, and summarize several state and local groundwater protection programs. The resulting report identifies those significant technical and institutional features that the Committee believes provide effective groundwater quality protection (NRC, 1986). These features are to be used as practical models for state and local officials who are attempting to develop or enhance their groundwater quality protection programs (NRC, 1986).

Criteria

The following eight criteria are considered necessary components of a comprehensive groundwater protection program, and are used for measuring program effectiveness (NRC, 1986).

1. *Goals and Objectives* The goals and objectives of a groundwater program should be clearly defined, reflect the entire groundwater resource problem, and be based on adequate legal authority. In addition, periodic program evaluation should take place to review program success in order to make needed modifications (NRC, 1986).

2. *Information* Effective protection programs should be based on the evaluation of information gathered from the following surveys:

(a) water resources and their location;

(b) groundwater basin characteristics with respect to potential contamination; and

(c) current and anticipated land and surface water uses that may affect groundwater quality (NRC, 1986).

3. *Technical Basis* An effective groundwater protection program should be based on sound technical information, such as:

(a) appropriate physical, social, and behavioral assumptions;

310 (b) physical, chemical engineering, and hydrologic principles; and

(c) sound relationships linking mandated actions with desired results (NRC, 1986).

4. *Source Elimination and Control* A major goal in all groundwater protection programs should be the elimination and reduction of groundwater pollution sources. This can be attained by:

(a) prohibition of certain harmful activities and products;

(b) siting groundwater threatening activities and facilities away from sensitive areas through land use controls, permits, and regulations;

(c) developing incentives for the use of products and technologies less threatening to groundwater quality; and

(d) developing incentives for the recycling and reuse of waste products (NRC, 1986).

5. *Intergovernmental and Interagency Linkages* A strong, coherent intergovernmental program, linking local, state and federal activities, is essential in protecting groundwater (NRC, 1986).

6. *Effective Implementation and Adequate Funding* Groundwater programs must have:

(a) adequate legal authority to take action,

(b) adequate long-term funding,

(c) sufficient personnel with adequate training, expertise, and skills, and an ongoing program for professional development, and

(d) funding mechanisms and strategies to sustain activity over time (NRC, 1986).

7. *Economic, Social, Political, and Environmental Impacts* Because prevention of groundwater contamination is the least costly alternative to groundwater contamination, protective actions should be evaluated in terms of economic, social, political, and environmental impacts (NRC, 1986).

8. *Public Support and Responsiveness* Groundwater programs

should:
- (a) encourage public understanding and support;
- (b) involve the public in program design and evalution;
- (c) exercise a balance of authority and consultation with affected parties; and
- (d) consider the equity of the distribution of benefits, costs, and burdens and the relative ability of the various responsible parties to bear them (NRC, 1986).

311

Recommendations

In addition to these eight components, the report outlines a number of recommendations which could have broad applicability in the development of state groundwater protection programs.

Information Base

1. *Hydrogeologic Information* The committee encourages state and local programs to obtain and collect hydrogeologic information over the long term in order to quantitatively understand the occurrence, quality, and dynamics of the resource along with the types, extent, and sources of potential contaminants. The resulting data should reflect program effectiveness (NRC, 1986).

2. *Types of Data and Data Management Systems* The committee recommends that both state and federal informational programs be carefully designed for the collection and storage of data in order to facilitate analysis of problems and long term trends. The management systems should be flexible and easily accessible, and should be reviewed and revised regularly to improve efficiency of data gathering (NRC, 1986).

Classification

The committee recommends that states consider the mapping and classifying of their groundwater along with the specification of areas for special protection. Classification criteria should be adopted through public process (NRC, 1986).

Groundwater Quality Standards

The Committee recommends that EPA should proceed with the promulgation of MCL's and MCL goals already proposed, and propose and set federal standards for all other chemicals commonly found in groundwater. EPA should also provide states with technical information on those chemicals found in groundwater for which there are no standards. Individual states should adopt ambient standards according to their goals and objectives, and should consider a multi-tiered standard setting approach for implementation (NRC, 1986).

Control of Contamination Sources

1. *Hazardous and Solid Waste Management* The committee recommends that states consider reducing waste quantities by such methods as incineration and resource recovery. These procedures should be accompanied by monitoring requirements, discharge or emission limits, and ambient environmental quality standards to ensure environmental safety. An essential element in each state's protection program should be a plan for treating, storing, and disposing of hazardous wastes, with waste minimization as a key element. Exportation of hazardous wastes should only be permitted in special circumstances. The program should, in addition, provide polluting activities with incentives for good housekeeping (NRC, 1986).

2. *Underground Storage Tanks* The committee recommends that each state consider developing a comprehensive program for chemical and petroleum storage tanks. Programs should include design standards for new tanks and requirements for monitoring, inspecting, and upgrading existing tanks (NRC, 1986).

3. *Non-Point Sources* The committee recommends that states maintain a data base on the spatial and temporal distribution of applied pesticides. They should flag those pesticides that have a potential for leaching into and contaminating groundwater, or if states are reluctant to cancel the registration of leachable pesticides, they should still cancel registration in those areas where leaching may pose a serious problem. States should consider funding pesticide monitoring through taxes on pesticides and their use. In addition they should encourage incentives, for pesticide and fertilizer reduction (NRC, 1986).

4. *Source Reduction* The committee believes that the best long term strategy for groundwater protection is source reduction and elimination. States should consider regulatory and economic incentives for source reduction by industry, government, commercial interests, and the public. State agencies, university-based groups, trade associations, and other institutions should develop educational programs on waste reduction technology for the general public and industry, and help assist in implementing these practices (NRC, 1986).

5. *Land Use Control* The committee recommends that land use controls be considered an essential part of a groundwater

protection program and be implemented at early stages of
development for the protection of vulnerable areas (NAS, 1986).

Implementation of Groundwater Protection Programs

The committee recommends that the federal government provide
states with financial support for the development and
implementation of self-supporting groundwater protection
programs. They should also provide technical support to state and
local governments on health and environmental effects of
groundwater contamination, fate and transport of pollutants, and
technologies and strategies for groundwater protection. They
should provide criteria, guidelines and standards for contaminants,
and also provide training to state and local officials on water
management and protection (NRC, 1986).

313

1. *Political Mobilization, Public Participation, and Support* The
committee recommends the consideration of the following at every
governmental level (NRC, 1986):

- Decision making processes that reflect public attitudes, and
  include active participation of public interest groups, industry
  and the general public.
- Development of high level political leadership to create
  comprehensive legislation, ensure commitment of continued
  funding, and implement an adequate groundwater protection
  program.
- Communication between groundwater program managers and
  the media.
- Exchange of information regarding groundwater protection
  problems and programs.
- Ongoing educational activities about groundwater protection
  in the school systems at every grade level.
- State funding provided to the universities and colleges in
  order to expand the numbers of trained hydrogeologists.
- A public-intravenor program, to provide the public with an
  avenue for legal action, so to ensure the incorporation of their
  interests into groundwater protection programs.

2. *Role of Economic Analysis* The committee believes that the
balance between the costs of pollution prevention and remediation,
and also between universal groundwater policies and problem-
specific measures should be analysed in order to evaluate the
effectiveness of groundwater protection programs (NRC, 1986).

3. *Comparison of Governmental Groundwater Responsibilities*
Table 10-1 shows the ways in which the needs identified in this
report can be satisfied (NRC, 1986).

Table 10-1. Roles of Various Governmental Units for Consideration in the Protection of Groundwater

Governmental Level	Information	Planning	Standards and Enforcement	Public Process
Federal USGS	National water quality data, surface/ground.	Conducts local studies or portions of studies jointly with states or local agencies.	Provides technical data on resource conditions to EPA and states.	Provides information from data storage and studies.
EPA	Assesses national ground water conditions and provides technical advice to states.	Provides grants conditioned on establishing self-sufficient continuing planning. Conducts basic national health research.	Establishes national standards and provides technical and scientific basis for state standards. Supports states in enforcement of standards. Initiates national source control.	Is national focus of scientific and public dialogue on ground water protection strategy.
State[a]	Maintains central record of local quality and quantity of ground water, chemical usage, and disposition and present and projected land use.	Conducts continuing statewide planning and provides assistance and grants to regional or local agencies preparing ground water management plans.	Sets state or regional standards, including prohibitions when required national standards have not been set, but based on national scientific	Provides information, education, and continuing forum for public dialogue on goals, standards, control strategies, and investments.

assessment. Primary enforcement agency for state and federal standards and source control regulations.

Regional[b]	Provides same information as state for specific ground water basin serving several local agencies.	Prepares and implements specific basin management plans.	Accepts state delegation for enforcement as appropriate, including standard setting source controls, class systems, and land use plans to reflect actual environmental conditions.	Is primary focus of public process, including citizens, committees, working groups, and information gathering and dissemination.
Local	Provides basic information to individual users.	Prepares and implements local ground water plans.	Enforces appropriate standards. Assists state or regional entities in specific local situations.	Is primary contact with individual citizens.

[a] Each state's structure varies; ground water protection responsibilities may also vary; clear central state coordination is needed.
[b] Interstate Basin Agency: coordinates roles in several states where ground and related surface water problems exist.

Source: After NRC, 1986

Eleven

Aquifer Classification

- Aquifer classification is used by several states as a groundwater management and protection strategy, and is being considered for implementation by other states as well.
- Selective protection of different aquifers to different quality levels would require a classification system. Preservation and restoration of all aquifers to drinking-water use or the prohibition of degradation of present water quality would not require such a system.
- The benefits and the disadvantages of classification systems are discussed.
- State aquifer classification systems in existence are outlined in this chapter and compared. They fall into distinct groups, ranging from simple narrative classification to sophisticated differentiated systems incorporating land use criteria and numerical water standards.

As groundwater becomes an increasingly valuable resource in the United States, strategies for its protection and management are becoming the focus of detailed study in several states. Aquifer classification is one management tool being used by some states and considered by several others. An analysis of the implications of adopting an aquifer-classification system is summarized below, followed by summaries of the Environmental Protection Agency's classification system and actual state aquifer classification systems. State strategies, classification systems, laws, and regulations are further discussed in Chapter 13 under state and local strategies for managing groundwater problems.

The purpose of aquifer classification is to establish water-quality goals for each aquifer and to identify standards or controls necessary to ensure that water quality meets those goals. It is one approach for implementing the state's water policy by formally designating use and water-quality goals for groundwater resources. A state's choice of policy largely determines the appropriateness of classification as part of its overall groundwater-management program. There are three major policy options: non-degradation policy, limited degradation policy, and differential protection policy. A non-degradation policy aims to protect all groundwaters at their existing quality, and sometimes even calls for their improvement. A limited-degradation policy aims to protect groundwater at as high a quality as possible and to prevent degradation beyond a given quality or standard. A differential protection policy aims to protect groundwaters of different levels of quality based on current needs, characteristics, and anticipated uses. The first two policies are blanket policies, whereas the last would require the classification of aquifers. Very rarely has a state adopted a policy that fits perfectly into one of these categories; however, this breakdown provides a frame of reference.

317

As a groundwater protection management strategy, aquifer classification has both benefits and drawbacks. The benefits provided by aquifer classification include:

- Legal protection for valuable aquifers and a basis for siting potential contamination sources in low-risk areas.
- A reduction in unnecessary economic burdens on waste-facility operators by requiring that stringent water-quality standards be met throughout the state.
- A common basis for critical regulatory decisions affecting future use of groundwater resources.
- An opportunity for public involvement in critical policy decisions.
- Guidance for planning and for programs to protect water quality at all levels of government.

The classification of aquifers has the following drawbacks:

- The difficulty and cost of putting the system into place and making it effective.
- The difficulty of delineating the boundaries of aquifers.
- The possibility of serious legal and public-policy limitations

on the feasibility of classification. Without the acknowledgment of some form of degradation zones, a classification system could offer very little flexibility in levels of protection given to groundwater areas.
- The difficulty in gaining public acceptance for formally designated waste-receiving zones.

The Process of Classification

No state has an aquifer-classification system in place today that could serve as a model for other states developing classification systems. The variability of geological and hydrological conditions across the country is too great, and the different programs and policies governing the use and protection of groundwater also vary from state to state. Therefore, in order for a state to adopt an effective classification system, many factors, both legal and environmental, have to be taken into consideration.

Definition of Aquifer Boundaries

Groundwater classes can be developed only after some sort of aquifer boundaries are defined. In some areas of the country, distinct aquifer units can be identified that are hydraulically isolated from any other aquifer systems, as, for example, the Cohansey Aquifer in the New Jersey Central Pine Barrens, the glacial aquifer of Cape Cod, the valley-fill deposits located in glaciated regions of New England and intermountain systems of the western United States, and the St. Peter Sandstone Aquifer of Texas. In most areas of the country, however, identification of isolated hydraulic systems is virtually impossible.

There are several ways to define boundaries. The identification of flow systems such as recharge areas, discharge areas, regional flow, and multi-aquifer systems is another aid in defining the boundaries of aquifers. Water quality may serve to delineate aquifer zones. Natural quality is controlled by precipitation, evaporation, geology, and the amount of time the water remains in the groundwater system. As a general rule, the total content of dissolved solids increases with the depth and length of time that the water has traveled through an aquifer from the point of discharge. High evaporation and low precipitation may result in poor quality groundwater. In areas of high precipitation, shallow aquifers may have water of very good quality, with a total dissolved solids (TDS) content of less than 500 mg/l.

From a review of the hydrogeologic and water quality data as described above, a state should be able to decide whether or not it can delineate distinct aquifers or parts of aquifers for classification. There may be no method for distinguishing aquifer regions, and in that event they can only be described rather than mapped. Narrative criteria should, however, be used with caution. They are most effectively applied if used to identify regional aquifer characteristics.

319

Groundwater Classification

Having determined the boundaries of its aquifers, a state can begin to devise a structure for classifying groundwater. Existing systems recognize from two to eight classes. Wyoming has the greatest number of classes—agricultural, livestock, industrial (two classes), mining, domestic uses, aquaculture, and "non-usable." The fundamental question to ask is whether or not an aquifer should be designated as being of drinking-water quality. Further breakdowns can be made to reflect land use, existing physical factors, sensitive environmental systems, and other factors. Connecticut is an example of a state where the drinking-water category is divided into three sub-categories based on existing uses and conditions.

Classification categories are structured according to combinations of the following factors:

- existing use
- water quality, primarily based on its total content of dissolved solids
- the determination to use either criteria or standards of quality (standards are more specific than criteria)
- land use
- other aquifer characteristics (e.g., soils and geology)
- yield and availability of water regardless of quality (less than 350 gallons per minute [gpm] is considered uneconomical for community water-supply development)
- ability of an aquifer to attenuate and assimilate wastes
- multi-aquifer flow systems (a plan for the management of groundwater resources based on flow systems has been developed for Nassau and Suffolk Counties on Long Island)

- an aquifer associated with mineral deposits and geothermal sources may be classified for production purposes
- contribution to surface waters
- socioeconomic factors
- depth to aquifer

Classified aquifers may be controlled by numerical standards for water quality, effluent-discharge limitations (based on pollutant toxicity or discharge volume), or non-numerical rules for land-use and waste disposal (Magnuson, 1981; Henderson et al., 1984).

320

EPA's report, "Drastic: A Standardized System for Evaluating Ground Water Pollution Potential Using Hydrologic Settings," provides a methodology for planners, managers, and administrators that systematically evaluates the groundwater pollution potential of any hydrologic setting from a variety of sources (Aller et al., 1985). This system is used by many states in designating classes for their groundwaters.

EPA Aquifer Classification System

In 1984 EPA established an aquifer classification system as part of the comprehensive groundwater-protection strategy. A central feature of the strategy is the establishment of a framework that accords differing levels of protection to groundwater based on its use, value to society, and vulnerability to contamination. EPA has developed guidelines for classifying groundwater that define the classes, concepts, and key terms related to the system and describe the procedures and informational needs for classifying.

EPA's groundwater classification proposes an extensive three-class system for the management of groundwater (U.S. EPA, 1986a). (A more detailed discussion of EPA's classification system is found in Chapter 10.) The first step in classification is defining the area to be evaluated, the Classification Review Area (CRA). The guidelines specify a CRA as the area within a two-mile radius of a source of groundwater contamination. This segment of groundwater near the source, EPA believes, is most likely to be affected should a release occur (U.S. EPA, 1986a). Information regarding public and private wells, demographics, hydrogeology, and surface waters and wetlands must be collected. A classification decision then can be made, based on the criteria for each class listed below (U.S. EPA, 1986a). Class I aquifers, or special groundwater, must be highly vulnerable to contamination because of sensitive hydrologic conditions; they must be an irreplaceable source of drinking water, where no practical alternate source is available; and they must be ecologically vital (i.e., they contribute to sensitive ecological systems that if polluted would destroy a unique habitat). Class II aquifers are all other groundwaters that are currently used or potentially available for drinking water and for other beneficial uses. Class III aquifers are those groundwaters that are not considered a potential source of drinking water and have limited beneficial use. They must be heavily saline, with a total dissolved solids content of greater than 10,000 mg/l; or they must have contamination levels above those that can be cleaned up using methods employed by public-water treatment. They must not migrate into Class I and II groundwater, or discharge into surface water and cause degradation (U.S. EPA, 1984a and 1986a).

321

The final draft of the Guidelines for Ground-Water Classification under the EPA Ground-Water Protection Strategy further defines the classes, concepts, and key terms related to classification systems outlined in the strategy. In addition, it defines steps to be taken and the information needed to classify aquifers according to EPA's system (U.S. EPA, 1986a).

State Aquifer Classification Systems

EPA's Overview of State Ground-Water Program Summaries, published in 1985, found 22 states with established classification systems. However, trying to pin down the exact number of systems at any point in time is very difficult. Currently, groundwater protection programs throughout the country are being reorganized or developed. Consequently, the study done for this report only identified 13 states with classification systems, with an additional 9 states developing systems. The states found to have classification systems are Connecticut, Florida, Maine, Massachusetts, Montana, New Jersey, New Mexico, New York, North Carolina, Rhode Island, South Carolina, Vermont, and Wyoming.

This summary is not intended to be comprehensive. Rather, it is intended to recognize a variety of protection policies being used or being considered by some states, reflecting their needs and philosophies on regulation. The 13 aquifer-classification systems identified in this report vary considerably. Maine has implemented a simple two-class system for its groundwaters. Florida, and New Mexico have classification systems based on total content of dissolved solids in the groundwater. Montana uses specific conductance to assess the quality of groundwater for classification. The remaining states have more complex systems that have varying quality requirements and take into account the fact that groundwater is used for several different purposes.

322 **Connecticut**

Connecticut is considered to be something of a model among the states having applied aquifer-classification systems. In September 1980 the state adopted a four-class system as follows (Connecticut Department of Environmental Protection, 1980. Table 11-1):

1. Class GAA groundwater is to be used for public or private drinking water supplies without treatment. The only allowable discharges into groundwater of this class are wastewaters of human or animal origin and other minor cooling and clean-water discharges.

2. Class GA is assigned to groundwater to be used for public or private drinking-water supplies without treatment. Discharges are restricted to those which pose no permanent threat to untreated drinking-water supplies.

3. Class GB groundwater may have to be treated to be potable because of existing or past land uses. Allowable discharges include certain treated industrial wastewaters that can be filtered by the soils but do not cause degradation threatening future potability without treatment.

4. Class GC groundwater may be suitable for some waste-disposal practices if land-use practices or hydrogeologic conditions render it more suitable for that purpose than for development as a potable-water supply. Down-gradient surface water must, however, be of medium-to-poor quality, and wastewater discharges must not result in the degradation of surface waters below the classification goals established by the state.

"It is the policy of the State to restore or maintain the quality of the ground water to a quality consistent with its use for drinking without treatment" (GAA or GA), except where it has been classified GB and is not needed as a potable water supply, or GC where restoration is not feasible (Connecticut Department of Environmental Protection, 1980). The criteria for chemical constituents are determined by EPA standards or the state

Table 11-1. Connecticut Groundwater Classification

Class GAA

Existing or proposed public or private drinking-water supplies.

1. Dissolved Oxygen	As naturally occurs.
2. Oils and Grease	None other than of natural origin.
3. Color and Turbidity	None other than of natural origin.
4. Coliform Bacteria (per 100 mi)	Not to exceed a monthly arithmetic mean of 1 or more than 4 in any individual sample collected.
5. Taste and Odor	None other than of natural origin.
6. pH	As naturally occurs or as may result from normal agricultural, horticultural, silvicultural, lawn maintenance or construction activity provided all reasonable controls are used.
7. Chemical Constituents	The waters shall be free from chemical constituents in concentrations or combinations which would be harmful to human, animal or aquatic life for the most sensitive and governing water use class. Criteria for chemical constituents contained in guidelines published by the U.S. Environmental Protection Agency shall be considered. In areas where fisheries are the governing consideration and numerical limits have not been established, bioassays may be necessary to establish limits on toxic substances. The recommendations for bioassay procedures contained in "Standard Methods for the Examination of Water and Wastewater" and the application factors contained in EPA water quality guidelines shall be considered. For groundwaters classified for use as public or private drinking water (Classes GAA and GA), the raw water sources must be maintained or restored at a quality as defined by criteria developed by the U.S. EPA or the State, whichever is more stringent, so that criteria for finished water can be met without treatment.

323

Class GA

May be suitable for public or private drinking water use without treatment.

1. Dissolved Oxygen	As naturally occurs.
2. Oils and Grease	None other than of natural origin.
3. Color and Turbidity	None other than of natural origin.
4. Coliform Bacteria (per 100 mi)	Not to exceed a monthly arithmetic mean of 1 or more than 4 in any individual sample collected.
5. Taste and Odor	None other than of natural origin.
6. pH	As naturally occurs or as may result from normal agricultural, horticultural, silvicultural, lawn maintenance or construction activity provided all reasonable controls are used.

Table 11-1. Connecticut Groundwater Classification (*continued*)

7. Chemical Constituents	The waters shall be free from chemical constituents in concentrations or combinations which would be harmful to human, animal or aquatic life for the most sensitive and governing water use class. Criteria for chemical constituents contained in guidelines published by the U.S. Environmental Protection Agency shall be considered. In areas where fisheries are the governing consideration and numerical limits have not been established, bioassays may be necessary to establish limits on toxic substances. The recommendations for bioassay procedures contained in "Standard Methods for the Examination of Water and Wastewater" and the application factors contained in EPA water quality guidelines shall be considered. For groundwaters classified for use as public or private drinking water (Classes GAA and GA), the raw water sources must be maintained or restored at a quality as defined by criteria developed by the U.S. EPA or the State, whichever is more stringent, so that criteria for finished water can be met without treatment.

Class GB
May not be suitable for public or private drinking water without treatment. No quantitative or qualitative limits apply, since the groundwaters specified as GB are known or presumed to be degraded.

Class GC
May be suitable for certain waste disposal practices because past land use or hydrogeologic conditions render these groundwaters more suitable for receiving permitted discharges than development for public or private water supply.
No qualitative or quantitative limits apply.

Source: After Connecticut Department of Environmental Protection, 1980.

standards, whichever are more stringent, and apply always to waters classified for use as public or private drinking water supplies (GAA or GA).

Florida

Florida currently has a four-class system that designates potable groundwaters as Classes G-I and G-II, and non-potable

Table 11-2. Florida Groundwater Classification Standards

Standards for Class G-I and G-II Groundwaters

(1) Minimum Criteria for Groundwater
All groundwater shall at all places and at all times be free from domestic, industrial, agricultural, or other man-induced, non-thermal components of discharges in concentrations which, alone or in combination with other substances, or components of discharges (thermal or non-thermal):
  (a) are harmful to plants, animals, or organisms that are native to soil and responsible for treatment or stabilization of discharges relied upon by Departmental permits;
  (b) are carcinogenic, mutagenic, teratogenic, or toxic to human beings;
  (c) are acutely toxic to indigenous species of significance to aquatic communities affected by groundwaters;
  (d) are a threat to public health, safety or welfare;
  (e) create or constitute a nuisance; or
  (f) impair the reasonable and beneficial use of adjacent waters.

(2) Primary Drinking Water Regulations—Maximum Contaminant Levels (MCL) (in mg/l unless otherwise noted)
  (a) Inorganics
      Arsenic—0.05
      Barium–1.0
      Cadmium—0.01
      Chromium—0.05
      Lead—0.05
      Mercury—0.002
      Nitrate (as N)—10.0
      Selenium—0.01
      Silver—0.05
      Sodium—160.0
      Fluoride—varies with temperature: from below 12°C (53.7°F)—2.4 to 32.5°C (90.5°F)—1.4*
  (b) Organics
      Chlorinated hydrocarbons:
      Endrin—0.0002
      Lindane—0.004
      Methoxychlor—0.1
      Toxaphene—0.005
      Chlorophenoxys:
      2,4,-D—0.1
      2,4,5-TP,Silvex—0.01
  (c) Turbidity—1 unit*
  (d) Microbiological (coliform bacteria)
      1) membrane filter technique: a) 1 per 100 ml as the arithmetic mean of all samples examined per month; b) 4 per 100 ml in more than one sample when less than 20 are examined per month; or c) 4 per 100 ml in more than 5% of the samples when 20 or more are examined per month.*
      2) fermentation tube method (10 ml standard portions): a) more than 60% of the portions in any month; b) 5 portions in more than 1 sample when less than 5 samples are examined per month; or c) 5 portions in more than 20% of the samples when 5 or more samples are examined per month.*

Table 11-2. Florida Groundwater Classification Standards (*continued*)

(e) Radionuclides
- combined radium-226 and radium-228—5pCi/l
- gross alpha particle activity (including radium-226, but excluding radon and uranium)—15 pCi/l
- beta particle and photon radioactivity from man-made radionuclides—average annual concentration from man-made radionuclides shall not produce an annual dose equivalent to the total body or any internal organ greater than 4 millirem/yr*

(f) Total trihalomethane—0.1 mg/l

(g) Volatile organics
Trichloroethylene—3 µg/l
Tetrachloroethylene—3 µg/l
Carbon tetrachloride—3 µg/l
Vinyl chloride—1 µg/l
1,1,1-Trichloroethane—200 µg/l
1,2-Dichloroethane—3 µg/l
Benzene—1 µg/l
Ethylene dibromide—0.02 µg/l

(3) Secondary Drinking Water Regulations
Chloride—250 mg/l
Color—15 units
Copper—1 mg/l
Corrosivity—neither corrosive nor scale forming
Foaming Agents—0.5 mg/l
Iron—0.3 mg/l
Manganese—0.05 mg/l
Odor—threshold number 3
pH—6.5 (min., no max. specified)
Sulfate—250 mg/l
TDS—500 mg/l (may be greater if no other MCL is exceeded)
Zinc—5 mg/l

If the concentration for any constituent listed in (2a) in the natural unaffected background quality of the groundwater is greater than the stated maximum, or in the case of pH is also less than the minimum, the representative background value shall be the prevailing standard for Class G-I and Class G-II groundwaters.

(4) Other Contaminants Without a Standard
It is prohibited to introduce into a public water system any contaminant which creates or has the potential to create an imminent and substantial danger to the public.

Standards for Class G-III Groundwaters
Minimum Criteria for Groundwaters (listed above)

Standards for Class G-IV Groundwaters
The Department shall apply standards on case-by-case basis for discharges to these groundwaters. Minimum criteria applied if danger to health, safety, or welfare.

* Conditions apply—look to FDER Public Drinking Water Systems Chapter 17-22 for more detailed information.

Source: After Florida Department of Environmental Regulation, 1985c, e.

groundwaters as Classes G-III and G-IV. According to designated uses, all groundwaters in the state are classified as follows:

Class G-I—Potable water use, groundwaters in single source aquifers having a TDS content of less than 3,000 mg/l. In order to designate an aquifer as a single source, the Environmental Regulatory Commission must find that (1) the aquifer or portion of the aquifer is the only reasonable available source of potable water to a significant segment of the population; and (2) the designated use is attainable upon consideration of the environmental, technological, water quality, institutional, societal, and economic factors (Florida Department of Environmental Regulation, 1985f).

327

Class G-II—Potable water use, groundwaters having a TDS of less than 10,000 mg/l, unless otherwise classified by the Environmental Regulatory Commission.

Class G-III—Non-potable water use, groundwater in unconfined aquifers having a TDS of greater than or equal to 10,000 mg/l, or a TDS of between 3,000 and 10,000 mg/l and has been classified by the Commission as having no reasonable potential use as a source of drinking water, or has been designated by the Department as an exempted aquifer.

Class G-IV—Non-potable water use, groundwater in confined aquifers having a TDS of greater than or equal to 10,000 mg/l.

In order to designate an aquifer as a single source, the Environmental Regulatory Commission must find that:

1) the aquifer or portion of the aquifer is the only reasonable available source of potable water to a significant segment of the population; and

2) the designated use is attainable, upon consideration of the environmental, technological, water quality, institutional, societal, and economic factors (Florida Department of Environmental Regulation, 1985f).

The set of criteria for discharges into classes G-I through G-IV aquifers is listed in Table 11-2 (Florida Department of Environmental Regulation, 1985c).

Maine

The state of Maine currently has two standards for the classification of groundwater (Maine Revised Statutes Annotated, 1979).

Class GW-A shall be the highest classification and shall be of such quality that it can be used for public water supplies. These waters shall be free of radioactive matter or any matter that imparts color, turbidity, taste, or odor that would impair usage of these waters, other than that occurring from natural phenomena. Federal drinking-water standards have been adopted for this class.

328

Class GW-B, the second highest classification, shall be suitable for all usages other than public water supplies. However, no groundwater has been designated as Class GW-B so far.

The groundwater classification subcommittee established by the Maine Land and Water Resources Council is currently assessing the practicality of forming a new three-tiered system for classification that would allow for the direction of land use activities (The New England Interstate Water Pollution Control Commission Newsletter, 1985).

Massachusetts

On October 15, 1983, the Massachusetts Ground-Water Discharge Permit Program and the Massachusetts Ground-Water Quality Standards were established by the Massachusetts Department of Environmental Quality Engineering to provide protection of all groundwaters in the Commonwealth (U.S. EPA, 1985d). One of the major features of the regulations was a statewide groundwater classification program. A three-class system was adopted based on the philosophy that classification should be a tool for guidance in future decision making on waste disposal sites rather than a course for definitive action.

Sub-basins are placed in one of three classes, ranging from high quality (Class I) to known contamination areas (Class III). Judgment for classification is "based upon the most sensitive uses for which the ground-water is to be maintained and protected" (Commonwealth of Massachusetts, 1983). The three classes are:

Class I—Fresh groundwaters found in the saturated zone of unconsolidated deposits or consolidated rock and bedrock, and designated as a source of potable water supply.

Class II—Saline groundwaters found in the saturated zone of the unconsolidated deposits or consolidated rock and bedrock, and designated as a source of potable mineral waters, for the conversion to fresh potable waters, or as raw material for the manufacture of sodium chloride or its derivatives or similar products.

Class III—Fresh or saline water found in the saturated zone of unconsolidated deposits or consolidated rock and bedrock, and designated for uses other than as a source of potable water supply. At a minimum, the most sensitive use of these waters is as a source of non-potable water which may come into contact with, but may not be ingested by, humans.

329

Currently, protection of groundwater from point sources is accomplished through a permit program. Discharges to Class I and II groundwater must meet the more stringent of either technology standards or National Primary Drinking Water Standards; where there are no standards, discharges must meet criteria laid out by the U.S. EPA Health Advisories (U.S. EPA, 1985d). See Table 11-3.

Montana

In 1982 a classification system was established to protect present and future most beneficial use of groundwater. "Most beneficial" here means the highest level of use into which groundwater could be placed based on existing quality. (Montana Department of Health and Environmental Sciences, 1982a). To assess the quality of groundwaters, Montana uses specific conductance, which measures the inorganic content of the water sample. Numerical and narrative standards for each of the four classes are defined below:

Class I—Groundwaters that are generally suitable for public and private water supplies, culinary and food-processing purposes, irrigation, livestock and wildlife watering, and commercial and industrial purposes with little to no treatment. Specific conductance of these waters is less than 1,000 micromhos/cm at 25°C.

Class II—Groundwaters that are generally suitable for public and private water supplies and culinary and food-processing uses, and are suitable for irrigation of some crops, for drinking water for most wildlife and livestock, and for most commercial and industrial purposes. These groundwaters may be used for municipal or domestic waters where better supplies are not readily available. Specific conductance of these waters ranges from 1,000 to 2,500 micromhos/cm at 25°C.

Table 11-3.   Massachusetts Groundwater Quality Criteria

Minimum Ground Water Quality Criteria

(1) Class I and Class II Ground Waters
The following minimum criteria are applicable to all Class I and Class II ground waters:

	Parameter	Criteria
(a)	Pathogenic Organisms	Shall not be in amounts sufficient to render the ground water detrimental to public health and welfare or impair the ground water for use as source of potable water.
(b)	Coliform Bacteria	Shall not exceed the maximum contaminant level as stated in the National Interim Primary Drinking Water Standards.
(c)	Arsenic	Shall not exceed 0.05 mg/l
(d)	Barium	Shall not exceed 1.0 mg/l
(e)	Cadmium	Shall not exceed 0.01 mg/l
(f)	Chromium	Shall not exceed 0.05 mg/l
(g)	Copper	Shall not exceed 1.0 mg/l
(h)	Fluoride	Shall not exceed 2.4 mg/l
(i)	Foaming Agents	Shall not exceed 0.5 mg/l
(j)	Iron	Shall not exceed 0.3 mg/l
(k)	Lead	Shall not exceed 0.05 mg/l
(l)	Manganese	Shall not exceed 0.05 mg/l
(m)	Mercury	Shall not exceed 0.002 mg/l
(n)	Nitrate-Nitrogen (as Nitrogen)	Shall not exceed 10.0 mg/l
(o)	Total Trihalomethanes	Shall not exceed 0.1 mg/l
(p)	Selenium	Shall not exceed 0.01 mg/l
(q)	Silver	Shall not exceed 0.05 mg/l
(r)	Sulfate	Shall not exceed 250 mg/l
(s)	Zinc	Shall not exceed 5.0 mg/l
(t)	Endrin (1,2,3,4,10,10-hexachloro-1,7-epoxy-1,4,4a,5,6,7,8,9a-octahydro-1,4-endo-5,8-dimethanonaphthalene)	Shall not exceed 0.0002 mg/l
(u)	Lindane (1,2,3,4,5,6-hexachlorocyclohexane, gamma isomer)	Shall not exceed 0.004 mg/l
(v)	Methoxychlor (1,1,1-trichloro-2,2-bis(p-methoxyphenyl) ethane)	Shall not exceed 0.1 mg/l
(w)	Toxaphene ($C_{10}H_{10}Cl_8$, technical chlorinated camphene, 67–69 percent chlorine)	Shall not exceed 0.005 mg/l
(x)	Chlorophenoxys: 2,4-D (2,4-	Shall not exceed 0.1 mg/l

Table 11-3.   Massachusetts Groundwater Quality Criteria (*continued*)

	dichlorophenoxyacetic acid)	
	2,4,5-TP,Silvex (2,4,5-tri-chlorophenoxypropionic acid)	Shall not exceed 0.01 mg/l
(y)	Radioactivity	Shall not exceed the maximum radionuclide contaminant levels as stated in the National Interim Primary Drinking Water Standards.
(z)	pH	Shall be in the range of 6.5–8.5 standard units or not more than 0.2 units outside of the naturally occurring range.
(aa)	All Other Pollutants	None in such concentrations which in the opinion of the Director would impair the waters for use as a source of potable water or to cause or contribute to a condition in contravention of standards for other classified waters of the Commonwealth.

(2) Class III Ground Waters.
The following minimum criteria are applicable to all Class III ground waters:

	Parameter	*Criteria*
(a)	Pathogenic Organisms	Shall not be in amounts sufficient to render the ground waters detrimental to public health, safety or welfare.
(b)	Radioactivity	Shall not exceed the maximum radionuclide contaminant levels as stated in the National Interim Primary Drinking Water Standards.
(c)	All Other Pollutants	None in concentrations or combinations which upon exposure to humans will cause death, disease, behavioral abnormalities, cancer, genetic mutations, physiological malfunctions or physical deformations or cause any significant adverse effects to the environment, or which would exceed the recommended limits on the most sensitive ground water use.

Source: Commonwealth of Massachusetts, 1983.

Table 11-4.   Montana Groundwater Quality Criteria

Quality Criteria for Classes I, II, and III (when III is Used for Drinking Water).

Maximum Inorganic Chemical Contaminant Levels

Constituent	Level (mg/l)
1) Arsenic	0.05
2) Barium	1.0
3) Cadmium	0.01
4) Chromium	0.05
5) Lead	0.05
6) Mercury	0.002
7) Nitrate (as N)	10.0
8) Selenium	0.01
9) Silver	0.05
10) Fluoride	2.4

Maximum Organic Chemical Contaminant Levels

Constituent	Level (mg/l)
1) Chlorinated hydrocarbons that include:	
a) Endrin	0.0002
b) Lindane	0.004
c) Methoxychlor	0.1
d) Toxaphene	0.005
2) Chlorophenoxys, including:	
a) 2,4-D	0.1
b) 2,4,5-TP Silvex	0.01
3) Total trihalomethanes	0.1

Maximum Radiological Contaminant Levels

Constituent	Level (pCi/l)
1) Combined radium-226 and radium-228	5
2) Gross alpha particle activity (including radium-226 but excluding radon and uranium)	15
3) Tritium	20,000
4) Strontium-90	8
5) Gross beta particle radioactivity	50
6) The average annual concentration of beta particle and photon radioactivity from man-made radionuclides in drinking water may not produce an annual dose equivalent in the total body or any internal organ greater than 4 millirem/yr.	

Table 11-4.   Montana Groundwater Quality Criteria
(*continued*)

Maximum Microbiological Contaminant Levels
1) When the membrane filter technique is used:
    a) 4 per 100 ml in more than one sample when less than 20 samples
        are examined per month;
    b) 4 per 100 ml in more than 5% of the samples when 20 or more sam-
        ples are examined a month; or
    c) 1 per 100 ml as the arithmetic mean of all validated samples ex-   **333**
        amined per month.
2) When the 10 ml fermentation tube is used:
    a) more than 10% of the portions in any month;
    b) 3 or more portions in more than one sample when less than 20 sam-
        ples are examined per month; or
    c) 3 or more portions in more than 5% of the samples when 20 or more
        samples are examined per month.
3) When coliform bacteria are found, daily samples from the same sampling
    point must be collected until the results obtained from at least 2 consec-
    utive samples are shown to be satisfactory bacteriological samples.

Source: Montana Department of Health and Environmental Science, 1982a and 1982b.

Class III—Groundwaters suitable for some industrial and commercial uses, as drinking water for some wildlife and livestock, and for irrigation of some salt-tolerant crops using special water management practices. Specific conductance of these waters ranges from 2,500 to 15,000 micromhos/cm at 25°C.

Class IV—Groundwaters suitable for some industrial, commercial and other uses, but unsuitable or (for practical purposes) untreatable for higher-class uses. Specific conductance of these waters is greater than 15,000 micromhos/cm at 25°C.

Concentrations may not exceed these levels to a point that would affect existing uses or any other use under its classification. Classes I and II, and Class III when it is used for drinking water, may not exceed the maximum contaminant levels set by the state. These standards are more stringent than those recommended by the EPA (Table 11-4). Other substances not covered in the standards must not exceed levels that may be harmful, detrimental, or injurious to public health (Montana Department of Health and Environmental Sciences, 1982a).

Degradation may occur only when necessary economic or social development is justifiable and will not prevent present or anticipated use. This non-degradation policy does not apply to non-point sources where all pollution-prevention actions have been applied (U.S. EPA, 1985d).

New Jersey

334

New Jersey has adopted a four-class system that has the total content of dissolved solids as its prime criteria for defining protection (New Jersey Department of Environmental Protection, 1981b). "It is the policy of this State to restore, enhance, and maintain the chemical, physical and biological integrity of its waters, to protect public health, to safeguard fish and aquatic life and scenic and ecological values and to enhance the domestic, municipal, recreational, industrial, and other uses of water" (New Jersey Department of Environmental Protection, 1981b).

"Existing and potential uses of ground water shall be maintained and protected." The State will upgrade the quality of any water that does not meet the standards for its intended use.

1. Class GW1 was established specifically to protect the groundwaters of the unique and fragile ecosystem of the Central Pine Barrens. The groundwater reservoir is of extremely high quality but also very vulnerable to contamination. No activity that would cause degradation is allowed in this area. Class GW1 groundwater "shall be suitable for potable water supply, agricultural water supply, continual replenishment of surface waters to maintain the existing quantity and high quality of the surface waters in the Central Pine Barrens, and other reasonable uses" (New Jersey Department of Environmental Protection, 1981b). Table 11-5 lists the quality criteria applicable to groundwater of this class.

2. Class GW2 groundwater has a natural concentration of total dissolved solids of 500 mg/l or less and shall be suitable for potable, industrial, or agricultural uses, or for the replenishment of quantity and quality of surface waters. Conventional water treatment may be necessary. Table 11-6 lists quality criteria for this class.

3. Class GW3 has a natural concentration of total dissolved solids of between 500 mg/l and 10,000 mg/l. It is suitable for conversion to fresh potable waters. Quality criteria are listed in Table 11-7.

Table 11-5. New Jersey Groundwater Quality Criteria for the Central Pine Barrens: Class GW1

Ground Water Quality Criteria for the Central Pine Barrens: Class GW1

Pollutant, Substance or Chemical	Ground Water Quality Criteria
1. Aldrin/Dieldrin	1. 0.003 μ/l
2. Arsenic and Compounds	2. 0.05 mg/l
3. Barium	3. 1.0 mg/l
4. Benzidine	4. 0.0001 mg/l
5. Cadmium	5. Natural background
6. Chromium (Hexavalent) and Compounds	6. Natural background
7. Cyanide	7. 0.2 mg/l
8. DDT and Metabolites	8. 0.001 μ/l
9. Endrin	9. 0.004 μ/l
10. Lead and Compounds	10. 0.05 mg/l
11. Mercury and Compounds	11. 0.002 mg/l
12. Nitrate-Nitrogen	12. 2.0 mg/l
13. Phenol	13. 0.3 mg/l
14. Polychlorinated Biphenyls	14. 0.001 μ/l
15. Radionuclides	15. Prevailing regulations adopted by the U.S. EPA pursuant to sections 1412, 1415 and 1450 of the Public Health Services Act as amended by the Safe Drinking Water Act (PL 93-523)
16. Selenium and Compounds	16. Natural background
17. Silver and Compounds	17. 0.05 mg/l
18. Toxaphene	18. 0.005 μ/l
19. Ammonia	19. 0.5 mg/l
20. BOD (5-day)	20. 3 mg/l
21. Chloride	21. 10 mg/l
22. Coliform Bacteria	22. a) by membrane filtration, not to exceed four per 100 ml in more than one sample when less than 20 are examined per month, or b) by fermentation tube, with a standard 10 ml portion, not to be present in three or more portions in more than one sample when less than 20 are examined per month, or c) prevailing criteria adopted

335

Table 11-5. New Jersey Groundwater Quality Criteria for the Central Pine Barrens: Class GW1 (continued)

Pollutant, Substance or Chemical	Ground Water Quality Criteria
	pursuant to the Federal Safe Drinking Water Act (PL 93-523)
23. Color	23. None noticeable
24. Copper	24. 1.0 mg/l
25. Fluoride	25. 2.0 mg/l
26. Foaming Agents	26. 0.5 mg/l
27. Iron	27. 0.3 mg/l
28. Manganese	28. 0.05 mg/l
29. Odor and Taste	29. None noticeable
30. Oil and Grease and Petroleum Hydrocarbons	30. None noticeable
31. pH	31. 4.2–5.8 standard units
32. Phosphate, Total	32. 0.7 mg/l
33. Sodium	33. 10 mg/l
34. Sulfate	34. 15 mg/l
35. Total Dissolved Solids	35. 100 mg/l
36. Zinc and Compounds	36. 5 mg/l

Source: New Jersey Department of Environmental Protection, 1981b.

Table 11-6. New Jersey Groundwater Quality Criteria for Class GW2

Ground Water Quality Criteria Statewide where the Total Dissolved Solids (TDS, Natural Background) Concentration is less than or equal to 500 mg/l: Class GW2

Pollutant, Substance or Chemical	Ground Water Quality Criteria
Primary Standards/Toxic Pollutants	
1. Aldrin/Dieldrin	1. 0.003 μ/l
2. Arsenic and Compounds	2. 0.05 mg/l
3. Barium	3. 1.0 mg/l
4. Benzidine	4. 0.0001 mg/l
5. Cadmium and Compounds	5. 0.01 mg/l
6. Chromium (Hexavalent) and Compounds	6. 0.05 mg/l
7. Cyanide	7. 0.2 mg/l
8. DDT and Metabolites	8. 0.001 μ/l
9. Endrin	9. 0.004 μ/l
10. Lead and Compounds	10. 0.05 mg/l
11. Mercury and Compounds	11. 0.002 mg/l
12. Nitrate-Nitrogen	12. 10 mg/l

Table 11-6. New Jersey Groundwater Quality Criteria for Class GW2 (*continued*)

Pollutant, Substance or Chemical	Ground Water Quality Criteria
13. Phenol	13. 3.5 mg/l
14. Polychlorinated Biphenyls	14. 0.001 µg/l
15. Radionuclides	15. Prevailing regulations adopted by the U.S. EPA pursuant to sections 1412, 1415 and 1450 of the Public Health Services Act as amended by the Safe Drinking Water Act (PL 93-523)
16. Selenium and Compounds	16. 0.01 mg/l
17. Silver and Compounds	17. 0.05 mg/l
18. Toxaphene	18. 0.005 µ/l

Secondary Standards

19. Ammonia	19. 0.5 mg/l
20. Chloride	20. 250 mg/l
21. Coliform Bacteria	21. a) by membrane filtration, not to exceed four per 100 ml in more than one sample when less than 20 are examined per month, or
	b) by fermentation tube, with a standard 10 ml portion, not to be present in three or more portions in more than one sample when less than 20 are examined per month, or
	c) Prevailing criteria adopted pursuant to the Federal Safe Drinking Water Act (PL 93-523)
22. Color	22. None noticeable
23. Copper	23. 1.0 mg/l
24. Fluoride	24. 2.0 mg/l
25. Foaming Agents	25. 0.5 mg/l
26. Iron	26. 0.3 mg/l
27. Manganese	27. 0.05 mg/l
28. Odor and Taste	28. None noticeable
29. Oil and Grease and Petroleum Hydrocarbons	29. None noticeable
30. pH	30. 5–9 standard units
31. Phenol	31. 0.3 mg/l
32. Sodium	32. 50 mg/l
33. Sulfate	33. 250 mg/l
34. Total Dissolved Solids	34. 500 mg/l
35. Zinc and Compounds	35. 5 mg/l

Source: New Jersey Department of Environmental Protection, 1981b.

Table 11-7. New Jersey Groundwater Quality Criteria for Class GW3

Ground Water Quality Criteria Statewide where the Total Dissolved Solids (TDS, Natural Background) Concentration is between 500 mg/l and 10,000 mg/l: Class GW3

Pollutant, Substance or Chemical	Ground-Water Quality Criteria
Primary Statewide/Toxic Pollutants	
1. Aldrin/Dieldrin	1. 0.003 µ/l
2. Arsenic and Compounds	2. 0.05 mg/l
3. Barium	3. 1.0 mg/l
4. Benzidine	4. 0.0001 mg/l
5. Cadmium and Compounds	5. 0.01 mg/l
6. Chromium (Hexavalent) and Compounds	6. 0.05 mg/l
7. Cyanide	7. 0.2 mg/l
8. DDT and Metabolites	8. 0.001 µ/l
9. Endrin	9. 0.004 µ/l
10. Lead and Compounds	10. 0.05 mg/l
11. Mercury and Compounds	11. 0.002 mg/l
12. Nitrate-Nitrogen	12. 10 mg/l
13. Phenol	12. 3.5 mg/l
14. Polychlorinated Biphenyls	14. 0.001 µ/l
15. Radionuclides	15. Prevailing regulations adopted by the U.S. EPA pursuant to sections 1412, 1415 and 1450 of the Public Health Services Act as amended by the Safe Drinking Water Act (PL 93-523)
16. Selenium and Compounds	16. 0.01 mg/l
17. Silver and Compounds	17. 0.05 mg/l
18. Toxaphene	18. 0.005 µ/l
Secondary Standards	
19. Ammonia	19. 0.5 mg/l
20. Chloride	20. Natural background
21. Coliform Bacteria	21. a) by membrane filtration, not to exceed four per 100 ml in more than one sample when less than 20 are examined per month, or b) by fermentation tube, with a standard 10 ml portion, not to be present in three or more portions in more

338

Table 11-7.   New Jersey Groundwater Quality Criteria for Class GW3 (*continued*)

Pollutant, Substance or Chemical	Ground-Water Quality Criteria
	than one sample when less than 20 are examined per month, or
	c) Prevailing criteria adopted pursuant to the Federal Safe Drinking Water Act (PL 93-523)
22. Color	22. None noticeable
23. Copper	23. 1.0 mg/l
24. Fluoride	24. 2.0 mg/l
25. Foaming Agents	25. 0.5 mg/l
26. Iron	26. 0.3 mg/l
27. Manganese	27. 0.05 mg/l
28. Odor and Taste	28. None noticeable
29. Oil and Grease and Petroleum Hydrocarbons	29. None noticeable
30. pH	30. 5–9 (standard units)
31. Phenol	31. 0.3 mg/l
32. Sodium	32. Natural background
33. Sulfate	33. Natural background
34. Total Dissolved Solids	34. Natural background
35. Zinc and Compounds	35. 5 mg/l

Source: New Jersey Department of Environmental Protection, 1981b.

4. Class GW4 has a natural concentration of total dissolved solids in excess of 10,000 mg/l. It is suitable for any reasonable beneficial use. Quality criteria are determined on a case-by-case basis.

New Mexico

New Mexico has a two-class system based on TDS. Groundwater with a TDS content greater than 10,000 mg/l is considered non-potable and water with a TDS content of less than 10,000 mg/l is protected for potable use (M. Goad, Environmental Improvement Division of New Mexico, personal communication). Thirty-five numerical standards have been adopted for groundwaters of 10,000 mg/l TDS concentration or less (Table 11-8).

Table 11-8. New Mexico Groundwater Standards

Standards for Ground Water of 10,000 mg/l TDS Concentration or Less

The following standards are the allowable pH range and the maximum allowable concentration in ground water for the contaminants specified unless the existing condition exceeds the standard or unless otherwise provided. When an existing pH or concentration of any water contaminant exceeds the standard specified in Subsection A, B, or C, the existing pH or concentration shall be the allowable limit, provided that the discharge at such concentrations will not result in concentrations at any place of withdrawal for present or reasonably foreseeable future use in excess of the standards of this section.

A. *Human Health Standards*—Ground water shall meet the standards of Sections A and B unless otherwise provided

Arsenic (As)	0.1 mg/l
Barium (Ba)	1.0 mg/l
Cadmium (Cd)	0.01 mg/l
Chromium (Cr)	0.05 mg/l
Cyanide (CN)	0.2 mg/l
Fluoride (F)	1.6 mg/l
Lead (Pb)	0.05 mg/l
Total Mercury (Hg)	0.002 mg/l
Nitrate (NO_3 as N)	10.0 mg/l
Selenium (Se)	0.05 mg/l
Silver (Ag)	0.05 mg/l
Uranium (U)	5.0 mg/l
Radioactivity: Combined Radium-226 and Radium-228	30.0 pCi/l
Benzene	0.01 mg/l
Polychlorinated Biphenyls (PCB's)	0.001 mg/l
Toluene	15.0 mg/l
Carbon Tetrachloride	0.01 mg/l
1,2-Dichloroethane (EDC)	0.02 mg/l
1,1-Dichloroethylene (1,1-DCE)	0.005 mg/l
1,1,2,2-Tetrachloroethylene (PCE)	0.02 mg/l
1,1,2-Trichloroethylene (TCE)	0.1 mg/l

B. *Other Standards for Domestic Water Supply*

Chloride (Cl)	250.0 mg/l
Copper (Cu)	1.0 mg/l
Iron (Fe)	1.0 mg/l
Manganese (Mn)	0.2 mg/l
Phenols	0.005 mg/l
Sulfate (SO_4)	600.0 mg/l
Total Dissolved Solids (TDS)	1000.0 mg/l
Zinc (Zn)	10.0 mg/l
pH	6–9 s.u.

Table 11-8.   New Mexico Groundwater Standards (*continued*)

C. Standards for Irrigation Use—Ground water shall meet the standards of
   subsections A, B, and C unless otherwise provided.

Aluminum (Al)	5.0 mg/l
Boron (B)	0.75 mg/l
Cobalt (Co)	0.05 mg/l
Molybdenum (Mo)	1.0 mg/l
Nickel (Ni)	0.2 mg/l

Source: New Mexico Water Quality Control Commission, 1983.   **341**

Regulations are implemented to "protect all groundwater of the
state of New Mexico which has an existing TDS concentration of
10,000 mg/l or less, for present and potential future use as
domestic and agricultural water supply, and to protect those
segments of surface waters which are gaining because of
groundwater inflow, for uses designated in the New Mexico Water
Quality Standards" (State of New Mexico Water Quality Control
Commission, 1983).

New Mexico not only protects its public water of present and
potential drinking-water use, but also its private wells because of
the state's large rural population. These measures are necessary in
such a water-poor state as New Mexico (U.S. EPA, 1985d).

New York

New York State adopted a three-class system in September 1978,
which was similar to North Carolina's (see below) but lacked the
classifications for recharge waters and specifications for depth of
occurrence of groundwater. "The purpose of these classes, quality
standards, and effluent standards and/or limitations is to prevent
pollution of groundwaters and to protect the groundwaters for use
as a potable water" (New York Department of Environmental
Conservation, 1978).

1. Class GA waters are fresh and potable and are protected by
   the most stringent quality standards determined by the state
   and EPA. The waters are found in the saturated zone of
   unconsolidated deposits and consolidated rock and bedrock
   (Tables 11-9 and 11-10).
2. Class GSA waters are saline waters containing between 250
   mg Cl/l and 1000 mg Cl/l that can be used for potable
   mineral water, converted to freshwater, or as a raw material in

Table 11-9. New York Groundwater Quality Standards for Class GA

The following quality standards shall be applicable to Class GA waters.

Items	Specifications
1. Sewage, industrial waste or other wastes, taste or odor producing substances, toxic pollutants, thermal discharges, radioactive substances or other deleterious matter.	1. None which may impair the quality of the ground waters to render them unsafe or unsuitable for a potable water supply or which may cause or contribute to a condition in contravention of standards for other classified waters of the State.
2. The concentration of the following substances or chemicals:	2. Shall not be greater than the limit specified, except where exceeded due to natural conditions:
(1) Arsenic (As)	(1) 0.025 mg/l
(2) Barium (Ba)	(2) 1.0 mg/l
(3) Cadmium (Cd)	(3) 0.01 mg/l
(4) Chloride (Cl)	(4) 250 mg/l
(5) Chromium (Cr) (Hexavalent)	(5) 0.05 mg/l
(6) Copper (Cu)	(6) 1.0 mg/l
(7) Cyanide (CN)	(7) 0.2 mg/l
(8) Fluoride (F)	(8) 1.5 mg/l
(9) Foaming Agents[1]	(9) 0.5 mg/l
(10) Iron (Fe)[2]	(10) 0.3 mg/l
(11) Lead (Pb)	(11) 0.025 mg/l
(12) Manganese (Mn)[2]	(12) 0.3 mg/l
(13) Mercury (Hg)	(13) 0.002 mg/l
(14) Nitrate (as N)	(14) 10.0 mg/l
(15) Phenols	(15) 0.001 mg/l
(16) Selenium (Se)	(16) 0.02 mg/l
(17) Silver (Ag)	(17) 0.05 mg/l
(18) Sulfate (SO_4)	(18) 250 mg/l
(19) Zinc (Zn)	(19) 5 mg/l
(20) pH Range	(20) 6.5–8.5
(21) Aldrin, or 1,2,3,4,10-hexa-chloro-1,4,4a,5,8,8a-hexa-hydro-*endo*-1,4-exc-5,8-dimethanonaphthalene.	(21) Not detectable[3]
(22) Chlordane, or 1,2,4,5,6,7,8,8-octachloro-2,3,3a,4,7,7a-hexahydro-4,7-methanoindene.	(22) 0.1 μ/l

Table 11-9.   New York Groundwater Quality Standards for Class GA (*continued*)

Items	Specifications
(23) DDT, or 2,2-bis-(*p*-chloro-phenyl)-1,1,1-trichloro-ethane and metabolites	(23) not detectable[3]
(24) Dieldrin, or 6,7-epoxy aldrin.	(24) not detectable[3]
(25) Endrin, or 1,2,3,4,10,10-hexachloro-6,7-epoxy-1,4,4a,5,6,7,8,8a-octahydro-*endo*-1,4-*endo*-5,8-di-methanonaphthalene.	(25) not detectable[3]
(26) Heptachlor, or 1,4,5,6,7,8,8-heptachloro-3a,4,7,7a-tetra-hydro-4,7-methanoindene and metabolites.	(26) not detectable[3]
(27) Lindane and other hexa-chlorocyclohexanes or mixed isomers of 1,2,3,4,5,6-hexachloro-cyclohexane	(27) not detectable[3]
(28) Methoxychlor, or 2,2-bis-(*p*-methoxyphenyl)-1,1,1-tri-chloroethane.	(28) 35.0 μ/l
(29) Toxaphene (a mixture of at least 175 chlorinated cam-phene derivatives).	(29) not detectable[3]
(30) 2,4-Dichlorophenoxyacetic acid (2,4-D)	(30) 4.4 μ/l
(31) 2,4,5-Trichlorophenoxy-propionic acid (2,4,5-TP) (Silvex)	(31) 0.26 μ/l
(32) Vinyl chloride (chloroeth-ene)	(32) 5.0 μ/l
(33) Benzene	(33) not detectable[3]
(34) Benzo(a)pyrene	(34) not detectable[3]
(35) Kepone or decachloro-octahydro-1,3,4-metheno-2H-cyclobuta(cd)pentalen-2-one (chlordeone).	(35) not detectable[3]
(36) Polychlorinated biphenyls (PCB) (Aroclor)	(36) 0.1 μ/l
(37) Ethylene thiourea (ETU)	(37) not detectable[3]
(38) Chloroform	(38) 100 μ/l

343

Table 11-9. New York Groundwater Quality Standards for Class GA (*continued*)

Items	Specifications
(39) Carbon tetrachloride (tetrachloromethane)	(39) 5 μ/l
(40) Pentachloronitrobenzene (PCNB)	(40) not detectable[3]
(41) Trichloroethylene	(41) 10 μ/l
(42) Diphenylhydrazine	(42) not detectable[3]
(43) bis(2-Chloroethyl) ether	(43) 1.0 μ/l
(44) 2,4,5-Trichlorophenoxy-acetic acid (2,4,5-T)	(44) 35 μ/l
(45) 2,3,7,8-Tetrachlorodibenzo-*p*-dioxin (TCDD)	(45) 3.5×10^{-5} μ/l
(46) 2-Methyl-4-chlorophenoxy-acetic acid (MCPA)	(46) 0.44 μ/l
(47) Amiben, or 3-amino-2,5-dichlorobenzoic acid (chloramben)	(47) 87.5 μ/l
(48) Dicamba, or 2-methoxy-3,6-dichlorobenzoic acid	(48) 0.44 μ/l
(49) Alachlor, or 2-chloro-2',6'-diethyl-*N*-(methoxymethyl)-acetanilide (Lasso)	(49) 35.0 μ/l
(50) Butachlor, or 2-chloro-2',6'-diethyl-*N*-(butoxymethyl)-acetanilide (Machete)	(50) 3.5 μ/l
(51) Propachlor, or 2-chloro-*N*-isopropyl-*N*-acetanilide (Ramrod)	(51) 35.0 μ/l
(52) Propanil, or 3',4'-dichloro-propionanilide	(52) 7.0 μ/l
(53) Aldicarb [2-methyl-2-(methylthio)propionaldehyde *O*-(methylcarbamoyl)oxime] and methomyl[1-methyl-thioacetaldhyde *O*-(methylcarbamoyl)oxime]	(53) 0.35 μ/l
(54) Bromacil, or 5-bromo-3-sec-butyl-6-methyluracil	(54) 4.4 μ/l
(55) Paraquat, or 1,1'-β-di-methyl-4,4'-dipyridylium	(55) 2.98 μ/l
(56) Trifluralin, or α,α,α-trifluoro-2,6-dinitro-*N*-dipropyl-*p*-toluidine (Treflan)	(56) 35.0 μ/l

344

Table 11-9.   New York Groundwater Quality Standards for Class GA (continued)

Items	Specifications
(57) Nitralin, or 4-(methylsulfo-nyl)-2,6-dinitro-*N*,*N*-dipropy-laniline (Planavin)	(57) 35.0 μ/l
(58) Benefin, or *N*-butyl-*N*-ethyl-α,α,α-trifluoro-2,6-dinitro-*p*-toluidine (Balan)	(58) 35.0 μ/l
(59) Azinphosmethyl, or *O,O*-di-methyl-*S*-4-oxo-1,2,3-benzo-triazin-3(4H)-ylmethylphos-phorodithioate (Guthion)	(59) 4.4 μ/l
(60) Diazinon, or *O,O*-diethyl *O*-(2-isopropyl-4-methyl-6-pyr-imidinyl)-phosphorothioate.	(60) 0.7 μ/l
(61) Phorate (also for disulfoton), or *O,O*-diethyl-*S*-[(ethyl-thio)methyl]-phosphorodi-thioate (Thimet R), and di-sulfoton, or *O,O*-di-ethyl-*S*-[(2-ethylthio)ethyl]-phosphorodithioate (Di-Sys-tem R)	(61) not detectable[3]
(62) Carbaryl, or 1-naphthyl-*N*-methylcarbamate	(62) 28.7 μ/l
(63) Ziram, or zinc salts of di-methyl-dithiocarbamic acid.	(63) 4.18 μ/l
(64) Ferbam, or iron salts of di-methyl-dithiocarbamic acid.	(64) 4.18 μ/l
(65) Captan, or *N*-trichloro-methylthio-4-cyclohexene-1,2-dicarboximide.	(65) 17.5 μ/l
(66) Folpet, or *N*-trichloro-methylthiophthalimide.	(66) 56.0 μ/l
(67) Hexachlorobenzene (HCB)	(67) 0.35 μ/l
(68) Paradichlorobenzene (PDB) (also orthodichlorobenzene)	(68) 4.7 μ/l
(69) Parathion (and methyl para-thion), or (*O,O*-diethyl-*O*-*p*-nitrophenylphosphoro-thioate, and methylpara-thion, or *O,O*-dimethyl-*O*-*p*-nitrophenylphosphoro-thioate.	(69) 1.5 μ/l

345

Table 11-9. New York Groundwater Quality Standards for Class GA (*continued*)

Items	Specifications
(70) Malathion, or S-1,2-bis(ethoxycarbonyl)ethyl-O,O-dimethylphosphorodithioate.	(70) 7.0 μ/l
(71) Maneb, or manganese salt of ethylene-bis-dithiocarbamic acid.	(71) 1.75 μ/l
(72) Zineb, or zinc salt of ethylene-bis-dithiocarbamic acid.	(72) 1.75 μ/l
(73) Dithane, or zincate of manganese ethylene-bis-dithiocarbamate.	(73) 1.75 μ/l
(74) Thiram, or tetramethylthiuramdisulfide	(74) 1.75 μ/l
(75) Atrazine, or 2-chloro-4-ethylamino-6-isopropyl-amino-S-triazine.	(75) 7.5 μ/l
(76) Propazine, or 2-chloro-4,6-diisopropyl-amino-S-triazine.	(76) 16.0 μ/l
(77) Simazine, or 2-chloro-4,6-diethylamino-S-triazine.	(77) 75.25 μ/l
(78) Di-n-butylphthalate	(78) 770 μ/l
(79) Di(2-ethylhexyl)phthalate (DEHP)	(79) 4.2 mg/l
(80) Hexachlorophene, or 2,2'-methylene-bis(3,4,6-trichlorophenol)	(80) 7 μ/l
(81) Methyl methacrylate	(81) 0.7 mg/l
(82) Pentachlorophenol (PCP)	(82) 21 μ/l
(83) Styrene	(83) 931 μ/l

1. Foaming agents determined as methylene blue active stubstances (MBAs) or other tests as specified by the Commissioner.
2. Combined concentration of iron and manganese shall not exceed 0.5 mg/l.
3. "Not detectable" means by tests or analytical determinations referenced in Section 703.4.

Source: New York Department of Environmental Conservation, 1978.

Table 11-10. New York Effluent Standards or Limitations on Class GA

Applicability. The following effluent standards or limitations shall apply to all Class GA waters in New York State.

Biological organisms. Coliform and/or pathogenic organisms shall not be discharged in amounts sufficient to render fresh ground waters detrimental to public health, safety or welfare.

Chemical Characteristics

Substance	Maximum Allowable Concentration (mg/l, unless otherwise noted)
(1) Aluminum	(1) 2.0
(2) Arsenic	(2) 0.05
(3) Barium	(3) 2.0
(4) Cadmium	(4) 0.02
(5) Chloride	(5) 500
(6) Chromium (Cr) (Hexavalent)	(6) 0.10
(7) Copper	(7) 1.0
(8) Cyanide	(8) 0.40
(9) Fluoride	(9) 3.0
(10) Foaming Agents[1]	(10) 1.0
(11) Iron[2]	(11) 0.6
(12) Lead	(12) 0.05
(13) Manganese[2]	(13) 0.6
(14) Mercury	(14) 0.004
(15) Nickel	(15) 2.0
(16) Nitrate (as N)	(16) 20
(17) Oil and Grease	(17) 15
(18) Phenols	(18) 0.002
(19) Selenium	(19) 0.04
(20) Silver	(20) 0.1
(21) Sulfate	(21) 500
(22) Sulfide	(22) 1.0
(23) Zinc	(23) 5.0
(24) pH Range[3]	(24) 6.5–8.5
(25) Aldrin, or 1,2,3,4,10,10-hexachloro-1,4,4a,5,8,8a-hexahydro-*endo*-1,4-exc-5,8-dimethanonaphthalene	(25) not detectable[4]
(26) Chlordane, or 1,2,4,5,6,7,8,8-octachloro-2,3,3a,4,7,7a-hexahydro-4,7-methanoindene	(26) 0.1 μ/l
(27) DDT, or 2,2-bis-(*p*-chlorophenyl)-1,1,1-tri-chloroethane and metabolites	(27) not detectable[4]
(28) Dieldrin, or 6,7-epoxy aldrin	(28) not detectable[4]

Table 11-10. New York Effluent Standards or Limitations on Class GA (*continued*)

Substance	Maximum Allowable Concentration (mg/l, unless otherwise noted)
(29) Endrin, or 1,2,3,4,10,10-hexachloro-6,7-epoxy-1,4,4a,5,6,7,8,8a-octahydro-*endo*-1,4-*endo*-5,8-dimethanonaphthalene	(29) not detectable[4]
(30) Heptachlor, or 1,4,5,6,7,8,8-heptachloro-3a,4,7,7a-tetrahydro-4,7-methanoindene and metabolites	(30) not detectable[4]
(31) Lindane and other hexachlorocyclohexanes or mixed isomers of 1,2,3,3,5,6-hexachloro-cyclohexane	(31) not detectable[4]
(32) Methoxychlor, or 2,2-bis-(*p*-methoxyphenyl)-1,1,1-trichloroethane	(32) 35 μ/l
(33) Toxaphene (a mixture of at least 175 chlorinated camphene derivatives)	(33) not detectable[4]
(34) 2,4-Dichlorophenoxyacetic acid (2,4-D)	(34) 4.4 μ/l
(35) 2,4,5-Trichlorophenoxypropionic acid (2,4,5-TP) (Silvex)	(35) 0.26 μ/l
(36) Vinyl chloride (chloroethene)	(36) 5.0 μ/l
(37) Benzene	(37) not detectable[4]
(38) Benzo(a)pyrene	(38) not detectable[4]
(39) Kepone, or decachlorooctahydro-1,3,4-metheno-2H-cyclobuta(cd)pentalen-2-one (chlordeone)	(39) not detectable[4]
(40) Polychlorinated biphenyls (PCB) (Aroclor)	(40) 0.1 μ/l
(41) Ethylene thiourea (ETU)	(41) not detectable[4]
(42) Chloroform	(42) 100 μ/l
(43) Carbon tetrachloride (tetrachloromethane)	(43) 5 μ/l
(44) Pentachloronitrobenzene (PCNB)	(44) not detectable[4]
(45) Trichloroethylene	(45) 10 μ/l
(46) Diphenylhydrazine	(46) not detectable[4]
(47) bis(2-chloroethyl)ether	(47) 1.0 μ/l
(48) 2,4,5-Trichlorophenoxyacetic acid (2,4,5-T)	(48) 35 μ/l
(49) 2,3,7,8-Tetrachlorodibenzo-*p*-dioxin (TCDD)	(49) 3.5×10^{-5} μ/l
(50) 2-Methyl-4-chlorophenoxyacetic acid (MCPA)	(50) 0.44 μ/l
(51) Amiben, or 3-amino-2,5-dichlorobenzoic acid (chloramben)	(51) 87.5 μ/l
(52) Dicamba, or 2-methoxy-3,6-dichlorobenzoic acid	(52) 0.44 μ/l
(53) Alachlor, or 2-chloro-2',6'-diethyl-*N*-(methoxymethyl)-acetanilide (Lasso)	(53) 35.0 μ/l
(54) Butachlor, or 2-chloro-2',6'-diethyl-*N*-(butoxymethyl)-acetanilide (Machete)	(54) 3.5 μ/l
(55) Propachlor, or 2-chloro-*N*-isopropyl-*N*-acetanilide (Ramrod)	(55) 35.0 μ/l

Substance	Maximum Allowable Concentration (mg/l, unless otherwise noted)
(56) Propanil, or 3′,4′-dichloropropionanilide	(56) 7.0 μ/l
(57) Aldicarb, [2-methyl-2-(methylthio)-propionaldehyde-*O*-(methylcarba-moyl)oxime] and methomyl[1-methylthio-acetaldehyde-*O*-(methyl-carbamoyl)oxime]	(57) 0.35 μ/l
(58) Bromacil, or 5-bromo-3-sec-butyl-5-methlur-acil	(58) 4.4 μ/l
(59) Paraquat, or 1,1′-dimethyl-4,4′-dipyridylium	(59) 2.98 μ/l
(60) Trifluralin, or α,α,α-trifluoro-2,6-dinitro-*N*-dipropyl-*p*-toluidine (Treflan)	(60) 35.0 μ/l
(61) Nitralin, or 4-(methylsulfonyl)-2,6-dinitro-*N*-*N*-dipropylaniline (Planavin)	(61) 35.0 μ/l
(62) Benefin, or *N*-butyl-*N*-ethyl-α,α,α-trifluoro-2,6-dinitro-*p*-toluidine (Balan)	(62) 35.0 μ/l
(63) Azinphosmethyl, or *O*,*O*-dimethyl-*S*-4-oxo-1,2,3-benzotriazin-3(4H)-ylmethylphospho-rodithioate (Guthion)	(63) 4.4 μ/l
(64) Diazinon, or *O*,*O*-diethyl-*O*-(2-isopropyl-4-methyl-6-pyrimidinyl)-phosphorothioate	(64) 0.7 μ/l
(65) Phorate (also for disulfoton), or *O*,*O*-diethyl-*S*-[(ethylthio)methyl]-phosphorodithioate (Thimet R), and disulfoton, or *O*,*O*-diethyl-*S*-[(2-ethylthio)ethyl]phosphorodithioate (Di-System R)	(65) not detectable[4]
(66) Carbaryl, or 1-naphthyl-*N*-methylcarbamate	(66) 28.7 μ/l
(67) Ziram, or zinc salts of dimethyldithiocar-bamic acid	(67) 4.18 μ/l
(68) Ferbam, or iron salts of dimethyldithiocar-bamic acid	(68) 4.18 μ/l
(69) Captan, or *N*-trichloromethylthio-4-cyclohex-ene-1,2-dicarboximide	(69) 17.5 μ/l
(70) Folpet, or *N*-trichloromethylthiophthalimide	(70) 56.0 μ/l
(71) Hexachlorobenzene (HCB)	(71) 0.35 μ/l
(72) Paradichlorobenzene (PDB) (also orthodi-chlorobenzene)	(72) 4.7 μ/l
(73) Parathion (and methyl parathion), or (*O*,*O*-diethyl-*O*-*p*-nitrophenylphosphorothioate, and methyl parathion, or *O*,*O*-dimethyl-*O*-*p*-nitrophenylphosphorothioate	(73) 1.5 μ/l
(74) Malathion, or *S*-1,2-bis(ethoxycarbonyl)-ethyl-*O*,*O*-dimethylphosphorodithioate	(74) 7.0 μ/l
(75) Maneb, or manganese salt of ethylene-bis-dithiocarbamic acid	(75) 1.75 μ/l

Table 11-10.   New York Effluent Standards or Limitations on Class GA (_continued_)

Substance	Maximum Allowable Concentration (mg/l, unless otherwise noted)
(76) Zineb, or zinc salt of ethylene-bis-dithiocarbamic acid	(76) 1.75 μ/l
(77) Dithane, or zincate of manganese ethylene-bis-dithiocarbamate	(77) 1.75 μ/l
(78) Thiram, or tetramethylthiuramdisulfide	(78) 1.75 μ/l
(79) Atrazine, or 2-chloro-4-ethylamino-6-isopropylamino-S-triazine	(79) 7.5 μ/l
(80) Propazine, or 2-chloro-4,6-diisopropyl-amino-S-triazine	(80) 16.0 μ/l
(81) Simazine, or 2-chloro-4,6-diethylamino-S-triazine	(81) 75.25 μ/l
(82) di-n-butylphthalate	(82) 770 μ/l
(83) Di(2-ethylhexyl)phthalate (DEHP)	(83) 4.2 mg/l
(84) Hexachlorophene, or 2,2'-methylene-bis(3,4,6-trichlorophenol)	(84) 7 μ/l
(85) Methyl methacrylate	(85) 0.7 mg/l
(86) Pentachlorophenol (PCP)	(86) 21 μ/l
(87) Styrene	(87) 931 μ/l

In addition to the effluent standards and/or limitations the following also apply in the counties of Nassau and Suffolk

Chemical Characteristics

Substance	Maximum Allowable Concentration in mg/l
Dissolved Solids, Total	1000
Nitrogen, Total (as N)	10

1. Foaming agents determined as methylene blue active substances (MBAs) or other tests specified by the Commissioner.
2. Combined concentration of iron and manganese shall not exceed 1.0 mg/l.
3. When natural ground waters have a pH outside the range indicated above, that natural pH may be one extreme of the allowable range.
4. Not detectable means by tests or analytical determination referenced in Section 703.4.

Source: New York Department of Environmental Conservation, 1978.

the manufacture of sodium chloride or its derivatives or similar products (Table 11-11).
3. Class GSB waters have chloride concentrations in excess of 1000 mg/l. They are found in the saturated zone and best used for the disposal of wastes. This class is only assigned if adjacent and tributary groundwaters will not be impaired (Table 11-11).

The New York State Department of Environmental Conservation has direct statutory responsibility for ambient groundwater standards, as listed in Tables 11-9, 11-10, and 11-11. However, the New York State Department of Health has statutory responsibility for drinking water standards, including those supplies provided by

351

Table 11-11. New York Groundwater Quality Standards for Classes GSA, GSB

The following quality standards shall be applicable to Class GSA waters.

Items	Specifications
1. Sewage, industrial wastes or other wastes, color, taste or odor producing substances, toxic pollutants, thermal discharges, radioactive substances or other deleterious matter.	1. None which may impair the waters for use as sources of saline waters for the best usage outlined above or as to cause or contribute to a condition in contravention of standards for other classified waters of the State.

The following quality standards shall be applicable to Class GSB waters.

Items	Specifications
1. Sewage, industrial wastes or other wastes, color, taste or odor producing substances, toxic pollutants, thermal discharges, radioactive substances or other deleterious matter.	1. None which may be deleterious, harmful, detrimental or injurious to the public health, safety or welfare or which may cause or contribute to a condition in contravention of standards for other classified waters of the State.

Class GSB shall not be assigned to any groundwaters of the State unless the Commissioner finds that adjacent and tributary ground waters and the best usage thereof will not be impaired by such classification.

Source: New York Department of Environmental Conservation, 1978.

groundwater. In other words, if groundwater supplies are to be distributed by a public water system, drinking-water standards set by the New York State Department of Health must be observed. In some instances these standards are more strict; furthermore, additional standards are being established for toxic organics (New York State Department of Environmental Conservation, 1983).

352

North Carolina

In June of 1979 North Carolina adopted a classification system in recognition of the fact that land uses were changing from agriculture and silviculture to housing, commerce, and industry, all of which would increase the potential and incidence of groundwater contamination (North Carolina Environmental Management Commission, 1979). North Carolina has a five-class system whose major parameters are the suitability of water for drinking, culinary use, and food processing; chloride concentration; and the depth at which the water occurs below the land surfaces (Table 11-12).

Table 11-12.   North Carolina Groundwater Classification Standards

1) Class GA Waters:*

   1) Where naturally occurring concentrations exceed the established standard, the standard will be the naturally occurring concentration as determined by the director.
   2) Total coliform—1 per 100 ml
   3) Endrin—0.0002 mg/l
   4) Lindane—0.004 mg/l
   5) Methoxychlor—0.1 mg/l
   6) Toxaphene—0.005 mg/l
   7) 2,4,-D—0.1 mg/l
   8) 2,4,6,-TP Silvex—0.01 mg/l
   9) Total trihalomethanes—0.1 mg/l
   10) Arsenic—0.05 mg/l
   11) Barium—1.0 mg/l
   12) Cadmium—0.01 mg/l
   13) Chromium—0.05 mg/l
   14) Lead—0.05 mg/l
   15) Mercury—0.002 mg/l
   16) Nitrate (as N)—10.0 mg/l
   17) Nitrite (as N)—1.0 mg/l

Table 11-12. North Carolina Groundwater Classification Standards (*continued*)

18) Selenium—0.01 mg/l
19) Silver—0.05 mg/l
20) Fluoride—1.5 mg/l
21) Combined radium-226 and radium-228—5 pCi/l
22) Gross alpha particle activity—15 pCi/l
23) Gross beta particle activity—50 pCi/l
24) Iron—0.03 mg/l
25) Manganese—0.05 mg/l
26) pH—no increase from naturally occurring pH values in acidity below or increase in alkalinity above 7.
27) Chloride—250 mg/l
28) Color—less than 15 units
29) Phenol—1.0 μg/l
30) TDS—500 mg/l
31) Thermal—not greater than 30°F variance from the naturally occurring level as determined by the director.

353

2) **Class GSA Waters:*

1) Where naturally occurring concentrations exceed the established standard, the standard will be the naturally occurring concentration as determined by the director.
2) Total coliform—1 per 100 ml
3) Endrin—0.0002 mg/l
4) Lindane—0.004 mg/l
5) Methoxychlor—0.1 mg/l
6) Toxaphene—0.005 mg/l
7) 2,4,-D—0.1 mg/l
8) 2,4,6,-TP Silvex—0.01 mg/l
9) Total trihalomethanes—0.1 mg/l
10) Arsenic—0.05 mg/l
11) Barium—1.0 mg/l
12) Cadmium—0.01 mg/l
13) Chromium—0.05 mg/l
14) Lead—0.05 mg/l
15) Mercury—0.002 mg/l
16) Nitrate (as N)—10.0 mg/l
17) Nitrite (as N)—1.0 mg/l
18) Selenium—0.01 mg/l
19) Silver—0.05 mg/l
20) Fluoride—1.5 mg/l
21) Combined radium-226 and radium-228—5 pCi/l
22) Gross alpha particle activity—15 pCi/l
23) Gross beta particle activity—50 pCi/l
24) Iron—0.3 mg/l
25) Manganese—0.05 mg/l
26) pH—no increase from naturally occurring pH values in acidity below or increase in alkalinity above 7.

Table 11-12. North Carolina Groundwater Classification Standards (*continued*)

27) Chloride—allowable increase not to exceed 100 percent of the naturally occurring chloride concentration.
28) Color—less than 15 units
29) Phenol—1.0 μg/l
30) TDS—1000 mg/l
31) Thermal—not greater than 30°F variance from the naturally occurring level as determined by the director.

3) Class GB and GSB Waters:

No increase above the naturally occurring concentration of any toxic or deleterious substance unless it can be shown, upon request, to the satisfaction of the director that the increase:
1) will not cause or contribute to the contravention of water-quality standards in adjoining waters of a different class;
2) will not accumulate in a manner such that unusual or different hydrological conditions may cause a threat to public health or the environment; and
3) will not cause an existing or potential water supply to become unsafe or unsuitable for its current use.

4) Class GC Waters:

All chemical, radioactive, biological, taste-producing, odor-producing, thermal, and other toxic or deleterious substances shall not exceed the concentration existing at the time of classification.

* For substances not specified, the standard is the naturally occurring concentration as determined by the director.
Synthetic, man-made, or other substances that do not occur naturally are prohibited.

Source: North Carolina Environmental Management Commission, 1985.

"The regulations . . . are intended to maintain and preserve the quality of the subsurface groundwaters, prevent and abate pollution and contamination, protect public health, and permit management of the groundwaters for their best usage by the citizens of North Carolina" (North Carolina Environmental Management Commission, 1985). These standards and classifications are considered the foundation of North Carolina's protection strategy.

1. Class GA waters are existing or potential sources of water supply for drinking, culinary use, and food processing without treatment, except where necessary to correct naturally occurring condition. Chloride concentrations are required to be less than or equal to 250 mg/l. These waters occur at

depths greater than 20 feet below the land surface and in the saturated zone above 20 feet, where they are a principal source of potable water supply.

2. Class GSA waters are similar to class GA waters but differ in that they are existing or potential sources of potable mineral water for culinary use or food processing, and that they can be converted to fresh waters through treatment. Naturally occurring chloride concentrations are greater than 250 mg/l and are considered safe for the uses listed.

3. Class GB waters are suitable for the recharge of groundwaters of class GA quality and occur in the saturated zone above a depth of 20 feet. This class was created in recognition of the fact that the upper 20 feet of the earth's surface is particularly vulnerable to contamination and should be considered a cycling zone for the removal of most or all contaminants from infiltrating water. In North Carolina most water in this saturated zone is of drinking-water quality.

4. Class GSB waters are similar to class GB waters but differ in that they are suitable for the recharge of class GSA waters.

5. Class GC waters are best used for activities other than drinking, culinary use, or food processing. They do not meet the quality standards of the higher classifications, and it is not possible to upgrade them for technical and economic reasons (North Carolina Environmental Management Commission, 1985).

Pollution prevention and detection requirements are included in the state permit program to protect groundwater standards. Any pollution discharged from facilities not requiring a permit may result in a state ordered cleanup or some other form of remedial action.

355

Rhode Island

During the spring of 1985, the state legislature of Rhode Island passed the Rhode Island Groundwater Protection Act of 1985. The act asserts that it is the policy of the state "to restore, enhance, and maintain the chemical, physical and biological integrity of its waters, to protect public health, to safeguard fish and aquatic life and scenic and ecological values and to enhance the domestic, municipal, recreational, industrial and other uses of water" (Rhode Island Department of Environmental Management, 1985). Among the provisions of this act was the establishment of a groundwater classification system similar to that of Connecticut "for the purposes of protecting existing, proposed or potential drinking water supplies" (Rhode Island Department of Environmental Management, 1985). A survey of all groundwater sources and related groundwater aquifers, watersheds, and deep flow recharge areas is to be conducted, and upon completion all groundwater sources will come under one of the following classifications:

GAA—Groundwater sources suitable for public drinking-water use without treatment;

356

GA—Groundwater sources that may be suitable for public or private drinking water without treatment;

GB—Groundwater sources that may not be suitable for public or private drinking water without treatment, due to known or presumed degradation;

GC—Groundwater sources that may be suitable for certain waste-disposal practices because past or present land use or hydrogeologic conditions render said groundwaters more suitable for receiving permitted discharges than for development as public or private water supply (Rhode Island Department of Environmental Management, 1985).

Also, water quality standards establishing maximum contaminant levels for each classification are to be developed. No degradation of groundwaters is permitted under this act unless the state finds it essential, desirable, and justifiable for economic, commercial, industrial, or social development (Rhode Island Department of Environmental Management, 1985). Water quality standards shall be used to promote restoration of groundwater to drinking water quality without treatment except where the groundwater is 1) in a zone of discharge otherwise permitted by the provisions of the general laws; 2) classified as GB and there exists no demonstrated present or future need to upgrade to GA; or 3) classified as GC.

South Carolina

A goal of the South Carolina Department of Health and Environmental Control is "to maintain the quality of ground waters consistent with the highest potential uses" (South Carolina Department of Health and Environmental Control, 1985). With this goal in mind, South Carolina established a three-tiered classification system of uses. On the date that this regulation became effective, all of the state's groundwaters were designated

Class GB. Class GB groundwaters are defined as groundwaters suitable for drinking water without treatment. Although most of South Carolina's groundwaters fell into this class, two additional classes were established for the unavoidable exceptions. Those aquifers of exceptional value were designated Class GA aquifers, and those having little potential as a drinking-water source are designated Class GC (South Carolina Department of Health and Environmental Control, 1985). Protection of these groundwaters is enforced through a permit system in which all discharges must be permitted by the South Carolina Department of Health and Environmental Control. All discharges must be treated or controlled to a degree that produces effluent that is consistent with this act, the Clean Water Act, and related regulations. These classes and their standards are defined in more detail in Tables 11-13 and 11-14.

357

Vermont

In 1985 Vermont's legislature passed a groundwater protection act that gave Vermont's Department of Water Resources the power it needed to protect its groundwaters for present and future use. This act called for the mapping and classification of groundwater into four clasces and the deVelopmenT of a framework for the regulation of above-ground activities according to the risks they pose on the groundwaters below (Vermont Agency of Environmental Conservation, 1985a). Vermont's classification is based on the vulnerability to contamination and acceptable degree of risk for existing and future groundwater use, rather than on established water-quality standards (U.S. EPA, 1985d). The state is now in the process of classifying its groundwaters, and as of 1985 the Aquifer Protection Area (APA) mapping program had charted 209 areas, with another 230 under study (Vermont Agency of Environmental Conservation, 1985a). The four aquifer classes are as follows:

Class I—Aquifers suitable for public (drinking) water supply. Character uniformly excellent. No exposure to activities which pose a risk to its current or potential use as a public water supply.

Class II—Aquifers suitable for public (drinking) water supply. Character uniformly excellent but exposed to activities which may pose a risk to its current or potential use as a public water supply.

Class III—Aquifers not suitable as a source of water for

Table 11-13. South Carolina Groundwater Classification

Class descriptions and Specific Standards for Ground Waters

Class GA—Those ground waters that are highly vulnerable to contamination because of the hydrological characteristics of the areas under which they occur and that are also characterized by either of the following two factors:

(a) Irreplaceable, in that no reasonable alternative source of drinking water is available to substantial populations;

or

(b) Ecologically vital, in that the aquifer provides the base flow for a particularly sensitive ecological system that, if polluted, would destroy a unique habitat.

Quality Standards for Class GA Ground Waters

Items	*Standards*
Treated wastes, toxic wastes, deletrious substances, or constituents thereof.	None allowed.

Class GB—All ground waters of the State, unless classified otherwise, which meet the definition of underground sources of drinking water* (USDW) as defined below.

Quality Standards for Class GB Ground Waters

Items	*Standards*
(a) Inorganic chemicals	As set forth in the State Primary Drinking Water Regulations.
(b) Organic chemicals	As set forth in the State Primary Drinking Water Regulations.
(c) Man-made radionuclides, priority pollutant volatile organic compounds, acid extractable organic compounds, pesticides, herbicides, polychlorinated biphenyls, any other synthetic organic compounds not specified above, treated wastes, thermal wastes, deleterious substances, colored wastes or other wastes or constituents thereof.	Not to exceed concentrations or amounts such as to interfere with use, actual or intended, as determined by the Department.

Table 11-13.   South Carolina Groundwater Classification (*continued*)

Class GC—Those groundwaters not considered potential sources of drinking water and of limited beneficial use (i.e., ground waters that exceed a concentration of 10,000 mg/l total dissolved solids or are otherwise contaminated beyond levels that allow cleanup using methods reasonably employed in public-water-system treatment. These ground waters also must not migrate to GA or GB ground waters or have a discharge to surface water that could cause degradation.

359

Quality Standards for Class GC Ground Waters

Items	*Standards*
Treated wastes, toxic wastes, deleterious substances, or other constituents thereof.	None which interfere with any existing use of an underground source of drinking water.*

* Underground source of drinking water refers to an aquifer or its portion:
   1) which supplies any public water system; or
   2) which contains a sufficient quantity of ground water to supply a public water system; and,
       (a) currently supplies drinking water for human consumption; or
       (b) contains water with fewer than ten thousand milligrams per liter total dissolved solids.

Source: South Carolina Department of Health and Environmental Control, 1985.

individual domestic (drinking) water supply, irrigation, agricultural use and general industrial and commercial use.

Class IV—Aquifers not suitable as a source of potable water but suitable for some agricultural, industrial and commercial use (Vermont Agency of Environmental Conservation, 1985b).

All groundwater is initially classified as Class III. Class I or II shall apply to aquifers in use as a public water supply source or which in the opinion of the secretary have a high probability for such use (Vermont Agency of Environmental Conservation, 1985b). The Department of Water Resources will submit recommendations for the regulation of activities over Class I, II, and IV aquifers by September 1987. (Michael Smith, personal communication).

Wyoming

Wyoming adopted groundwater classification in April 1980 "in order to apply standards to protect water quality. Groundwaters of the State are classified by use, and by ambient water quality"

Table 11-14. South Carolina State Primary Drinking Water Regulations

MAXIMUM CONTAMINANT LEVELS IN DRINKING WATER
(1) Maximum Contaminant Levels for Inorganic Chemicals
The maximum contaminant level for nitrate is applicable to both community water systems and non-community water systems except as provided below. The levels for the other inorganic chemicals apply only to community water systems. If it is determined that a non-community water system exceeds a maximum contaminant level for any inorganic constituent, other than nitrates, an investigation of the possible health effects will be made. If it is determined that a health risk does exist, the maximum contaminant level for that particular chemical will apply.

Contaminant	Level (mg/l)
Arsenic	0.05
Barium	1.0
Cadmium	0.001
Chromium	0.05
Fluoride	1.6
Lead	0.05
Mercury	0.002
Nitrate (as N)	10.0
Selenium	0.01
Silver	0.05

At the discretion of the Department, nitrate levels not to exceed twenty milligrams per liter may be allowed in a non-community water system if the supplier of water demonstrates to the satisfaction of the Department that:
a) Such water will not be available to children under six months of age; and,
b) There will be continuous posting of the fact that nitrate levels exceed ten milligrams per liter and the potential health effects of exposure; and,
c) No adverse health effects shall result from the consumption of this water.

(2) Maximum Contaminant Levels for Organic Chemicals
The maximum contaminant levels for organic chemicals apply to community water systems. If it is determined that a non-community water system exceeds a maximum contaminant level for any organic constituent, an investigation of the possible health effects will be made. If it is determined that a health risk does exist, the maximum contaminant level for that particular organic chemical will apply.

Contaminant	Level (mg/l)
(a) Chlorinated hydrocarbons:	
Endrin	0.0002
Lindane	0.004
Methoxychlor	0.1
Toxaphene	0.005
(b) Chlorophenoxys:	
2,4-D	0.1
2,4,5-TP Silvex	0.01

Table 11-14. South Carolina State Primary Drinking Water Regulations (*continued*)

(3) Maximum Contaminant Levels for Turbidity
The maximum contaminant levels for turbidity are applicable to both community and non-community water systems using surface water sources in whole or in part. The maximum contaminant levels for turbidity are applicable to water systems using groundwater sources where turbidity problems occur, as determined by the Department.

    (a) One turbidity unit, as determined by a monthly average, except that **361** five or fewer turbidity units may be allowed if the supplier of water can demonstrate to the State that the higher turbidity does not do any of the following:
- Interfere with disinfection
- Prevent maintenance of an effective disinfectant agent throughout the distribution system
- Interfere with microbiological determinations.

    (b) Five turbidity units based on an average for two consecutive days.

(4) Maximum Microbiological Contaminant Levels
These maximum contaminant levels shall apply to community and non-community water systems.

    (a) When the membrane filter technique is used, the maximum contaminant levels for coliform bacteria are as follows:*
- One per 100 milliliters as the arithmetic mean of all samples examined per month
- Four per 100 milliliters in more than one sample when less than 20 samples are examined per month
- Four per 100 milliliters in more than 5 percent of the samples when 20 or more are examined per month.

    (b) When the fermentation tube method and 10-milliliter standard portions are used, coliform bacteria shall not be present in any of the following:
- More than 10 percent of the portions in any month;
- Three or more portions in more than one sample when less than 20 samples are examined per month; or
- Three or more portions in more than 5 percent of the samples when 20 or more samples are examined per month.

    (c) When the fermentation tube method and 100-milliliter standard portions are used, coliform bacteria shall not be present in any of the following:
- More than 60 percent of the portions in any month
- Five portions in more than one sample when less than 5 samples are examined per month; or
- Five portions in more than 20 percent of the samples when 5 or more samples are examined per month.

(5) Maximum Contaminant Levels for Naturally Occurring Radionuclides
The maximum contaminant levels for naturally occurring radionuclides are applicable to community water systems.

Table 11-14. South Carolina State Primary Drinking Water Regulations (*continued*)

Contaminant	*Level (pCi/l)*
Radium-226 and radium-228	5
Gross alpha particle activity (including radium-226 but excluding radon and uranium)	15

362

(6) Maximum Contaminant Levels for Man-Made Radionuclides
The maximum contaminant levels for man-made radionuclides are applicable to community water systems.
   (a) The average annual concentration of beta particle and photon radioactivity from man-made radionuclides in drinking water shall not produce an annual dose equivalent to the total body or any internal organ greater than 4 millirem/year.
   (b) Except for the radionuclides listed below, the concentration of man-made radionuclides causing 4 mrem total body or organ dose equivalents shall be calculated on the basis of a 2-liter-per-day drinking-water intake using 168-hour data listed in "Maximum Permissible Body Burdens and Maximum Permissible Concentration of Radionuclides in Air or Water for Occupational Exposure," NBS handbook 69, as amended August 1963, U.S. Department of Commerce. If two or more radionuclides are present, the sum of their annual dose equivalent to the total body or to any organ shall not exceed 4 millirem/year.

Radionuclide	*Critical Organ*	*pCi/l*
Tritium	Total body	20,000
Strontium-90	Bone marrow	8

(7) Maximum Contaminant Levels for Total Trihalomethane Concentration
The maximum contaminant for total trihalomethane concentration shall apply to community water systems serving populations of 10,000 or more which add a disinfectant (oxidant) to the water in any part of the treatment process.

The MCL for total trihalomethane concentrations shall be 0.1 mg/l.

SECONDARY MAXIMUM CONTAMINANT LEVELS
The secondary maximum contaminant levels are applicable to both community and non-community water systems.

Contaminant	*Level*
Chloride	250 mg/l
Color	15 color units
Copper	1 mg/l
Corrosivity	noncorrosive
Foaming Agents	0.5 mg/l
Iron	0.3 mg/l

Table 11-14.   South Carolina State Primary Drinking Water Regulations (*continued*)

Manganese	0.05 mg/l
Odor	3 threshold odor number
pH	6.5–8.5
Sulfate	250 mg/l
Total dissolved solids	500 mg/l
Zinc	5 mg/l

The Department may establish higher or lower levels which may be appropriate depending upon local conditions provided the supplier of water is able to demonstrate that use of the water will not adversely affect the public health and welfare. In evaluating the affect to the public health and welfare, the supplier of water may evaluate the unavailability of alternate water sources; the economic evaluation of necessary treatment or other compelling factors that may prevent compliance.

363

* Additional conditions apply—look to South Carolina's State Primary Drinking Water Regulations for more detailed information.

Source: South Carolina Department of Health and Environmental Control, 1981.

(Wyoming Department of Environmental Quality, 1980). The Wyoming system is an intricate one and appears to accommodate all the major competitors for groundwater use in the state: people, crops, livestock, fish and aquatic life, industry, excavation of hydrocarbons and minerals, geothermal energy. The four classes are:
1. Class I—suitable for domestic use
2. Class II—suitable for agricultural use where all other conditions (i.e., soil) are adequate
3. Class III—suitable for livestock
4. Class Special (A)—suitable for fish and aquatic life

Each of the above classes has a set of parameters limiting its chemical constituents (Table 11-15). In addition, those waters may not contain biological, hazardous, toxic, or potentially toxic materials or substances in concentrations greater than those determined by the EPA under the National Primary Drinking Water Regulations. Discharges into water of these four classes are only permitted if the water can be returned to its original quality.
5. Class IV—suitable for industry; quality standards vary with type of industry
    a) Class IV (A)—TDS ≤ 10,000 mg/l.
    b) Class IV (B)—TDS > 10,000 mg/l.

Table 11-15. Wyoming Groundwater Quality Standards

Constituent or Parameter	Underground Water Class Use Suitability and Concentration*		
	I Domestic	II Agriculture	III Livestock
Aluminum (Al)	—	5.0	5.0
Ammonia (NH-N)	0.5[1]	—	—
Arsenic (As)	0.05	0.1	0.2
Barium (Ba)	1.0	—	—
Beryllium (Be)	—	0.1	—
Boron (B)	0.75	0.75	5.0
Cadmium (Cd)	0.01	0.01	0.05
Chloride (Cl)	250.0	100.0	2000.0
Chromium (Cr)	0.05	0.1	0.05
Cobalt (Co)	—	0.5	1.0
Copper (Cu)	1.0	0.2	0.5
Cyanide (CN)	0.2	—	—
Fluoride (F)	1.4–2.4[2]	—	—
Hydrogen Sulfide (H_2S)	0.05	—	—
Iron (Fe)	0.3	5.0	—
Lead (Pb)	0.05	5.0	0.1
Lithium (Li)	—	2.5	—
Manganese (Mn)	0.05	0.2	—
Mercury (Hg)	0.002	—	0.00005
Nickel (Ni)	—	0.2	—
Nitrate (NO_3-N)	10.0	—	—
Nitrite (NO_2-N)	1.0	—	10.0
($NO_3 + NO_2$)-N	—	—	100.0
Oil & Grease	Virtually free	10.0	10.0
Phenol	0.001	—	—
Selenium (Se)	0.01	0.02	0.05
Silver (Ag)	0.05	—	—
Sulfate (SO_4)	250.0	200.0	3000.0
Total Dissolved Solids (TDS)	500.0	2000.0	5000.0
Uranium (U)	5.0	5.0	5.0
Vanadium (V)	—	0.1	0.1
Zinc (Zn)	5.0	2.0	25.0
pH	6.5–9.0 s.u.	4.5–9.0 s.u.	6.5–8.5 s.u.
SAR	—	8	—
RSC	—	1.25 meq/l	—
Combined Total Radium-226 and Radium-228	5 pCi/l	5 pCi/l	5 pCi/l
Total Stontium-90	8 pCi/l	8 pCi/l	8 pCi/l

Table 11-15. Wyoming Groundwater Quality Standards (*continued*)

Constituent or Parameter	Underground Water Class Use Suitability and Concentration*		
	I Domestic	II Agriculture	III Livestock
Gross alpha particle radioactivity (including Radium-226 but excluding Radon and Uranium)	15 pCi/l	15 pCi/l	15 pCi/l

365

Constituent or Parameter	Underground Water Class Use Suitability and Concentration
	Special (A) Fish/Aquatic Life
Aluminum (Al)	0.1
Ammonia (NH_3)[3]	0.02
Arsenic (As)	0.05
Barium (Ba)	5.0
Beryllium (Be)	0.011–1.1[4]
Boron (B)	—
Cadmium (Cd)	0.0004–0.015[4]
Chloride (Cl)	—
Chromium (Cr)	0.05
Cobalt (Co)	—
Copper (Cu)	0.01–0.04[4]
Cyanide (CN)	0.005
Fluoride (F)	—
Hydrogen Sulfide (H_2S)[5]	0.0025
Iron (Fe)	0.5
Lead (Pb)	0.004–0.15[4]
Lithium (Li)	—
Manganese (Mn)	1.0
Mercury (Hg)	0.00005
Nickel (Ni)	0.05–0.4[4]
Nitrate $(NO_3\text{-N})$	—
Nitrite $(NO_2\text{-N})$	—
$(NO_3 + NO_2)\text{-N}$	—
Oil & Grease	Virtually free
Phenol	0.001
Selenium (Se)	0.05
Silver (Ag)	0.0001–0.00025[4]

Table 11-15. Wyoming Groundwater Quality Standards (*continued*)

Constituent or Parameter	Underground Water Class Use Suitability and Concentration*		
	I Domestic	II Agriculture	III Livestock
Sulfate (So$_4$)		—	
Total Dissolved Solids (TDS)		500.0–1000.0–2000.0	
Uranium (U)		0.03–1.4	
Vanadium (V)		—	
Zinc (Zn)		0.05–0.6[4]	
pH		6.5 s.u.–9.0 s.u.	
Combined Total Radium-226 and Radium-228		5 pCi/l	
Total Strontium-90		8 pCi/l	
Gross alpha particle radioactivity (including Radium-226 but excluding Radon and Uranium)		15 pCi/l	

[1] Total ammonia nitrogen
[2] Dependent on the annual average of the maximum daily air temperature
[3] Ionized ammonia
[4] Dependent on hardness
[5] Undissociated H_2S
* mg/l, unless otherwise indicated

Source: Wyoming Department of Environmental Quality Division, 1980.

Discharges into industrial-use aquifers are allowed as long as the water remains fit for its intended use. Oil and grease concentrations cannot be greater than 10 mg/l. Radiation concentrations may not exceed the limits of the first four classes. EPA standards for maximum chemical constituents may not be exceeded.

6. Class V—groundwater closely associated with hydrocarbon deposits or other minerals, or groundwater considered a geothermal resource. Discharges into these waters are for the purpose of recovering the resources.

7. Class VI—groundwater that is unsuitable for any use.

Summary of State Aquifer Classification Systems

Many states have adopted some form of classification, and as discussed, these vary from state to state. They range from simple narrative systems to sophisticated differentiating systems that incorporate both land-use criteria and numerical standards. Maine currently has a very simple two-class system of classification. Florida and New Mexico classify their aquifers into potable and non-potable water supplies based on the total content of dissolved solids in the water. Montana uses specific conductance to assess the quality of groundwater for classification. Those aquifers classified as potable are carefully protected to maintain drinking-water quality. The remaining states—Connecticut, New Jersey, New **367** York, North Carolina, Massachusetts, South Carolina, Vermont, and Wyoming—have a variety of more complex classification systems. For the most part, these systems are based on the understanding that different land uses require different levels of water quality, and thus, protection. The states have set up standards of water quality for varying uses such as housing, agriculture, and industry. Most of these states have classification systems that permit wastes to be discharged into low-quality aquifers.

Twelve

Federal Statutes Relevant to the Protection of Groundwater

- There is no federal groundwater protection statute. Several federal laws provide for the prevention of groundwater contamination, but they focus on a narrow range of polluting activities.
- At the federal level, the predominant focus has been on groundwater contamination from the disposal of industrial hazardous wastes pursuant to the Resource Conservation and Recovery Act (RCRA). The Hazardous and Solid Waste Amendments of 1984 (HSWA) strengthen RCRA in this regard.
- Other federal statutes directly or indirectly impacting groundwater protection are the Safe Drinking Water Act (SDWA), the Clean Water Act (CWA), the Comprehensive Environmental Response, Compensation and Liability Act (CERCLA or "Superfund Law"), the Toxic Substances Control Act (TSCA), the Surface Mining Control and Reclamation Act (SMCRA), the Federal Insecticide, Fungicide and Rodenticide Act (FIFRA), and the National Environmental Policy Act (NEPA).

The federal legal framework outlined above reflects an imperfect approach to deal with the diverse problems of groundwater contamination described in earlier chapters. In examining this legal framework, it is important to consider the relationship between federal regulations and state and local regulations, since in some instances the respective bodies of law cover quite different areas of concern. It should also be noted that some activities which may give rise to groundwater contamination do not lend themselves easily to control through traditional "command and control" regulatory mechanisms, thus far, these activities have tended to receive less attention from lawmakers.

Another feature of most of these federal environmental laws is that they are not focused exclusively, or in many instances even primarily, on groundwater. Instead, they are directed at particular activities or practices likely to pose a variety of threats to the environment, with groundwater being one of the several environmental media warranting protection. Thus, it is particularly important to consider whether the various statutes, in combination, provide a comprehensive framework for protecting the quality of groundwater, or whether there are gaps or weaknesses that need to be addressed. The Environmental Protection Agency (EPA) established an Office of Ground-Water Protection in 1984 to examine this question and to develop an overall groundwater strategy for the country. Progress toward integrating this strategy with the sometimes competing goals of the various federal environmental statutes and the difficult technical issues implicit in groundwater protection has been slow.

369

At the federal level, the predominant concern over groundwater quality has been contamination resulting from land disposal of industrial hazardous wastes. The Resource Conservation and Recovery Act of 1976 (RCRA), together with the sweeping Hazardous and Solid Waste Amendments (HSWA) to RCRA enacted by Congress in 1984, establish a comprehensive framework for the management of municipal and industrial solid and hazardous wastes. These laws manifest Congress's recognition of a need to regulate waste disposal practices in such a way as to minimize the threat of soil and groundwater contamination and to shift this nation's reliance on land disposal to less polluting methods for dealing with waste.

Other important federal statutes affecting groundwater protection are: the Safe Drinking Water Act (SDWA), which was enacted to provide for sanitary drinking water supplies; the Clean Water Act (CWA), which establishes a program to control the discharge of pollutants into navigable waters; and the Comprehensive Environmental Response, Compensation and Liability Act of 1980 (CERCLA or "Superfund Law"), which authorizes the federal government to respond to releases or threatened releases of hazardous substances into the environment which may present an imminent and substantial danger to public health. Other statutes bearing less directly on contamination of groundwater, although relevant to the overall national effort to protect the resource, are the Toxic Substances Control Act (TSCA), the Surface Mining Control and Reclamation Act (SMCRA), the Federal Insecticide, Fungicide and Rodenticide Act (FIFRA), and the National Environmental Policy Act. In addition to the statutes listed above, there are a **370** number of other Federal laws that apply to groundwater only marginally or under specific situations. They are the Atomic Energy Act of 1954 (42 U.S.C. 2011), the Coastal Zone Management Act of 1976 (16 U.S.C. 1451), the Federal Land Policy and Management Act of 1976 (43 U.S.C. 1701), the Hazardous Liquid Pipeline Safety Act of 1979 (49 U.S.C. 2001), the Hazardous Materials Transportation Act of 1975 (49 U.S.C. 1801), the Reclamation Act of 1902 (43 U.S.C. 390(b)), the Uranium Mill Tailings Radiation Control Act of 1978 (42 U.S.C. 7901), and the Water Research and Development Act of 1978 (42 U.S.C. 7801).

Many of these statutes also contain broad "imminent hazard" provisions that enable the federal government to take immediate action to enjoin activities that pose threats to public health or the environment, and thus provide additional authority to protect groundwater. In fact, hundreds of cases have been filed by federal and state governments pursuant to these laws to abate ground and surface water pollution attributable to the migration of toxic wastes in underground aquifers.

When it enacted RCRA, SDWA and CWA, Congress envisioned that the states would play a major role in administering these programs once the regulatory framework was in place. The role of states in groundwater protection efforts has thus been significantly augmented by several features of federal environmental statutes that enable the states to obtain EPA authorization to administer and enforce these environmental programs and that provide for federal financial and technical assistance for such activities. In order to obtain authorization to administer one of these environmental programs, a state must propose a program for EPA approval that is substantially equivalent to the federal program and obtain the opinion of the State Attorney General that the state has sufficient legal authority from its state legislature to enforce the program. This process has taken place under RCRA, CWA and SDWA to varying extents, resulting in state primacy in the implementation of federal environmental programs. State programs, in some instances, are stricter than federal law requires.

Resource Conservation and Recovery Act of 1976, 42 U.S.C. §6901 *et seq.,* Pub. L. No. 94–580, as amended by Act of Nov. 8, 1984, Pub. L. No. 98–616

Statutory Goals

The Resource Conservation and Recovery Act of 1976 (RCRA) was enacted to regulate the generation, transportation and disposal of hazardous and other solid wastes. Major provisions of the law are directed toward minimizing groundwater contamination. After extremely limited federal involvement in the area of hazardous waste management, the enactment of RCRA in 1976 signaled major federal concern over the significant health and environmental impacts from disposal of wastes.

It is significant that, prior to RCRA the enactment of pollution control programs such as the Clean Air Act and Clean Water Act substantially increased waste management problems by creating massive amounts of wastes from pollution control activities. The original Solid Waste Disposal Act of 1965, which RCRA amended, did little to alleviate these problems. Many unregulated land disposal practices for these and other wastes created situations where toxic leachate might migrate into groundwater, causing potentially serious and widespread contamination. One of the original purposes of RCRA was to ameliorate the problems of waste disposal that existed after the implementation of other pollution-control programs. As stated by Congress in section 1002 of RCRA:

> As a result of the Clean Air Act, the Water Pollution Control Act and other Federal and State Laws respecting public health and the environment, greater amounts of solid waste (in the form of sludge and other pollution treatment residues) have been created. Similarly inadequate and environmentally unsound practices for disposal or use of solid waste have created greater amounts of air and water pollution and other problems for the environment and health.

Amendment to RCRA by the Hazardous and Solid Waste Amendments of 1984 reflects an appreciation that in many regions of the United States it is difficult, if not impossible to dispose of wastes safely by traditional land disposal techniques. The Amendments also acknowledge that corrective measures for removing contamination are likely to be expensive, complex and time consuming. Their goal, therefore, is to force new planning and management practices in the treatment, storage, and disposal of hazardous waste.

Overview of Waste Management Programs Impacting Groundwater

Subtitles C and D of RCRA set forth the primary federal regulatory programs for controlling groundwater contamination from waste disposal. Subtitle C, which has been extensively implemented by EPA, applies to industrial hazardous waste management. Subtitle D, which covers municipal solid waste disposal, directs the states to prepare state solid waste plans and also leaves permitting and enforcement programs to the states themselves. A new provision, Subtitle I of RCRA, enacted by the HWSA of 1984, is directed at the problem of leaking underground storage tanks.

RCRA provides for the regulation of only certain types of groundwater contaminants, as a result, in part, from the way Congress defined exclusions to the definition of the term "solid waste." Under Section 1004(27) of RCRA, "solid waste" is defined as any garbage, refuse, sludge, or other discarded material. "Hazardous waste", a subset of solid waste, is defined as any solid waste that may have specified properties such as ignitability, corrosivity, reactivity, or toxicity or is a waste listed by EPA that poses harm to human health and the environment. Specifically excluded from both definitions, and thus from regulation under RCRA, are solids or dissolved materials in domestic sewage, irrigation return flows, industrial point source discharges subject to the Federal Water Pollution Control Act Amendments of 1972, nuclear wastes, exploration and production waste, and certain wastes associated with mining activities. Household wastes and hazardous wastes generated in small quantities by certain businesses are exempted from Subtitle C regulation by EPA. Some of these materials do result in groundwater contamination in certain areas of the country.

It is clear that RCRA was intended to regulate disposal activities affecting groundwater. Section 1004(3) defines the term "disposal" as "the discharge, deposit, injection, dumping, spilling, leaking, or placing of any solid waste or hazardous waste into or on any land or water so that such solid waste or hazardous waste or any constituent thereof may enter the environment or be emitted into the air or discharged into any waters, including ground waters." Unfortunately, many significant activities that result in groundwater contamination in many parts of the country, such as the application of pesticides and deicing salts, and the release of septic tank leachates, are not reached by RCRA, since they do not fall within the definition of "disposal."

372

Subtitle C—Industrial Hazardous Waste

The heart of the Subtitle C RCRA program is the tracking of hazardous wastes from "cradle to grave." EPA has promulgated regulations imposing manifesting, record keeping, and reporting requirements on generators, transporters, and disposers of hazardous waste. Generators and transporters must obtain hazardous waste identification numbers and meet EPA standards for the storage and transport of hazardous wastes. Hazardous wastes can only be sent to RCRA-permitted disposal facilities. Under Section 3005 of the Act, effective November 19, 1980, it is unlawful to treat, store, or dispose of any hazardous waste without a RCRA permit. Finally, the 1984 HSWA prohibits certain land disposal practices within the next ten years and creates incentives for innovative disposal technologies.

373

Key to the protection of groundwater from hazardous wastes is Section 3004 of RCRA, which directs EPA to establish by federal regulation strict design and operating standards for all hazardous waste treatment, storage, or disposal facilities (TSD facilities) that are necessary to protect human health and the environment. These standards are incorporated into RCRA permits for individual hazardous waste management facilities. In order to obtain permits pursuant to RCRA, owners and operators of hazardous waste TSD facilities must install groundwater monitoring wells to detect contamination from the facility, apply corrective measures if contamination occurs, and obtain financial assurances to cover corrective measures and liability releases if they occur.

Because the more than 10,000 hazardous waste facilities in existence on the effective date of RCRA, November 19, 1980, could not be permitted overnight, RCRA provided interim status to existing operating facilities that submitted brief "Part A" permit applications until such time as lengthy "Part B" permit applications could be completed, reviewed, and approved or disapproved for these hazardous waste treatment, storage, and disposal (TSD) units. "Interim status" allowed existing facilities to be treated as though they had received individual RCRA permits and to continue to operate pending final action on their permit applications if they complied with the notification procedures required to obtain interim status and met certain federal standards for operating practices.

In 1985 the vast majority of the hazardous waste management land disposal facilities in the U.S. were still operating under interim status. Because of the hazard they posed to groundwater, the 1984 Amendments cancelled interim status for land disposal facilities that failed to submit a final Part B application under Section 3005 of the statute by November 8, 1985. The 1984 HSWA Amendments not only required the submission of Part B RCRA permits, but also required such facilities to certify that they had installed groundwater monitoring and met financial responsibility requirements by November 8, 1985 or else cease operating. As a result of the new HSWA requirements, and largely owing to the difficulty in obtaining commercial pollution liability insurance, EPA estimates that over 1,000 or two-thirds of the interim status land disposal facilities in 1985 were forced to close down operations.

Different performance standards and other operating requirements apply to hazardous waste facilities, depending on whether they are in interim status or have received a final RCRA "Part B" permit from EPA. Interim status standards are set out in the Code of Federal Regulations (40 C.F.R., Part 265). Standards set out in 40 C.F.R., Part 264 ("Part 264 standards") apply to both new and existing facilities when they receive their Part B permit. Facilities that cannot meet these permitting standards must cease handling hazardous waste following a denial of their permit applications.

RCRA Operating Standards and Groundwater Protection

Operating standards for both interim status and "Part B" permitted facilities include: (1) general standards that apply to all hazardous waste treatment, storage, or disposal units; and (2) design and construction standards applicable to specific types of facilities, e.g., landfills, land treatment units, surface impoundments, waste-piles, tanks, or incinerators. The "Part B" standards are stricter and more site-specific than the "interim status" standards.

General facility standards applicable to all regulated hazardous waste management units require emergency preparedness and prevention measures; contingency plans and emergency procedures; systems for maintenance of manifests and other record keeping and reports; groundwater monitoring; corrective action plans to be implemented if contaminant migration occurs; plans for closure; post-closure maintenance and groundwater monitoring; and financial responsibility guarantees.

The performance standards applicable to different categories of TSD units impose specific design, operating, and construction requirements depending on the type of hazardous waste facility and its potential threat to the environment. Many of these requirements are directly aimed at protecting groundwater; for example, owners and operators of land treatment units are required to demonstrate, for each waste treated, that hazardous constituents are completely degraded, transformed or immobilized in the treatment zone. As another example, waste piles must be placed in or on liners that prevent precipitation run-off or leachate.

Two basic groundwater protection elements embodied in EPA's RCRA permitting standards are a "liquids-management strategy" and a "groundwater-monitoring and response program." The liquids-management strategy is intended to minimize the generation of leachate and to remove it from the waste management unit before it enters surrounding soils and groundwater. This strategy reflects EPA's awareness that when hazardous waste is in liquid form or in a mixture with other liquids it presents the greatest threat to groundwater, because of the potential for migration below the surface. The most significant feature of the liquids-management strategy mandated by Part 264 was the use of a single liner and leachate collection systems on new surface impoundments, waste piles, and landfills. Design and operating standards for surface impoundments and landfills permitted after passage of the 1984 Amendments to RCRA require these facilities to have *double* liners and a leachate collection and removal system above and between the liners to prevent migration of wastes to adjacent subsurface soil, groundwater, or surface water. (Exemptions are allowed if EPA determines that alternative design or operation practices or the location of the unit will prevent the migration of wastes.) EPA must also promulgate standards for leak detection systems applicable to new landfill units, waste piles, surface impoundments, land treatment units and underground tanks by May, 1987.

The groundwater monitoring and response program is the other pivotal feature of the Part 264 permitting standards whose principal goal is to protect groundwater. Parts 264 and 265 establish that the groundwater protection standards applicable to each facility are background levels of contaminants in the uppermost aquifer upgradient of the unit, or Maximum Contaminant Levels (MCLs) for contaminants promulgated under the Safe Drinking Water Act, whichever level is lower. (A regulatory mechanism allows EPA to set alternate concentration levels [ACLs] on a case by case basis, but EPA has thus far not set such a groundwater standard.) Compliance with these standards is measured through the placement of monitoring wells upgradient and downgradient of the facility. If standards are exceeded downgradient of a TSD unit, RCRA requires that a compliance monitoring program be established; if violations continue, corrective action must be instituted to remedy the contamination.

Bans on Liquids in Landfills and Land Disposal

Section 3004, which provides authority for EPA's promulgation of performance and operating standards, was amended by the 1984 HSWA Amendments. These amendments should have an enormous effect on protecting groundwater. They establish a comprehensive, phased approach to limiting land disposal of hazardous wastes. Land disposal includes any placement of hazardous wastes in a landfill, surface impoundment, waste pile, land treatment unit, mine, cave, injection well, salt well or salt bed. The Amendments also immediately banned placement of non-containerized or bulk liquid hazardous waste in salt domes or beds, mines, or caves until EPA determined whether such disposal was acceptable and issued permits incorporating standards for these practices. Placement of other hazardous wastes in these specified types of disposal units was also banned until the facility received its Part B permit.

The Amendments also banned disposal of bulk liquid wastes in *landfills* in 1985 and the land *disposal* of dioxin-containing hazardous wastes and certain solvents in November 1986, except where EPA determined that a particular disposal practice was safe. All land *disposal* of liquid hazardous wastes and free liquids containing concentrations of toxic materials such as arsenic, cadmium, lead, and PCBs, is banned on November 8, 1987 except for wastes that EPA determines can be disposed of safely by specific land disposal techniques.

EPA must also review one-third of all listed and characteristic hazardous wastes by August 8, 1988, two-thirds by June 8, 1989 and all listed wastes by May 8, 1990, to determine whether they should be banned from land disposal. The 1984 law would have imposed a complete ban on all landfilling and surface impoundment disposal, and a ban on the storage of hazardous waste banned from land disposal had EPA missed the initial rulemaking deadline. Similarly, if EPA misses the May 8, 1990 rulemaking deadline, the Amendments contain a hammer automatically banning all land disposal of hazardous wastes.

EPA can only grant variances from the land disposal bans

imposed by the 1984 amendments on a case by case basis for up to one year if an applicant can show a binding contractual commitment to construct some alternative means of treatment or disposal and demonstrate that circumstances beyond his control have prevented construction from being completed. The Agency may also determine by regulation that a land disposal ban should not be imposed for certain hazardous wastes because sufficient alternative treatment, recovery, or disposal capacity does not exist; in such a case EPA can extend the effective date of a ban for up to **377** two years after promulgation of a ban.

It is important to note that the land disposal ban(s) do not prevent the disposal of regulated materials if they are suitably treated. When EPA promulgates a land disposal ban, it must specify a type and level of treatment that would render the waste acceptable for land disposal by the banned method.

Imminent Hazard Authority

Section 7003 of RCRA empowers the Administrator, upon receipt of evidence that the handling of any solid waste or hazardous waste is presenting or may present an "imminent and substantial endangerment to health or the environment," to bring suit in the federal district court in which the activity occurs to restrain such handling of solid waste or hazardous waste. Alternatively, the Agency may issue such administrative orders "as may be necessary to protect public health and the environment." Willful violation of an administrative order subjects the violator to additional civil penalties, which may be collected in a separate action in federal district court. EPA's imminent hazard authority has been used to compel owners and operators of the hazardous waste management facilities to correct groundwater contamination.

State Role

Congress designed the Subtitle C program so that it also could be administered by the states. The state-authorization process takes place in two phases pursuant to Section 3006. First, in order to obtain interim authorization, the state must submit a description of the program it proposes to administer, including a description of the agency that will administer it, the staff, the estimated cost, and the type of forms and papers that will be used. After EPA review and opportunity for public notice, hearing, and comment, EPA will grant interim authorization to the state if the program is substantially equivalent to the federal program. A state with interim authorization must receive final authorization by 26 January 1985. If no final authorization has been obtained by that date, the authority of the state agency to issue RCRA permits will terminate, and EPA will have to reinstitute a federal program in that state. (States were allowed additional time to adopt regulations in conformance to the requirements of the Hazardous and Solid Waste Amendments of 1984.) In order to obtain final authorization, a state program must be equivalent to the federal program, be consistent with the federal program and other state programs, and provide for adequate enforcement.

378

As of January 1987, 42 states had received final authorization. Only 3 states had failed to submit requests to EPA to administer their own programs.

Solid Waste Disposal Program

Subtitle D of RCRA provides for somewhat limited federal supervision of state and municipal solid waste disposal. EPA is required to provide technical and financial assistance to the states in the development of methods of solid waste disposal that are environmentally sound and that encourage resource conservation. Pursuant to Sections 1008(a) and 4010(c) of RCRA, EPA is to issue guidelines for solid waste management facilities, including methods and controls for the protection of public health, ground waters, and surface waters. Under Section 4006 of RCRA, states are required to develop plans for the management of solid wastes and the conservation of resources in conformance with EPA guidelines. The plans must be approved by EPA. In 1984, Congress required states to implement permit programs by November 1987 for facilities that may receive hazardous wastes from households and small generators of hazardous wastes exempted from Title C. It also requires states to assure that such facilities meet criteria promulgated by EPA for sanitary landfills.

The solid waste program envisaged by Subtitle D, in contrast to the Subtitle C program, does not impose a heavy federal regulatory burden on individual waste management facilities. Instead, it shifts primary responsibility for implementation of the regulatory features of managing wastes to the states.

The key to the Subtitle D solid waste program is Section 4004 of RCRA, which requires EPA to establish guidelines for classifying facilities for the disposal of solid waste as either "open dumps" or "sanitary landfills." The Act provides that at a minimum, a facility may be classified as a sanitary landfill and not as an open dump "if there is no reasonable probability of adverse effects on health or the environment from disposal of solid waste at such facility." States with approved Subtitle D plans then use the EPA criteria to classify their facilities for solid waste disposal and take appropriate action to upgrade or close any that are classified as open dumps.

EPA has established eight guidelines to be used by the states in classifying and managing solid waste facilities (40 C.F.R. §257.3 [1987]). One guideline deals specifically with effects on groundwater, while the others address other types of environmental concerns (40 C.F.R. § 257.3–4 [1987]). It states that a facility for solid waste disposal is an open dump if it contaminates groundwater that either is being used as a source of drinking water or contains less than 10,000 mg/l total dissolved solids. Contamination is deemed to occur whenever the concentration of a contaminant in the groundwater at the "solid waste boundary" is increased so as to exceed any of the maximum contaminant levels (MCLs) specified by the Safe Drinking Water Act for nine inorganic chemicals and six pesticides. (MCLs for volatile organic chemicals in drinking water were proposed by EPA in 1986, but together these MCLs represent only a small fraction of the contaminants found in groundwater.) If the groundwater at the "solid waste boundary" already exceeds any of these standards, then any further increase in the concentration of that contaminant is also deemed to be contamination.

Although the EPA groundwater criteria measure contamination at the "solid waste boundary" (i.e., at the property line of the solid waste facility), the states are allowed to establish alternative boundaries for particular facilities based upon a number of site specific factors. These include the nature of the wastes involved, the present quality and quantity of the groundwater, and the availability of alternate supplies of drinking water.

Once a state with an approved Subtitle D plan classifies a facility as an open dump, Section 4005 of RCRA requires that the state must then provide for either the upgrading or the closure of that facility. The state may issue a compliance schedule to a facility that plans to upgrade, thereby insulating the facility from possible citizen suits to enjoin the prohibited act of "open dumping" under 4005(a). There is however, no mechanism in

Subtitle D that requires states or others to clean up contamination in any groundwater aquifer from such a facility.

The 1984 Amendments to RCRA provided little additional federal authority over potential groundwater contamination from municipal sanitary landfills. Congress, however, charged EPA to conduct a study of the extent to which the guidelines and criteria issued under Subtitle C are adequate to protect human health and the environment from groundwater contamination. A report of this study must be submitted to Congress in late 1987. It is likely that such a study will begin to reveal the magnitude of groundwater contamination from municipal sanitary waste facilities and that it may prompt consideration of further legislation.

380

Leaking Underground Storage Tanks

The Hazardous and Solid Waste Amendments of 1984 provided EPA authority to regulate underground storage tanks that can cause contamination of groundwaters. Owners of underground storage tanks containing RCRA-regulated substances had to notify state agencies of the existence of the tank by May 8, 1986, and specify its age, size, type location and use. New Section 9003 of RCRA, Subtitle I, requires EPA to promulgate release detection, prevention, and correction regulations along with monitoring, recordkeeping, and reporting requirements by May 8, 1987. These will be applicable to owners and operators of regulated underground tanks. In promulgating these regulations, EPA is allowed to distinguish among categories of tanks on the basis of age, soil, climate, history of maintenance, hydrogeology, water table and other factors. The Administration must also issue new tank performance standards for regulated underground storage tanks containing petroleum and crude oil by November 8, 1987. Financial responsibility requirements for leaking tanks must be finalized by November 8, 1988. Until EPA completes these rulemakings, the statute prohibits the installation of any underground storage tank for the purpose of storing regulated substances, unless the tank will prevent groundwater releases due to corrosion and the material the tank is constructed of is compatible with the substance to be stored. States that submit approvable plans may receive delegated authority from EPA to administer the leaking underground tank program.

Since leaking underground storage tanks are a major source of groundwater contamination, Subtitle I will be integral in preventing and detecting groundwater contamination. In addition, EPA has authority to issue compliance orders under the new program and may assess civil penalties of up to $25,000 a day for failure to comply with such an order. Civil penalties of up to $10,000 are also allowed if the owner or operator fails to comply with the Act's interim requirements or knowingly submits false information when it notifies states of existing underground storage tanks.

381

Groundwater Commission

Finally, the 1984 RCRA Amendments authorized a National Groundwater Commission to report to Congress concerning the nation's groundwaters. Such a commission was not funded by Congress, and therefore no Commission was appointed. This gap has been filled to varying degrees by the Committee on Groundwater established by the National Academy of Sciences (see Chapter 11) and several other ad hoc groups such as the National Groundwater Forum convened in 1985–1986 by the Conservation Foundation.

Safe Drinking Water Act of 1974, 42 U.S.C. §§300f–300j-11, Pub. L. No. 93–523, as amended by Act of June 19, 1986, Pub. L. No. 99–339

The Safe Drinking Water Act (SDWA) was enacted by Congress to provide for sanitary drinking water supplies, to protect sole source aquifers, and to establish a program to control underground injection in order to prevent endangerment of subsurface waters. 1986 Amendments to the Act require states to adopt programs to protect wellhead areas that supply public drinking water supplies.

Drinking Water Standards

The SDWA program provides EPA's authority to promulgate primary and secondary drinking water standards under Section 1412. These standards apply to every public water system in each state. The primary drinking water standards specify a maximum contaminant level (MCL) for each such contaminant and contain criteria and procedures to assure that supplies of drinking water are in compliance with the prescribed MCLs. The secondary drinking water standards specify MCLs for contaminants that may adversely affect the odor or appearance of drinking water in such a way as to cause a substantial number of persons to discontinue its use. The secondary standards also prescribe MCLs for any contaminant that may in any other way adversely affect public welfare. Both standards are based on technological and economic feasibility. The SDWA also gives EPA authority to promulgate maximum contaminant level goals (MCL goals) under the SDWA. In contrast to the MCLs, the MCL goals are based only on health criteria and are unenforceable regulatory goals.

382

Primary and secondary drinking water standards have only been set for a small fraction of the contaminants now found in surface waters and groundwater. The 1986 Amendments to the Act require EPA to establish MCLs for 83 contaminants by June 12, 1989. These standards and the MCL goals will become increasingly important as they are applied to groundwater protection and remediation efforts under other state and federal environmental programs, such as RCRA corrective actions and "Superfund" cleanups.

Underground Injection Wells

The underground injection control (UIC) program under the Safe Drinking Water Act was the first federal effort directly aimed at the prevention of groundwater contamination. Under the UIC program, EPA asserts direct authority over underground waste disposal wells. Although additional statutory authority was supplied to EPA by RCRA to regulate underground waste disposal wells, the UIC program continues to be managed under the Safe Drinking Water Act. RCRA amendments, however, did mandate a study of disposal practices involving UIC wells. More significantly, the 1984 Hazardous and Solid Waste Amendments banned the disposal of hazardous wastes by underground injection into or above formations which contain (within one quarter mile of the well used for such underground injection) an underground source of drinking water. Limited exceptions are provided for waters reinjected into formations as part of Superfund or RCRA-mandated remedial or corrective actions, if such wastes have been treated and the remedial response or corrective will protect human health and the environments.

Under the UIC program established under Section 1421 of the SDWA, underground injection is prohibited without an authorized state permit. In order to obtain a state permit, the applicant must satisfy the state that the underground injection will not endanger drinking water sources. The provision also imposes inspection, monitoring, record keeping, and reporting requirements on operators of underground injection wells. The 1986 Amendments to **383** SDWA gave EPA enhanced authority to enforce UIC regulations, including the power to issue administrative orders and impose administrative penalties for their violation.

Sole Source Aquifers

Another aspect of the SDWA program specifically aimed at groundwater protection is the "sole source aquifer" program under Section 1424, enabling EPA to take special measures to protect those areas which have only one aquifer as their principal source of drinking water. States must apply to EPA for designation of sole source aquifers. Once such an aquifer is designated, no new underground injection wells can be drilled in these areas without a permit. In addition, if EPA determines that a sole source aquifer, if contaminated, "would create a significant hazard to public health," no commitment of federal financial assistance through grants, contracts, or loan guarantees may be given to any project which EPA determines may contaminate such an aquifer so as to create a significant hazard to human health.

Wellhead Protection Areas

New provisions of the 1986 SDWA Amendments were designed to provide for state management of sources and activities of contaminants that might result in pollution of groundwater supplies. In 1986 legislation was offered in the House of Representatives (1986 HR 1650) that would have mandated more stringent state groundwater protection efforts tied to federal financial assistance for Superfund programs. The compromise legislation passed as Senate bill S 124 provides far less fiscal incentive for states to enact these programs. It is significant, though, in providing federal grantsharing assistance to states to develop groundwater management programs. Such programs identify and control sources of groundwater contamination within the surface and subsurface area affecting a water well or well field, called a "well head protection area" by the Amendments, which provides public drinking water supplies.

While the House legislation would have mandated broad state groundwater protection programs, Section 205 of the Amendments, codified in Section 1428, requires states to adopt and submit well head protection area plans by June 1989 to protect wellhead areas from contaminants that may have an adverse affect on public health in order to qualify for federal matching grants.

These programs must at a minimum:

1. Specify the duties of state and local agencies and public water supply systems with respect to the development and implementation of plans.

2. Determine the wellhead protection areas, based on all reasonably available hydrogeologic information on groundwater flow, recharge and discharge.

3. Identify within each wellhead protection area all potential man-made sources of contaminants that may have an adverse effect on public health.

4. Describe a program that contains, as appropriate, technical assistance, financial assistance, implementation of control measures, education, training, and demonstration projects to protect the water supply within the wellhead protection area from such contaminants.

5. Include contingency plans for the location and provision of alternative drinking water supplies for each public water system in the event that the wellhead is contaminated.

6. Include a requirement that consideration be given to all potential sources of such contaminants within the expected wellhead area of a new water well serving the public drinking water supply system.

State Primacy

When Congress enacted the SDWA, it envisioned that the states would have primary responsibility for assuring compliance with SDWA regulations. Under Section 1413, EPA is authorized to give the states primary enforcement responsibility for assuring compliance with the primary and secondary drinking water standards. In order to obtain such authorization, the state must adopt drinking water standards in conformance with the federal standards and adopt implementing procedures for the enforcement of the state standards. Similarly, the SDWA enables states to have primary enforcement responsibility for programs to control underground injection. Under Section 1422, EPA is required to list all states that should have such programs, and all states that are listed by EPA must adopt such a program which meets the **385** requirements specified by EPA. EPA is authorized to take over these programs if a state fails to ensure their enforcement.

Imminent Hazard Authority

SDWA also has an imminent hazard provision that is relevant to the implementation of groundwater protection measures. In order to protect drinking water supplies, Section 1431 of SDWA allows EPA to institute a civil action or issue administrative orders to prevent contamination of a drinking water source. Enforcement penalties of up to $25,000 a day are now authorized by the Act.

Clean Water Act of 1977, 33 U.S.C. §1251 *et seq.,* Pub. L. No. 95–217, as amended by Act of Jan. 30, 1987, Pub. L. No. 100–4

The Clean Water Act (CWA) is a comprehensive federal water pollution control program which has as its primary goal the reduction and control of the discharge of pollutants into the nation's navigable waters. While the CWA does not explicitly specify control of groundwater pollution as one of its objectives, it nevertheless offers limited statutory authority to implement groundwater protection measures. In some instances the reduction of pollution in surface waters that recharge an aquifer will protect the quality of groundwater. As part of the overall program to prevent and eliminate sources of water pollution, Section 104 of the CWA requires EPA to "establish, equip and maintain a water quality surveillance system for the purpose of monitoring the quality of the navigable waters and ground waters. . . ." This section has been implemented most completely with regard to surface waters, with only limited surveillance of groundwater.

Effluent Limitations and Water Quality Standards

The goals of CWA are accomplished by setting and enforcing effluent limitations and water quality standards that apply to all point sources (i.e., pipes and outfalls) that discharge pollutants into navigable waters. These features of the Clean Water Act apply exclusively to surface water and are relevant to groundwater protection to the extent that there are situations where surface water may be recharging groundwater aquifers. Section 304 of CWA directs the Administrator to provide guidelines for effluent limitations in the form of regulations that identify the degree of effluent reduction attainable through the application of varying levels of pollution control technology applicable to any point source. Section 304 also requires EPA to establish water quality standards consistent with the purposes of the Act.

Effluent limitations established under the CWA for categories of sources are enforced by EPA or the states through National Pollution Discharge Elimination System (NPDES) under Section 402. This section provides for permits for the discharge of any pollutant or combination of pollutants if certain specified conditions are met and requires compliance with specific standards in order to release certain types of industrial and municipal wastes into the nation's surface waters. If groundwater is leaching from a facility into a surface water through some conduit which is not permitted, it would arguably be a violation of Section 402.

Section 303 also gives EPA the authority to require states to promulgate groundwater quality standards where a clear "hydrologic nexus" has been established between ground and surface waters. EPA has interpreted this provision of the law very narrowly.

Non-Point Source Pollution

Section 208 of CWA provides for the development and implementation of area-wide plans for the management of non-point sources of pollution, such as "run-off" from fields and parking lots and stormwater discharges to both surface and ground waters. Until 1987, Section 208 plans were not an enforceable element of CWA. This statutory authority is potentially the most effective means of controlling groundwater pollution as part of the CWA program.

Under Section 208, in coordination with state authorities, EPA is required to publish guidelines for "the identification of those areas which, as a result of urban-industrial concentrations or other factors, have substantial water quality control problems." The state must develop plans that identify the type of treatment facility necessary to meet municipal and industrial waste-treatment needs. The plan should also identify agriculturally and silviculturally related non-point sources of pollution, including irrigation return flows and their cumulative effects, and set procedures to control these non-point sources of pollution. The plan is also required to identify methods of controlling salt-water intrusion into rivers and lakes as a result of the reduction of fresh-water flow due to the migration, obstruction, or extraction of groundwater. The Section 208 plan should also contain similar remedial measures to control pollution from mining and construction activities.

The programs implemented under Section 208, unfortunately, have received lower priority and achieved much more limited results than have other parts of the Clean Water Act; these programs nevertheless possess the potential for having a significant bearing on the protection of groundwater. In some areas, for instance, they have led to the reduction in use of septic tanks, which have been known to contribute to groundwater contamination in urban environments.

The 1987 CWA Amendments *require* governors to submit state non-point source management plans to EPA for approval. These plans must identify: a) best management practices and measures to reduce pollutant loadings from categories of nonpoint sources, taking into account their impact on groundwater quality; programs for enforcement, technical and financial assistance, and training necessary to implement best management practices and measures; schedules for utilizing these measures; adequate state authority for enforcing a management program; sources of other state and federal funding outside of CWA moneys; and, federal financial development and development projects for which the state will review consistency of the project with these non-point source management plans.

Information and Guidelines

The information and guideline requirements of Section 304(e) of CWA are also relevant to groundwater issues, since they provide for education. Section 304(f) requires EPA to publish information including guidelines for identifying and evaluating the nature and extent of non-point sources of pollutants, and processes, procedures, and methods to control pollution resulting from agriculture and silviculture, mining and construction, excavation, salt-water intrusion, and changes in surface or groundwater flow.

Imminent Hazard Authority

388 The CWA also includes an imminent hazard authority allowing EPA to bring suit or to take any other action that may be necessary to restrain persons or activities allegedly causing the discharge of pollutants. Section 504 states that where a pollutant source or combination of sources presents an imminent and substantial endangerment to human health or where the endangerment is to the livelihood of such persons, EPA may go to court to restrain any persons causing such pollution. Because this provision gives EPA broad authority to protect against environmental as well as economic injury, i.e., injury to livelihood or welfare, it could be effective in restraining activities threatening to contaminate groundwater supplies.

Studies and Grants

Section 316(i) of the 1987 Amendments adds new §319(i), which provides for modest 50% matching grants to states for groundwater protection research and other activities. These will not exceed $150,000 in any year. The section also authorized EPA technical assistance to states who request it with respect to their groundwater protection programs. The new law also provides for studies of groundwater contamination in seven of the nation's aquifers, with report is to be submitted to Congress by January 1989.

Comprehensive Environmental Response, Compensation and Liability Act of 1980, 42 U.S.C. §4611 *et seq.,* Pub. L. No. 96–510, as amended by Act of Oct. 17, 1986, Pub. L. No. 99–499

The Comprehensive Environmental Response, Compensation and Liability Act (CERCLA or the "Superfund Law"), was enacted in

1980 to allow the federal government to respond immediately to the release or threatened release of hazardous substances into the environment which pose a threat of imminent and substantial danger to public health. Since many such incidents represent threats to groundwater quality, CERCLA is an extremely important statute for groundwater protection and remediation.

Groundwater Response Actions

Section 104 of CERCLA provides the federal government with authority to take direct action to remove and provide remedial measures whenever hazardous substances are released into the environment or there is the threat of such a release which poses a serious and imminent risk to public health or the environment. The authority to respond to releases or threatened releases includes conducting investigations, testing, and monitoring. It also includes the ability of the government to implement remedial measures to remove contaminants from groundwater, clean up a site, or institute mitigation measures. Under Section 106 of the Superfund law the government can also order private parties to implement such response actions themselves. These response activities are to be carried out under EPA regulations set forth in the National Contingency Plan (NCP). The Superfund is a powerful mechanism for protecting and cleaning up groundwater.

The remedial measures carried out by the federal government under CERCLA are financed by an $8.5 billion Hazardous Substance Response Trust Fund, a five year financing mechanism in the Act commonly referred to as the "Superfund." Seven-eighths of the Trust Fund money is provided by industry through feedstock and corporate environmental taxes, and one-eighth through appropriations from general revenues.

Liability

The government may seek reimbursement of the Superfund for response costs it incurs under Section 107 of the Act. In cases where the responsibility for the wastes causing contamination can be traced to companies with financial resources, CERCLA requires that the financial responsibility for the cleanup be placed on those companies. Potentially responsible parties under CERCLA are persons who owned or operated a site where hazardous wastes were disposed, who generated wastes found at the disposal site, or who transported wastes to the site. Liability can be imposed on such individuals or companies even though they were never directly involved in the ownership or operation of the facilities where the wastes were disposed or were non-negligent in handling the wastes.

390   Liability under CERCLA attaches without regard to fault or negligence and is subject only to a few narrowly drawn defenses. Liability can be avoided by a showing that the release of hazardous substances resulted solely from an act of God; an act of war; or an act or omission of a third party, other than an employee or a person with a direct or indirect contractual relationship with the defendant, provided that the defendant exercised due care and took precautions against foreseeable acts and omissions of the third party. The latter defense, for instance, has been held by Courts generally not to apply where generators contracted with owners and operators of disposal facilities for disposition of wastes, despite the companies' ability to demonstrate that they exercised due care when the wastes were within their own control.

National Contingency Plan

EPA published the National Contingency Plan (NCP) to carry out response actions for releases of Hazardous Substances and Oil Spills under both CERCLA and the Clean Water Act. The NCP is of particular importance to groundwater protection because it outlines how the government will investigate releases and threatened releases to see if removal or other remedial actions to cleanup the release are necessary. This assessment procedure is outlined in seven phases:

Phase I—Discovery and Notification
Phase II—Preliminary Assessment
Phase III—Immediate Removal
Phase IV—Evaluation and Determination of Appropriate
          Response
Phase V—Planned Removal
Phase VI—Remedial Action
Phase VII—Documentation and Cost Recovery

The NCP also includes criteria to be applied by the government for determining priorities for remedial actions among releases.

Based upon these criteria EPA has developed a Hazard Ranking System (HRS). The HRS "score" establishes the foundation for determining whether further response actions are warranted at a site, and it has been used by EPA to list sites which should be priorities for further investigation. This list, called the National Priority List (NPL), contains 703 sites, and EPA had proposed to add another 248 sites to the list as of January 1987.

The HRS takes into account the population at risk, the hazard potential of hazardous substances at such facilities, the potential **391** for contamination of drinking water supplies, the potential for direct human contact, and the potential for destruction of sensitive ecosystems. With these considerations in mind, EPA looked at five potential "pathways" of exposure to the human population or a sensitive environment and other factors to arrive at an HRS mathematical score for relative ranking of the priority list. The mathematical computations are subject to some uncertainty, and the actual selection of sites for remedial action depends not only on the relative-risk rankings but also on the availability of state cost sharing and other financial assurances.

Groundwater Cleanup

One critical question in deciding on the appropriate remedial response when there has been a release or threatened release of hazardous substances into groundwater is the standard that cleanup should attain. EPA attempted to answer this question when it revised the NCP in 1985 by suggesting that other relevant and appropriate state laws should be considered in determining the appropriate extent of cleanup. The 1986 Amendments to Superfund mandate such an approach at Section 121, requiring that remedial actions attain "applicable or relevant and appropriate requirements" (ARARs) of federal and/or state environmental laws. For groundwater remedial actions, the new law and its legislative history concludes that this will often require the application of the Maximum Contaminant Level Goals published by EPA under the Safe Drinking Water Act, if the groundwater is or will be used in the future for drinking water. Since the MCLGs for many contaminants is zero or are those for which EPA cannot determine a threshhold health effect, and such a standard may not be feasible to attain (or measure) because of technological and/or economic reasons, it will be important to see how ARARs are applied to groundwater remedial actions.

Toxic Substances Control Act, 15 U.S.C. 2601 *et seq.*, Pub. L. No. 94–469

The Toxic Substances Control Act (TSCA) was enacted in 1976 to regulate the manufacture, use, and disposal of chemical substances and mixtures that pose a significant risk of injury to health and the environment. Although groundwater protection is not a specific statutory objective of TSCA, to the extent that TSCA can be relied on by the government to regulate the disposal of toxic chemicals, it is relevant to a discussion of groundwater protection.

TSCA provides a comprehensive framework for the regulation of the manufacture and use of chemical substances and mixtures. It requires extensive testing, notification, labeling, and record keeping. EPA has the authority under TSCA to prohibit or limit the quantity of the manufacture, processing, distribution, and use of chemicals. But most importantly for groundwater concerns, TSCA gives EPA the authority to prohibit or regulate the manner or method of disposal of chemicals or of any article containing a chemical. This authority extends to the manufacturer and processor as well as to any person who uses or disposes of a chemical substance for commercial purposes. EPA has used this authority under Section 6 to regulate closely the disposal of polychlorinated biphenyls (PCBs), an environmentally persistent synthetic compound used as insulant in electronic equipment, and PCB-containing equipment. EPA also used this authority prior to RCRA's passage, to regulate the disposal of dioxins in Missouri.

TSCA also has an imminent-hazard provision. Pursuant to Section 7, EPA has emergency powers to commence civil action to seize or demand such relief as is necessary to protect the public or the environment from imminently hazardous chemical substances and mixtures. This authority has never been relied on, however, for any threat to groundwater posed by a toxic substance.

Surface Mining Control and Reclamation Act of 1977, 30 U.S.C. 1201 *et seq.*, Pub. L. No. 95–87

In 1977 Congress enacted the Surface Mining Control and Reclamation Act (SMCRA) to protect the environment from the adverse effects of surface mining. Generally SMCRA enables the Secretary of the Interior or authorized states to take measures to ensure that surface mining is conducted in an environmentally sound manner, to ensure that steps are promptly taken to reclaim areas where surface mining has taken place, and to prevent surface mining where reclamation is not feasible.

Under Section 1265 of the Act, all permitted surface coal mining and reclamation operations must be in compliance with environmental protection performance standards, which specify the restoration and reclamation activities that must be conducted contemporaneously with production operations. Included among these measures are the requirements under Section 1265(10) to take special measures to minimize disturbances to the prevailing hydrologic balance of the mine site and surrounding areas and to minimize disturbances to the quality and quantity of surface and groundwater systems. Section 1265(b)(10)(A) lists special design and operating requirements that must be observed by all permitted surface mining operations in order to prevent toxic drainage liquids from entering groundwater.

Similarly, under Section 1258, in order to obtain a permit to conduct a surface mining operation, the permit applicant must include, among other things, a detailed description of measures that will be taken during the mining and reclamation process to protect groundwater.

Section 1237 gives the Secretary of the Interior the authority to enter upon land adversely affected by past coal mining practices and do whatever is necessary to control or abate any adverse effects upon land or water resources.

393

Federal Insecticide, Fungicide and Rodenticide Act, 7 U.S.C. §136 *et seq.,* Pub. L. No. 92–516

The Federal Insecticide, Fungicide and Rodenticide Act (FIFRA) was enacted to provide for the establishment of procedures for the registration, classification, sale, use, research, monitoring, and disposal of pesticides. Because pesticides have been known to be the source of groundwater contamination, it is appropriate to consider the effect FIFRA's provisions have on the protection of groundwater.

FIFRA gives EPA broad powers to regulate all pesticides. Under Section 3(a), no pesticide may be bought, sold, distributed, or otherwise handled if it is not registered. Pursuant to Section 3(c)(2)(A), EPA is required to make the stringency of standards which must be met in order to register or reregister a pesticide commensurate with the anticipated pattern of use and the degree of potential exposure of man and the environment to the pesticide. Pursuant to this registration authority EPA can and does require batteries of studies of pesticide effects on humans, animals and the environment to determine whether the pesticide causes unreasonably adverse effects on the environment. ("Environment" is defined by the statute to include man.)

One of the most significant features of FIFRA for groundwater protection is Section 3(d), which enables EPA, as part of the registration process, to classify pesticides for either general or restricted use. A pesticide classified for general use has been determined by EPA generally not to cause unreasonable adverse effects on the environment. For pesticides classified for restricted use, it has been determined that their use may cause unreasonable adverse effects on the environment without additional regulatory restrictions. Under Section 3(d) restricted-use pesticides may only be used or applied by a certified applicator, a person authorized by EPA to use or supervise the use of any pesticide.

EPA has broad power under Section 6(b) to change the classification or to cancel the registration of a pesticide if it appears generally to cause unreasonable adverse effects on the environment. A hearing must be held before such action can be taken. Thus, in order to take immediate steps in the case of an emergency, Section 6(c) authorizes EPA to suspend the registration of a pesticide if necessary to prevent an imminent hazard during the time required for cancellation or change in classification.

National Environmental Policy Act of 1969, 42 USC §4321 *et seq.*, Pub. L. No. 91–190

Finally, in reviewing federal statutes affecting groundwater, it is appropriate to consider the impact of the National Environmental Policy Act of 1969 (NEPA). NEPA's goal is to require that federal agencies consider the effects on the environment of any major federal action, including legislation, regulation, or project. Section 102 of NEPA requires that as part of any recommendation or report on a proposal for legislation or any other major federal action significantly affecting the quality of the environment, the federal agency must prepare an environmental impact statement (EIS) assessing the environmental effect of the proposed action, the unavoidable environmental effects if the proposal were implemented, alternatives to the proposed action, the short-term uses of man's environment in comparison with the maintenance and enhancement of long-term productivity, and any irreversible and irretrievable commitments of resources which would be involved if the proposal were implemented. Thus, NEPA causes major projects sponsored by or permitted by the various federal agencies to be evaluated and studied for their potential adverse effects on the environment, including groundwater. If a major federal action is determined to have an adverse environmental effect, mitigation measures must be identified and described in the EIS. Many of these measures become required components of any federal program.

395

Conclusion

It is clear from this brief summary of the federal environmental statutes that Congress has responded to various environmental concerns by enacting legislation to limit specific practices which threaten health and the environment. The desire to prevent contamination of groundwater is only one of many enunciated environmental objectives in these federal laws. Although there is no one single statute specifically aimed at protecting groundwater, there are more than a half dozen federal statutes which, in combination with each other, are directed at accomplishing that objective. Most significant among these are RCRA, SDWA, CWA, and CERCLA. A review of these and other environmental statutes discussed in this chapter indicates that there is a legal framework which can significantly aid in the implementation of a national groundwater protection strategy.

With increasing reports of incidents of groundwater contamination across the United States, Congress has been actively debating whether additional federal groundwater legislation is necessary to protect this valuable resource. One major issue in this debate is whether states, and not the federal government, should have primary control over groundwater protection and how to insure that states will institute programs to address their own unique groundwater problems. Another major issue is whether the nation should adopt a nondegradation goal for a groundwater protection program or whether differential protection of the resource should be allowed. Comprehensive groundwater legislation was offered in Congress in both the Senate and House in 1986 and 1987. Passage of a federal groundwater law could therefore occur in the 100th Congress, and many think that passage of such a law is inevitable by 1990.

Thirteen

State and Local Groundwater Quality Protection Programs

The states have given a great deal of attention to groundwater contamination since 1980.

- Most states, in cooperation with USGS, have mapped their aquifers and determined various characteristics of the aquifers.
- Several states have developed classification systems of aquifers based on their natural characteristics and use.
- Many states have water quality standards for groundwater.
- Most states have regulations regarding construction and use of wells.
- Increasing attention is given to underground storage tanks and non-point sources.

Policies and strategies for management of groundwater contamination vary greatly from state to state, with the result that state laws and regulations differ. This variation is in part because of variability in the quantity and quality of local groundwater and the ease with which it is contaminated (unconfined versus confined aquifers), and also because of varying land use and population density and distribution. This review will not attempt to describe state and local laws, regulations, and institutional arrangements in any detail, but will summarize them according to what is operative in the various states.

Bases for State Groundwater Quality Protection Programs

In order to have an effective overall groundwater protection program, one should have a step by step process for adopting: (1) an overall groundwater policy; (2) groundwater management strategies; and (3) specific techniques for obtaining groundwater protection. The bases for such a well-founded program have been summarized for the Environmental Law Institute by Henderson et al. (1984), as described below.

397

The first step in developing an overall groundwater protection policy involves the reconciling of other competing state interests and competing goals such as economic development. For this reason groundwater protection policies will vary from state to state.

There are typically three major categories of state groundwater policies: (1) non-degradation, which calls for maintaining existing groundwater at its highest and best use for human consumption; (2) limited degradation, which acknowledges that some groundwater degradation will or has already occurred, but tries to protect groundwater quality as much as possible; and (3) differential protection, which is aimed at satisfying present and anticipated future uses by singling out critical and/or high quality aquifers. The groundwater policies focus mainly on use as opposed to ambient quality.

The second step in developing an overall program is the establishment of management strategies. Typically, a state management strategy includes one or more of the following types of management: (1) Aquifer classification—wherein aquifers or portions of aquifers are placed in different categories, generally according to present and future uses or to the natural conditions of the aquifer. Aquifers in each category are then protected to different degrees so that each aquifer or portion of an aquifer can be used for a designated purpose. (2) Contamination and source classification—in which the causes of serious groundwater pollution problems are singled out for priority attention and control. (3) Uniform management—where groundwater management controls are applied equally across the state. (4) Recharge zone protection—special protection by the states for areas important for groundwater replenishment or recharge, e.g., the Pine Barrens of New Jersey. This is a subset of aquifer classifications and is typically implemented by local agencies who design the recharge protection programs.

Specific techniques for correcting or controlling groundwater contamination are the primary basis of a groundwater program. These techniques usually include one or a combination of the following. (1) Groundwater quality standards—these standards are usually set to define the maximum concentrations of contaminants allowed in order to maintain a described, desired level of groundwater quality. Applied alone these standards are reactive and serve no preventative function, therefore, they are often linked to other protection techniques. (2) Source oriented controls—these apply directly to the sources of groundwater contamination, and are aimed at preventing contamination. They may include permits, effluent limitations, discharge zones, facility design standards, and prescribed management practices for non-point sources. (3) Groundwater use regulations—these regulations are applied to those who use the groundwater, they include well drilling and siting permits, groundwater withdrawal permits, and groundwater recycling rules. (4) Land use controls—these are controls used to govern the land overlying or affecting groundwater, particularly the recharge zones. Examples of land use control restrictions are zoning, siting requirements, and public acquisition of lands. (5) Contamination and cleanup requirements—these require the cleanup of contaminants (such as hazardous and non-hazardous waste sites) that are a threat to groundwater. They supplement the federal Superfund hazardous waste cleanup program, and are either publicly or privately funded.

There are several other factors necessary for an effective, comprehensive groundwater program. Various states have considered as supplemental measures: (1) consolidation of available information and creating a system for collecting, and storing new data on state groundwater resources; (2) coordination of various state agencies with groundwater responsibilities; (3) development of mechanisms for funding groundwater protection programs; (4) adoption of adequate groundwater monitoring and enforcement procedures.

It should also be pointed out that federal and local participation and assistance are central ingredients for a successful state groundwater program.

Federal regulatory programs often dictate or provide guidance to states regarding how certain aspects of groundwater protection efforts should be structured, but the federal government should also provide the states with specially targeted information and technical assistance on groundwater protection, and should aid in the financing of the states' groundwater programs. Local governments should encourage and actively promote alternative community practices that mitigate groundwater pollution threats, e.g., new land use ordinances, improved pesticide application methods, safer procedures for cleaning septic tanks, and more effective uses of fertilizers (Henderson et al., 1984).

Specific Sources of Groundwater Contamination

Because of the difficulty of cleaning up groundwater contamination **399** once it has occurred, many states focus on prevention of groundwater contamination through the restriction of activities that generate it. Regulations applicable to specific sources of contamination typically require a discharger to obtain a permit from the state's Environmental Protection Agency. Before receiving a permit, the applicant may be required to meet various design, construction, monitoring, and reporting specifications.

Industrial Liquid Waste Impoundments

These impoundments include ponds for the treatment of wastes, the evaporation of wastes, or, in some cases, the place of final waste disposal. It has been estimated that there are 50,000 such ponds in the United States leaking over 100 billion gallons of wastewater per year into the ground (Henderson et al., 1984). EPA (1983) has inventoried 27,912 sites that can threaten ground water quality.

Many of these structures are not lined and are not monitored. The largest volumes of impounded liquid wastes are generated by the following industries: paper, petroleum and coal products, primary metals, and chemicals and allied products (Henderson et al., 1984). Such wastes include materials such as solvents, heavy metals, organic materials, acids, and cyanide.

Landfills

Landfills may be permitted to receive non-toxic wastes only (i.e., no priority pollutants), or they may be permitted to include both industrial and municipal wastes. Typically, they contain residual and commercial garbage as well as liquid and solid industrial waste such as sludges, scraps, solvents, and filter cakes. It has been estimated that there are over 18,500 municipal facilities in the

United States. Most of these are not lined or monitored; many of these are very close to water supply wells and thus bring about contamination of groundwater. It has been estimated that landfills leach over 90 billion gallons per year into groundwater (Henderson et al., 1984). Industrial non-hazardous waste landfills are estimated to be 75,700 (OTA, 1984).

400   Septic Tanks

Septic tanks are widely distributed over the United States, not only in rural areas but also in many suburban and small town areas. Approximately one-third of the population relies on these systems for sewage disposal; there are approximately 22 million septic tanks or cesspools in the United States, most of which are connected to individual homes (EPA, 1986b).

Septic systems, if they are properly constructed and managed, can be an effective way of disposing of domestic sewage in low population density areas. Poorly designed septic systems that are sited too close together can contaminate groundwater with many different types of chemicals and organisms. The most common chemical contaminants are nitrates and phosphates, but complex organic molecules are also often found. The most common organisms are bacteria, fungi, and viruses (Henderson et al., 1984). These septic tanks and cesspools discharge approximately one trillion gallons of sewage per year. They are very important sources of groundwater contamination (EPA, 1986a).

In most states landowners are required to obtain a permit before installing a septic system, and as a precondition are required to furnish soil reports, detailed plans of the system design, and equipment specifications. In addition, some states restrict the density of septic systems in a given area so that attenuation and dilution processes can operate effectively. Usually the administration of septic tank regulations is handled primarily through local health departments. To date, very little effort has been made by the states to monitor the management of these septic tanks after they are installed.

Municipal Wastewater Systems

There are many thousands of miles of leaking sanitary and stormwater sewers in the United States. In fact, many of them are designed or expected to leak. Other routes of leakage come from municipal wastewater holding ponds for the treatment of sewage, and for the prevention of excessive rates of land spreading of wastewater.

In 1977 EPA estimated that 15 billion gallons per day of these wastes across the United States threatened to contaminate groundwater.

Land Spreading of Pollutants

Wastes from agricultural activities and from industry may be sprayed as liquids or spread as sludges on the landscape. This is an effective way of getting rid of organic wastes if it is carefully managed; however, often it is not and thus the practice poses a risk of contaminating groundwater. This sludge, if spread at excessive rates, may cause nitrate contamination of groundwater.

Municipal sludges may contain not only nitrogen and phosphorus, but also heavy metals and salts and many complex organic molecules which are used as pesticides and for cleaning in homes.

Industrial sludges are usually produced from coal-fired utilities, and the textile, canning, petroleum refining, and paper industries. These wastes contain a great many types of chemicals, particularly many of the complex organics. The spreading of these types of wastes is projected to increase in the future (U.S. EPA, 1977; Henderson et al., 1984).

Only a few states specifically regulate land spreading of such pollutants. For example, the Pennsylvania Spray Irrigation Manual contains comprehensive regulations; fixes guidelines for locating and evaluating sites; and sets standards for treatment, storage, screening controls, piping, sprinklers, spacing, and application rates (Pennsylvania Department of Environmental Resources, 1972). Maryland issues permits for spray irrigation of pollutants as part of its program to enforce water quality standards (Lehr et al., 1976), and New York requires that the design specifications for spraying systems be submitted for approval (Lehr et al., 1976).

The land spreading of municipal and industrial sludge is regulated in approximately twenty states. Some states, such as Minnesota and Oregon, have separate regulations for municipal sludge, while other states deal with it as part of a general regulation of municipal waste treatment plants. Colorado's criteria for the review of wastewater facilities, for example, require that submitted plans include a description of the process for stabilizing sludge (Colorado Department of Health, 1973). Illinois waste treatment regulations require that a permit be obtained from the Division of Land Pollution Control to dispose of non-liquid sludges, or from the Division of Water Pollution Control to dispose of liquid sludges (Illinois Environmental Protection Agency, 1971).

402
To regulate industrial sludges most states rely on the general provisions of statutes governing water pollution. Under such statutes effective enforcement is hampered by the fact that the state has the burden of showing that sludge disposal has actually resulted in pollution. A few states, including New York, have avoided the burden of proof problem by invoking laws governing solid wastes or hazardous waste disposal as a basis for regulation of land spreading of pollution (U.S. EPA, 1977).

Mining Wastes

Active and abandoned mines have caused groundwater contamination through seepage of tailing ponds, runoff of waste piles, and discharge of mine drainage into soils and injection wells. The most common contaminants from these sources are acids, dissolved solids, metals, and radioactive materials. Surface and underground mining may contaminate groundwater, and particularly significant is the excess acidity of groundwater caused by mining (Henderson et al., 1984).

Mining takes place in every state in the United States, and therefore, is a nationwide source of groundwater contamination. The potential of large scale coal mining operations in midwestern states such as North Dakota, Wyoming, and Montana, will cause these states to be particularly vulnerable to groundwater contamination from this source (Henderson et al., 1984). Several states regulate these activities; for instance, Illinois and Pennsylvania require permits for all mining operations (U.S. EPA, 1977). Applications must describe the proposed method of mining and the disposal of refuse, and must meet the requirements for plugging holes, monitoring, and reporting.

Leaks and Spills from Pipelines, Storage Tanks, and Vehicles

Pipelines. There are many thousands of miles of pipelines in the United States carrying coal slurries and various petroleum products as well as gas and other products. Leaks from these pipes may or may not be quickly detected. In most cases they are fairly well monitored, but serious leaks that have resulted in groundwater contamination have occurred in various parts of the country.

Storage tanks. Storage tanks are scattered throughout the country; they contain many different kinds of products from industry and from agriculture. Attention has recently been given to leaking underground storage tanks, one of the major unregulated sources of groundwater contamination in the United States. These tanks are ubiquitous. A recent Congressional report estimated that 2.5 million underground storage tanks store non-hazardous liquids. Approximately 81% of these tanks are made of steel with no corrosion protection and were buried over twenty year ago (OTA, 1984).

403

Petroleum experts have estimated that 75,000 to 100,000 of these steel tanks leak, and that the problem will increase two- to threefold during the next five years (Henderson et al., 1984). No real estimate has been made on the percentage of leaking pipelines containing gasoline, diesel fuel, various chemicals, etc.

Accidental spills have become more common in recent years. They often contain liquid wastes, toxic fuels, and gasoline. All of these may travel through the soil and eventually reach aquifers. Many state programs designed to handle spills do not take into account the impact of spills on groundwater. Shallow, unconfined aquifers are particularly vulnerable to the chemicals in such spills. Some states have adopted spill regulations that attempt, among other things, to protect groundwater quality. In Pennsylvania, for instance, anyone responsible for a spill must immediately notify the Department of Environmental Resources (U.S. EPA, 1977). A geologist from the Groundwater Section will survey the scene of the spill and suggest procedures for its removal that will be least detrimental to groundwater quality.

Deicing Salts

Highway and airstrip deicing salts have been found to be contaminating groundwater in many areas of the United States, particularly in the snow belt. The main cause of groundwater contamination is runoff from salt storage areas, but dissolved de-icing salts trickling into the ground along roadsides also contribute to the elevated levels of chlorides in groundwater. More recently, chemicals other than chlorides have been used in deicing, and these are sometimes more toxic. Some states are aware of this danger and have begun to reduce the quantities of salts spread or to enclose salt storage piles to avoid dispersal through precipitation.

Saline Intrusion into Groundwater

404

Improper installation, poor use, and abandonment of wells used for the extraction of water, oil, and gas can cause groundwater contamination. However, one of the most common causes of saline intrusion into freshwater aquifers is excessive pumping. Large scale pollution has occurred in the coastal northeastern states, California, and Florida. Several northeastern states have reduced coastal saltwater intrusion by imposing strict controls on pumpage, so as to prevent excessive depletion. California has been successful in reversing the intrusion of saltwater by placing "barrier wells" parallel to the shoreline and injecting fresh water into depleted aquifers.

Saline intrusion may also occur where wells, such as oil or gas wells, are poorly cased or improperly cared for when they are abandoned. These wells may act as conduits for saline water to travel into freshwater aquifers. Reinjection of highly mineralized water used for extraction during oil and gas drilling can also cause excessive groundwater salinity or hardness.

Improperly cased or abandoned wells of all kinds may be channels for contaminants such as oil and gas, and may cause them to leak into groundwater from the surface or from shallow water contained aquifers into deeper clean aquifers (Wilson, 1982).

Construction and Operation of Wells

Wells are generally of two types: (1) for obtaining domestic and agricultural water and sometimes municipal waters, (2) for waste disposal. Water wells, if poorly designed or constructed, can be wasteful of groundwater and may be a conduit for polluting groundwater. Many states have undertaken to regulate the installation and abandonment of such wells. These regulations vary in detail from state to state and contain requirements for location, design, construction, drilling equipment, pumps, water testing, and record keeping. It is estimated that 500,000 or more new wells are drilled each year and, therefore, the enforcement of construction standards is not an easy task, but the licensing of drilling contractors appears to have been somewhat effective in improving well construction practices (Pye, Patrick, and Quarles, 1983).

Waste disposal wells are very common throughout the United States. There are tens of thousands of these wells nationwide used for the disposal of sewage, industrial waste, stormwater, oil field brines, and irrigation return flows. A few of these wells inject wastes into saline aquifers at depths greater than 1,000 ft, but most wells allow waste to seep into shallow fresh water aquifers. Contaminants from these sources are many, ranging from agriculture to domestic and industrial wastes (Henderson et al., 1984; EPA, 1977).

405

Agricultural Activities

Agricultural activities such as irrigation, the spreading of chemical fertilizers and pesticides, the disposal of animal feedlot wastes, and the problems associated with dry land farming may cause groundwater contamination.

Typically, excessive irrigation processes most often lead to the contamination of groundwater from dissolved salts and other agricultural chemicals. These percolate through the soil; for example, in California's Central Valley the leaching of pesticides into groundwater has caused the closing of drinking water wells across the valley (Henderson, et al., 1984).

Nitrates from nitrogen fertilizers and pesticides are common agricultural chemicals that have produced contamination of groundwater, particularly in the agricultural areas of the United States. No-till agriculture has increased these problems.

Another important agricultural source of groundwater contamination, particularly in the midwest, is runoff from concentrated livestock feeding operations which overload the natural capacity of the soil to assimilate organic materials. As a result, organisms, viruses, and various organic chemicals may enter the groundwater. Since animal feedlots are considered point sources under Section 502 (14) of the Federal Water Pollution Control Act Amendments, they are subject to the requirements of the NPDES permit program. States have adopted regulations for feedlots in which the ratio of animals to land area is deemed high enough to threaten groundwater quality. Usually regulations demand that water be diverted above the feedlot and that a settling pond or lagoon be provided below it. Some states require a permit and operators may be required to submit information concerning building and lot areas, lagoons, and direction of surface drainage. Some of the more detailed rules contain specifications for the operation of facilities and for the storage and disposal of wastes.

Dry land farming, which is the raising of crops in arid regions without irrigation, may cause groundwater contamination. When this cropland is not planted, precipitation that normally would be absorbed and transpired in the atmosphere by the native vegetation percolates into the soil, dissolving salts and carrying them into the groundwater (Henderson et al., 1984).

406

State Groundwater Protection Programs, Policies, Strategies, Laws, Regulations, and Institutional Arrangements

This section will present summaries of groundwater protection programs now in effect in various states. Effort has been made to obtain program data on all of the states. Detailed descriptions of classification systems and standards are given in Chapter 11.

Alabama

The state of Alabama has implemented very little legislation for the protection of groundwater quality. The Alabama Department of Environmental Management (ADEM) regulates public water supplies and is mainly concerned with the potability of water and the adequacy of the water supply system to meet demands. It demands certificates from well-drillers and has developed standards, but generally does not participate in the selection of well sites or regulate the spacing of wells. Permits are required for any well within the Coastal Area Zone that produces 50 gallons per minute or more. The ADEM investigates reports of groundwater contamination and has the authority to close wells that produce water that is unfit for human consumption. Self-supplied industrial, commercial, irrigation, and other agricultural uses of groundwater are not regulated in Alabama. The Geological Survey of Alabama maintains a statewide water data network and conducts investigations of Alabama's water resources (USGS, 1985).

Arizona

The Arizona Groundwater Management Act enacted on June 12, 1980, is the first comprehensive legislative framework for managing the groundwater resources of Arizona. The 1980 act made the Department of Water Resources responsible for administrating the law's complex provisions. The Act establishes four active management areas (AMAs). These are areas in which intensive groundwater management is needed because of the large and continuous groundwater overdraft. Within the AMAs the law requires a 45-year water conservation and water management program. The management goal is safe yield by the year 2025. Safe yield is a concept whereby long-term groundwater discharge is equal to groundwater recharge (USGS, 1985).

407

Arkansas

Groundwater management in Arkansas is currently in a data collection and planning stage, although several state and local agencies have limited or implied jurisdiction over groundwater. The responsibility of the Arkansas Department of Health is the protection of municipal and rural drinking water and the regulation of construction and use of septic tanks. The Arkansas Soil and Water Conservation Commission is responsible for the Arkansas State Water Plan, which evaluates groundwater resources problems and management strategies. It is the leading proponent of the Arkansas Water Code Bill, which if enacted, will require registration of groundwater withdrawals. The Arkansas Geological Commission provides geologic and hydrologic data on the state's water resources. The Department of Pollution Control and Ecology is responsible for the control of groundwater quality and the execution of federally delegated programs such as the underground injection control program, the Resource Conservation and Recovery Act, the Clean Water Act, and construction-grant programs. The Arkansas Oil and Gas Commission shares responsibility with the Arkansas Department of Pollution Control and Ecology over the underground injection control program. Well driller licensing and well construction standards are regulated by the Water Well Committee. Reports are filed on the completion of well construction. The Arkansas Plant Board, the Forestry Commission, and the cooperative extension service also have responsibilities that indirectly affect groundwater (USGS, 1985).

Groundwater monitoring is done by several agencies; including the U.S. Geological Survey and the Arkansas Department of Health.

California

California does not have a statewide comprehensive groundwater management law. Groundwater management is practiced largely by local agencies. The California Department of Water Resources is the state's principal water protection agency. It provides advice and technical support to local agencies, collecting data and conducting investigations. Standards are established and enforced by the State Water Resources Control Board and nine regional boards. The Department of Health Services monitors the quality of drinking water supplies. The USGS has cooperative programs for data collection and hydrologic investigations with several state and local agencies (USGS, 1985).

Proposition 65 was put on the ballot and received support by the voters in 1986. This proposition requires that by March 1, 1987, California's governor must publish a list of chemicals known to the state to cause cancer or reproductive toxicity. This list will be updated yearly. In addition, a separate list must indicate chemicals which have not been adequately tested for carcinogenesis or reproductive toxicity. It prohibits a person in the course of doing business from knowingly discharging a detectable amount of a covered chemical where it may reach a potential source of drinking water (which may well be interpreted by the State Water Resources Control Board to mean all state waters). A person may not knowingly and intentionally in the course of doing business expose anyone to any detectable amount of a chemical on the governor's list without providing a clear and reasonable warning which has yet to be identified.

Once on the governor's list a chemical is "guilty until proven innocent". The shift of burden of truth goes from the enforcer to the business which knowingly releases listed chemicals in violation of the Discharge Prohibition or causes a known and intentional exposure without providing a warning. This proposition will probably help to control the contamination of groundwater.

Conflicts over water rights have arisen among users in many water districts in southern California. The Water Resources Control Board has authority to file actions to restrict pumping or any other practice that may degrade the quality of groundwater. Major groundwater issues include *groundwater overdraft, seawater intrusion, land subsidence, artificial recharge,* and *conjunctive* use of groundwater (Peters, 1982). The California Department of Water Resources has identified 42 groundwater basins in overdraft; of these, 11 are in a critical condition of overdraft. Between 1945 and 1965, seawater intrusion was most intensive in the coastal basins of Los Angeles, Orange, and Ventura County, the Pajaro and Salinas Valleys, and the Fremont area of Alameda County. It is now under control in most of these areas as a result of management programs that include injection-well barriers, control on withdrawals, artificial recharge, and imported water. Intensive pumping of aquifers in the San Joaquin and Santa Clara valleys has caused *land subsidence* over large areas; as much as 29 feet in the Los Banos—Kettleman City area. Little subsidence has occurred since imported water became available in the later 1960s, except for a slight resumption during the drought of 1977–78. Artificial recharge and conjunctive use of surface and groundwater are major elements of groundwater management in California.

Artificial recharge was first used in southern California in the 1920s. It is widely used there now and in the Central and Santa Clara Valleys as well. Imported water is available in all of these areas. An interesting variation on artificial recharge is "in-lieu replenishment," whereby imported water is delivered directly to users in return for reduction of groundwater withdrawal by an equivalent amount (USGS, 1985).

409

Colorado

In 1965 the Groundwater Management Act (CRS 37-90-101 to 104), commonly referred to as HB 1066, was enacted; in 1969 the Water Rights Determination and Administration Act (CRS 37-92-101 to 602) was enacted. The latter controlled well drilling more effectively, particularly the effect that pumping groundwater would be allowed to have on surface water hydrologically connected to an aquifer. Groundwater that is part of a stream—aquifer system is classified as tributary groundwater. The withdrawal of this class of groundwater is administered within the priority system by the state engineer to minimize the effect of withdrawal on surface-water supplies. The Colorado Ground Water Commission and local management districts control designated groundwater sources. Outside of these designated basins and in areas where groundwater is considered not tributary to surface water, the groundwater is classified as non-tributary groundwater and is administered by the state engineer. In these areas groundwater cannot be withdrawn at an annual rate of greater than 1% of the volume of water stored beneath the well owner's property boundaries. Much of the water in the Denver Basin aquifers is classified as non-tributary (USGS, 1985).

410

Connecticut

Connecticut has a comprehensive program for managing groundwater resources that originated after the passage of the Clean Water Act of 1967 (Connecticut Public Act 57). This act has been strengthened by subsequent passage of federal clean-water legislation. It is jointly managed by the Connecticut Departments of Environmental Protection (DEP) and Health Services (DOHS) and the Office of Policy and Management. Water quality standards for groundwater that were adopted in 1980 (Connecticut DEP, 1980) provide the framework for basinwide plans that specify actions to eliminate water quality problems and the framework for such regulatory programs as waste discharge permits and enforcement actions. Permits for drilling wells and submission of well records have been required by the state since 1955.

The Connecticut Water Policy Diversion Act (Connecticut General Statutes, Section 22a-365) gives the DEP authority to regulate groundwater withdrawals that exceed 50,000 gallons per day. The process for permitting withdrawals addresses issues of water quality and quantity. The DEP is also responsible for investigating pollution incidents, for providing technical assistance to municipalities, and for conducting inventories and investigations of the state's water resources. The DOHS under Section 19-13 of the Connecticut General Statutes has the major role in managing groundwater resources used for drinking water. Responsibilities include protection and location of private and public supply wells, well construction requirements, and development and enforcement of standards for the quality of drinking water. Public water supply utilities are required by Connecticut Public Act 84-502 to submit long range water supply plans to DOHS to aid in identifying aquifers to be protected for future public supply. Most of the programs that have been submitted include regulations that prohibit uses or activities that could adversely affect groundwater quality (USGS, 1985).

Connecticut has a system of classification for groundwater uses and permissible discharges. Class GAA is public and private drinking water supplies without treatment. Class GA is private drinking water supplies without treatment. Class GB may not be suitable for potable use unless treated because of existing or past land uses. Classes GA-GC and GB-GC may be suitable for certain waste disposal practices due to past land use or hydrological conditions that render these groundwaters more suitable for receiving permitted discharges than for development for public and private water supplies. The downgradient surface water quality classification must be Class B or Class SB.

411

Through various regulations, Connecticut has developed the following controls that affect groundwater basin plans (Section 303e, Clean Water Act) and water utility plans (Public Act 84-502). The latter plans are implemented by a permit system.

Groundwater monitoring (Public Act 84-502, Section 22a-431) may be required if groundwater pollution may potentially affect water supply wells or is suspected of affecting them, and in the vicinity of permitted waste discharges.

The regulations regarding permits are as follows: wastewater discharges, Section 22a-430, as amended by Public Act 84-219; sewage, Section 22a-430; solid waste landfills, Section 22a-207 to 251; hazardous materials, Section 22a-114 to 133, Section 22a-449(c); underground fuel storage tanks, Public Act 83-142; diversions of more than 50,000 gallons per day of water from any surface or source, requires a permit (Section 22a-365 through 378); agricultural operations, require a permit (Sections 22a-46, 66Z, 430).

Enforcement against the pollution of groundwater is managed under the following acts: irrigation, Public Act 83-237; pollution abatement, Sections 22a-429, 431, 432; a Pollutor may be compelled to provide potable water to persons with contaminated supplies, Section 22a-471.

Connecticut has designated a hierarchy of land uses, as follows. Category A: land uses providing maximum protection to regionally significant aquifers; Category B: land uses posing minimal risk to regionally significant aquifers; Category C: land uses posing slight to moderate risk to groundwater; Category D: land uses considered to pose a substantial risk to groundwater; Category E: land uses posing a major threat to groundwater.

Connecticut also has technical guidelines for planning and inspecting commercial and industrial operations (Harrison and Dickinson, 1984).

Delaware

The goal of the state of Delaware is to ensure sufficient groundwater quality for protection of public health and for beneficial uses such as may be desired, including the preservation of significant ecological systems now and in the future.

Groundwater use is regulated by the Department of Natural Resources and Environmental Control (DNREC) under the Delaware Environmental Act (7 Delaware Code Chapter 60). The Water Supply Section of DNREC licenses well drillers, issues permits for construction of wells, requires reports on completion of wells, and issues allocations for the use of ground and surface waters. The DNREC issues permits for on-site wastewater treatment installations and monitors National Pollution Discharge Elimination Systems wastewater return flow data. The Delaware Department of Health and Social Services Division of Public Health (DPH) regulates the quality and adequacy of public water-supply systems (16 Delaware Code 122). Through the law and under 16 Delaware Code 1244 it can regulate any activity within one mile of the source of a public water supply. Public water supplies are also regulated by the Public Service Commission (PSC). This group can function as an enforcement arm of the Department of Health or other state agencies. The Delaware River Basin Commission, by agreement with the various states in the Delaware River Basin, regulates the use of surface waters and groundwaters, and projects that have an impact on them, in that part of Delaware within the basin boundary. Nonregulatory agencies involved in Delaware groundwater issues include the Water Resources Agency for New Castle County and the Delaware Geological Survey (DGS). The DGS, in cooperation with the USGS, maintains a statewide water-data network and investigates the groundwater resources of the state (USGS, 1985).

Florida

The Florida Water Resources Act of 1972 established authority for management of the state's water resources through five water management districts under the Florida Department of Natural Resources. Since 1975 the water management districts have functioned within the Department of Environmental Regulation. This department is concerned primarily with quality-rated aspects of water management. Permitting regulations control the construction of wells two inches or more in diameter, and the withdrawal of groundwater for all uses except private domestic use and other minor uses, through consumptive-use permitting. Permitting regulations pertaining to waste disposal or other activities that affect groundwater quality are administered directly by the Department of Environmental Regulation. The Florida Water Quality Assurance Act of 1983 made the department responsible for establishing a statewide groundwater-quality monitoring network and a centralized data base for the acquired information. Each of the districts has a cooperative study with the U.S. Geological Survey. They provide hydrological data and interpretive information needed to manage the quality and quantity of Florida's groundwater (USGS, 1985).

Florida has a classification system not only for aquifers but also for injection wells. The Florida Department of Environmental Regulation controls the permitting, operating, and monitoring of these wells. It also controls the plugging and abandonment of such wells. Classification of injection wells is as follows:

Class 1: wells used by generators of hazardous wastes or owners or operators of hazardous waste management facilities to inject hazardous wastes beneath the lowermost formation containing within a quarter mile of the well boring an underground source of drinking water. Other industrial and municipal disposal wells that inject fluids beneath the lowermost formation containing within a quarter mile of the well bore an underground source of drinking water are also considered Class 1 injection wells.

Class 2: wells inject fluids that are brought to the surface in connection with conventional oil or natural gas production and that may be comingled with wastewaters from gas plants, which are an integral part of the production operation unless these waters are classified as hazardous wastes at the time of injection. Also included in Class 2 are injection wells for enhanced recovery of natural gas and for the storage of hydrocarbons that are liquid at standard temperatures and pressure.

Class 3: wells that inject for extraction of minerals, including: the mining of sulfur by the Frasche process and solution mining of minerals.

Class 4: wells used by generators of hazardous or radioactive wastes, by owners or operators of hazardous waste management facilities or by owners or operators of radioactive waste disposal sites to dispose of hazardous waste or radioactive waste into or above a formation that is within a quarter mile of wells containing either an underground source of drinking water or an exempted aquifer.

Class 5: Injection wells not included in classes 1, 2, 3, or 4. *Group 1A*—air conditioning return-flow wells used to return to any aquifer the water used for heating or cooling (an air conditioning supply well, heat pump, and return-flow well used to inject water containing no additives into the same permeable zone from which it was withdrawn constitutes a closed loop system; *B*—cooling water return flow wells used to inject water previously used as cooling water. *Group 2A*—recharge wells used to replenish, augment, or store water in an aquifer to prevent salt intrusion; *B*—recharge wells or saltwater intrusion barriers; *C*—subsidence control wells; *D*—connector wells used to connect two aquifers to allow interchange of water between those aquifers. *Group 3A*—wells that are part of a domestic waste treatment system; *B*—swimming pool drainage wells; *C*—devices receiving wastes that have an open bottom and sometimes have perforated sides (this rule does not apply to single family residential waste disposal systems); *D*—wells used to inject spent brine into the same formation from which it was withdrawn or extracted; *E*—injection wells used in experimental technologies. *Group 4A*—dry wells used for the injection of wastes into a subsurface formation; *B*—sand backfill wells used to inject a mixture of water and sand, tailings, or other solids into mined out portions of subsurface mines; *C*—wells other than Class 4 used to inject radioactive wastes; *D*—injection wells used for in situ recovery of phosphates, uraniferous sandstone, clay, sand, and other minerals extracted by the borehole slurry mining method. *Group 5*—drainage wells used to drain surface fluids, primarily storm runoff or lake level. *Group 6A*—injection wells associated with recovery of geothermal energy for heating, agriculture, and production of electric power; *B*—other wells.

414

Florida has regulations for well construction for these various types of wells, as well as operating, monitoring, and reporting requirements. Permits are used to control the construction, operation, and plugging and abandonment of wells.

Florida also has regulations concerning aboveground and underground storage tanks. These regulations govern the construction, operation, and repair standards for such tanks (Florida Department of Environmental Regulation, Fact Sheet, 1985a).

Georgia

Georgia has a comprehensive set of rules governing the quality and use of groundwater. The Ground-Water Use Act of 1972

provided for the permitting of withdrawals for industrial and
municipal use that exceed 100,000 gallons per day and authorized
the Georgia Environmental Protective Division to issue regulations
about reporting, timing of withdrawals, abatement of saltwater
encroachment, well depth and spacing, and pumping levels or
rates. Amendments to the act in 1982 required that irrigation
withdrawals in excess of 100,000 gallons per day be reported to
the state, although permits are not required. The Oil and Gas Deep
Drilling Act of 1975 authorized the Board of Natural Resources to **415**
regulate drilling and use of oil, gas, and other types of wells, for
the purpose of protecting fresh groundwater supplies. The Georgia
Safe Drinking Water Act of 1977 provides for regulation of water
quality in public water supplies. The Georgia Environmental
Protective Division (EPD) is responsible for enforcing all surface
water, groundwater, and water quality laws. In 1984 a groundwater
management plan for Georgia was implemented to identify key
activities performed by EPA management to control and regulate
potential pollution sources and to develop a monitoring program to
provide water quality and water quantity data on the state's
principal aquifers. The Water Resources Management Branch
issues permits for groundwater withdrawals that exceed 100,000
gallons per day by industrial and municipal users and oversees the
reporting of groundwater use for irrigation in excess of 100,000
gallons per day. The Ground-Water Program of the Water Protection
Branch provides permits to public water suppliers that use
groundwater and monitors water quality for compliance with
drinking water standards.

The Industrial and Hazardous Waste Management Program of the
Land Protection Branch monitors groundwater at hazardous waste
sites. The Geological Survey Branch provides technical support for
other branches and cooperates with the USGS to provide most of
the basic data and interpretive information needed to manage
quality and quantity of groundwater in the state (USGS, 1985).

Georgia's groundwater management plan provides for (1)
protection of groundwater, (2) management of groundwater
quantity, and (3) monitoring of groundwater quality and quantity.

Protection There are three levels of groundwater quality
protection in Georgia: (1) non-degradation—groundwater resources
are maintained at their present levels of quality or are improved; (2)
limited degradation—aquifers are allowed to degrade to a specific
level designed to protect health or both health and welfare; and (3)
use orientation protection—groundwater resources are prevented
from deteriorating significantly, preserving them for present and
future uses. Selecting this last goal means that aquifers are

protected to a greater or lesser degree according to their existing quality, current use, and potential for future use.

Management of Groundwater Quantity EPD has a permitting program that includes proper siting, construction, and operation of various types of wells. For management of groundwater quantity, permits are required for new withdrawals, before a permit is issued it must be demonstrated that the groundwater resource is available to sustain the requested withdrawal on a permanent basis. EPD does not issue any industrial withdrawal permits for those aquifers where total withdrawal may exceed recharge (e.g., the Clayton Aquifer) or where the aquifer is threatened by salt water. In other areas where there are large cones of depression, withdrawal for industrial purposes may be limited or even prohibited.

416

Monitoring EPD performs two types of monitoring: aquifer monitoring and facility surveillence. Aquifer monitoring is designed to assess the state's groundwater resources, including their susceptibility to pollution. Aquifer monitoring includes: (1) the Georgia groundwater monitoring network, which is a network that involves measurements of both water quality and quantity; (2) aquifer (water level) measurements, which are performed cooperatively with the U.S. Geological Survey and the Georgia Geological Surveys.

Facilities surveillance monitoring includes: (1) water quality monitoring of both active and abandoned hazardous waste facilities on at least a yearly basis; (2) groundwater quality monitoring in the vicinity of many sanitary solid waste landfills on at least a yearly basis; (3) water quality monitoring of public water systems at least once every three years; (4) water quality monitoring at municipal and industrial wastewater land disposal operations at least once a year; (5) water quality monitoring of selected wells and/or systems in sinkhole areas on a yearly basis.

The Facilities Surveillence Program is carried out at selected locations as part of several ongoing regulatory programs such as the Safe Drinking Water Program, the Industrial Wastewater Program, the Municipal Compliance and Technical Support Programs, the Groundwater Programs (wells in sinkhole areas), the Municipal Solid Waste Control Program, and the Industrial Hazardous Waste Management Program (Georgia Department of Natural Resources, 1984).

Idaho

The Idaho Department of Water Resources and the Idaho Water Resource Board are responsible for the management of groundwater resources and their protection from waste and contamination. The Idaho Department of Health and Welfare, Division of Environment, is responsible for protecting groundwater quality in the state.

Extensive pumping for irrigation has caused declining groundwater levels. The state has therefore moved to curtail additional agricultural development in some areas, and it can declare an area as a Groundwater Management Area under Idaho Code 42-233b. In those areas, permits for new well construction must be approved by the Idaho Department of Water Resources to ensure that the rights of existing water users are not affected. If the water levels decline at a rate that will threaten a reasonably safe supply for existing users, the state can declare the area a critical groundwater area under Idaho's Code 42-233a. In such areas, no new permits are issued. Groundwater withdrawals are reduced to levels determined by the Idaho Water Resources Board. Currently five areas have been declared groundwater management areas, and eight have been declared critical groundwater areas.

417

Data collection programs are carried out by the Idaho Department of Water Resources and the Idaho Department of Health and Welfare in cooperation with USGS. This data forms an information base on which groundwater management decisions in Idaho are made (USGS, 1985).

Idaho has enforced or will shortly have enforced the following activities which make up their management program.

Underground storage tanks. Idaho recognizes that underground storage tanks constitute one of its greatest potential sources of groundwater contamination. In 1986 it inventoried all existing underground storage tanks, particularly those relating to petroleum handling and storage.

Feedlots and dairies. The Division of Environment prepared guidelines for feedlots and dairies during 1986 under the Clean Water Act, Section 205. The guidelines sought to address best management practices for mitigating feedlot and dairy impacts on groundwater.

Land application and wastewaters. Wastewaters derived from both municipal and industrial sources and applied to land have been implicated in local groundwater contamination. Guidelines for Land Application of Wastewaters were adopted in Idaho in 1983. The Division of Environment is now developing new rules and regulations for these wastes. A permitting system is anticipated along with substantially improved guidelines that were scheduled to be completed late in 1986.

Hazardous substances. Hazardous substances are regulated by

RCRA regulations. Other hazardous substances that may threaten groundwater are regulated by the Idaho Water Quality Standards and Wastewater Treatment Requirements. Only one hazardous waste disposal site is active in Idaho. This facility is regulated under Idaho's rules, regulations, and standards for hazardous wastes (IDHW 1985).

Pesticides. The Division of Environment is preparing regulations for handling and disposal of pesticides.

Pits, ponds, and lagoons. These structures include animal waste contaminants, surface runoff basins, chemical storage ponds, industrial and municipal wastewater treatment lagoons, and agricultural ponds. Idaho's Water Quality Standard and Waste Water Treatment Requirements (IDHW, 1985) provide limited criteria for the management of these facilities.

Radioactive substances. The Department of Energy controls the wastes from the Idaho National Engineering Laboratory and associated mining activities. Other users of man-made radionuclides such as hospitals, schools, and industries are licensed by the Bureau of Hazardous Materials.

Septic tank systems. Idaho's septic tank system regulations govern permitting, design and construction, and operation of such systems and have recently been updated (IDHW, 1985). A project to assay groundwater impact from large community drain fields will be conducted by IDHW in 1986–87.

Well drilling. Drinking water wells are governed by rules and regulations for public drinking water supplies (IDHW, 1985). Underground injection wells are controlled by the Department of Water Resources. The underground injection control program is administered by the Department of Water Resources.

Mining. The extraction of gold and silver is a major industry in Idaho. There is potential contamination of groundwater from cyanide and also from heavy metals such as mercury and cadmium. The Division of Environment is developing regulations for cyanide leaching operations under Clean Water Act 205 (J) funding. Oil and gas drilling is becoming increasingly important. The Department of Lands issues permits for such drilling. The Division of Environment has developed staff guidelines for handling drilling fluids at disposal sites.

Silviculture activities. These are the lowest-ranked potential contaminants of groundwater. The State Department of Land administers rules and regulations of the Forest Practices Act, which mitigates sedimentation and contamination from equipment and supplies, including pesticides, herbicides, fertilizers, and chemical road treatments.

Specific sole source aquifers have been designated for special protection. They are the Rathdrum Prairie–Spokane Aquifer, the Snake Plain Aquifer, and particularly sensitive aquifers such as the Boise and Marsh Creek–Lower Portneuf Aquifer (Idaho Department of Health and Welfare, 1985).

Illinois

Several state agencies have regulatory authority over activities that affect groundwater quality. The Illinois Environmental Protection Act grants extensive regulatory powers to the Illinois Pollution Control Board. The board is authorized to promulgate regulations to prevent groundwater pollution and has the authority to act for the state with regard to establishing standards for federal laws concerning environmental protection. The Illinois EPA enforces regulations of the IEP Act. The charge includes evaluation, surveillance, and inspection of discharges from contamination sources; monitoring of environmental quality of public water supplies and waste disposal sites, as well as some classes of subsurface and waste injection wells; and investigations of violations of regulations or permits issued.

The Illinois Department of Public Health has regulatory authority over a variety of activities that can affect groundwater quality. These activities include sanitation investigation and inspection of public recreational and tourist facilities and licensing of private sewage disposal contractors, water well contractors, and pump installers. The Illinois Department of Mines and Minerals has the responsibility of permitting and regulating activities in coal mining, oil and gas exploration, and subsurface waste injection of oil and other wastes, some of which might adversely affect groundwater quality. The Illinois State Water Survey and the Illinois State Geological Survey have the authority to study but not regulate groundwater. The USGS works cooperatively with these two agencies and with the Illinois EPA to maintain a statewide water-data network and to investigate the state's water resources. The Illinois Water Use Act of 1983 (Public Act 83-700) established a mechanism to identify withdrawal conflicts. (USGS, 1985).

The Illinois General Assembly passed Public Act 83-1268 in response to the need to monitor, assess, and resolve groundwater problems and to develop a unified statewide groundwater protection plan. The Illinois Environmental Protective Agencies recommended a five-year plan with the following points: (1) clarify the goals and objectives for groundwater protection in Illinois; (2) operate appropriate groundwater monitoring programs; (3) continue to address suspected or confirmed groundwater contamination problems; (4) implement technology-based programs for land pollution control; and (5) establish a statewide program for groundwater source protection (Illinois Environmental Protection Agency, 1986).

420

Indiana

In 1983, Indiana enacted the Water-Resources Management Act, which established a water management branch within the Division of Water in the Indiana Department of Natural Resources. According to this act, registration for water-withdrawal facilities must be established. In addition, this branch assesses the availability of water; maintains an inventory on the significant uses of water withdrawn; and plans for development, conservation, and use of water for beneficial uses. The Water Resources Management Act considers Indiana's water resource as unitary and makes no distinction between groundwater and surface water. The act has been codified as I.C. 13-2-6.1, which was fully implemented on July 1, 1984 (USGS, 1985).

The Department of Environmental Management, the Department of Natural Resources, and the State Board of Health manage the groundwater protection programs. The Department of Environmental Management regulates hazardous- and solid-waste management as well as wastewater treatment and discharge. The Department of Natural Resources collects information on groundwater availability, use, and ambient quality. It also protects the rights of small users from the effects of large water withdrawals through legislation passed in 1985. The State Board of Health administers a program for public water supplies (Indiana Department of Environmental Management, 1984-1985).

Iowa

The laws regarding groundwater management are found in the Code of Iowa, Chapter 455B; the rules regarding groundwater management, in the Iowa Administration Code, Chapter 900. These laws and rules are administered by the Iowa Department of Water,

Air, and Waste Management. Under the authority of this agency, groundwater withdrawal permits are granted, restrictions on withdrawals are enforced, and groundwater injection is regulated. The Iowa Geological Survey is the state's manager of water-resources information and supports various activities that assess the groundwater condition in the state. The University of Iowa Hygienic Laboratory System is responsible for analysis of the quality of community supplies (USGS, 1985).

The state of Iowa is in the process of developing a groundwater **421** protection strategy. For the development of such a strategy, the following recommendations have been made by the Environmental Protection Commission: (1) that non-degradation be maintained as a goal for Iowa's groundwater policies and programs; (2) that three policies guide Iowa toward its goals of non-degradation— prevention of contamination, improvement of groundwater quality when contamination is present, and enhancement and maintenance of public confidence in groundwater quality through public awareness; (3) that the Department of Natural Resources place emphasis on improving programs at this time and on developing guidelines and standards in the future if such guidelines become desirable or necessary. The Environmental Protection Commission further recommends that the State of Iowa direct its resources toward programs in groundwater protection according to the following priority issues:

- Priority 1—agricultural uses of nitrogen fertilizer and pesticides and abandoned hazardous waste disposal sites, abandoned dumps, and unpermitted land disposal.
- Priority 2—underground tanks and pipelines, landfills, and storage, handling, and transportation of hazardous substances.
- Priority 3—agricultural drainage wells, abandoned wells, and sinkholes.
- Priority 4—land application of solid and liquid wastes, lagoons and pits, septic tanks, urban use of nitrogen fertilizers and pesticides.
- Priority 5—the development of interagency cooperation in order to make the program successful.
- Priority 6—the development of methods to significantly increase the involvement of local governments in groundwater protection efforts.
- Priority 7—the significant improvement by the Department of Natural Resources of the resource information base.
- Priority 8—the apprisal of the U.S. EPA when any contaminant

is found in Iowa's groundwater and a concomitant request that the EPA develop a health-based risk assessment of the contaminant for drinking water.

- Priority 9—the significant increase by the Department of Natural Resources in educational programs on Iowa's groundwater resource in order to have better public awareness and understanding of key issues and of the application of preventive measures to daily activities.

422

- Priority 10—the acceleration by the Department of Natural Resources of its assessment of all known potentially hazardous waste-disposal sites and the conducting of preliminary on-site assessments on all sites by a special Iowa Department of Natural Resources team by 1991.
- Priority 11—the development of alternatives to conventional county landfills as a means of protecting Iowa's groundwaters.
- Priority 12—the full administration of an underground-storage-tank program by 1987.
- Priority 13—the special assessment of the possible effects on groundwater of pipelines, land application of solid and liquid wastes, lagoons and pits, septic systems, and urban use of nitrogen fertilizers and pesticides.

The Environmental Protection Commission has developed 12 recommendations that need general assembly action:

1. Legislation should acknowledge the importance of the groundwater resource, define contamination, establish a goal of non-degradation, establish individual and corporate responsibilities to prevent contamination, and grant increased authority to act in protection of groundwater resources.
2. The state of Iowa should act immediately to initiate a 10-year multiple agency, nonregulatory program directed at protecting groundwater from contamination resulting from the use of agricultural fertilizers and pesticides.
3. If the Iowa secretary of agriculture finds that contamination of groundwater from pesticides is widespread and represents a significant risk to human health and the environment, certain restrictions should be placed on the use of pesticides.
4. The State of Iowa should develop a facility to safely store and/or process hazardous wastes by 1992.
5. A statewide hazardous waste collection and transportation plan should be developed.
6. Pertinent health information should be collected and

evaluated as part of the continued development of a groundwater protection policy.

7. A program should be developed to establish permanent conservation easements within watersheds that drain to sinkholes or the areas that become wetlands after closure of agriculture drainage wells.
8. Agriculture drainage wells should be recognized as hazardous to groundwater resources, declared illegal, and phased out and permanently sealed.
9. All water wells should be required to be properly plugged when they are abandoned or can no longer be of service.
10. Programs should be developed to plug all drainage wells and abandoned water wells.
11. A program should be developed that targets soil conservation in watersheds that drain to sinkholes.
12. A program should be developed to help construct manure-storage facilities so that existing organic fertilizer sources may be more effectively used.

423

Methods of financing. The Environmental Protection Commission also recommends methods of financing such a program, including increases in long term financial resources to support the state's groundwater protection programs. The commission further recommends that the 37 million dollars from the oil-overcharge settlement fund be dedicated to support the proposed strategy (Iowa Department of Natural Resources, 1987).

Kansas

Kansas has five state agencies and one type of local state government unit with major responsibilities for managing groundwater. The Kansas Water Office is a water planning, policy, and coordination agency (K.S.A. 74-2605 *et seq.*). It prepares state plans for water resource management, conservation, and development. The Kansas Water Authority, a part of the Kansas Water Office (K.S.A. 74-2605 *et seq.*), is responsible for advising the governor, the legislature, and the director of the Kansas Water Office on water policy issues. The Kansas State Board of Agriculture, Division of Water Resources administers laws (K.S.A. 82A-701 *et seq.*) and is the authority for the appropriation of groundwater and assists in the organization of Groundwater Management Districts. The Kansas Department of Health, Division of Environment, has regulatory authority over matters dealing with water pollution (K.S.A. 65-161 *et seq.,* K.S.A. 55-1003 *et seq.,* K.S.A. 82A-1035–1038 and K.S.A. 82A-1201 *et seq.*). This agency is responsible for collecting, analyzing, and interpreting groundwater quality data, developing water quality management plans, and responding to emergency water pollution problems. The Kansas Corporation Commission has a mandate (K.S.A. 55-115 *et seq.*) to protect fresh groundwater supplies from adverse effects of mineral development activities. The Kansas Geological Survey conducts groundwater research, including the collection, analysis, and interpretation of data on groundwater quantity and quality (K.S.A. 76-322, 76-2610, 82A-903, 55-128). Groundwater management districts are locally managed political subdivisions of the state, and were formed as a result of the Groundwater Management District Act of 1972 (K.S.A. 82A-1020, *et seq.*). There are five such districts in Kansas; each is charged with managing groundwater resources within its boundaries (USGS, 1985).

The Kansas Water Quality Program seeks to do the following:
1. Effectively manage municipal sewage systems.
2. Utilize existing permit systems to regulate discharges of industrial wastes in accordance with minimum federal standings; maintain and control state requirements for the administration of federally mandated programs for the pretreatment of selected industrial wastes to prevent the discharge of toxic and hazardous substances into municipal sewage systems; prohibit the use of deep wells for the disposal of industrial wastes, salt brines, or other non-toxic materials incompatible with the injection formation.
3. Manage residual wastes by utilizing the established effluent- or hazardous-waste program for regulating the disposal of sludge from municipal- and industrial-waste treatment.
4. Control petroleum activities through implementation of additional controls as required by the underground injection control provisions of the Safe Drinking Water Act; require that the spillers of oil or hazardous materials assume responsibility for cleanup, and require installation of observation capabilities at new or replacement underground storage facilities.
5. Control coal mining, particularly as it relates to the existing Mine and Reclamation Act; accelerate water-quality monitoring programs, particularly in the southeast counties to help establish priorities for reclamation plans; utilize the existing effluent permit system to regulate discharges from operative facilities and, where practicable, from non-

operating facilities from the mining of materials other than coal, such as lead and zinc.

6. Take steps to address the problems associated with agricultural runoff, including acceleration of the present voluntary soil conservation program, continuing evaluation of the program's effectiveness, utilization of the existing pesticide-applicators-certificate program as a principal mechanism for avoiding excess pesticide concentrations in surface waters, and provision for additional studies and extension of existing monitoring systems to determine the effect of biological activity on groundwater. The plan also includes management of feedlot wastes which may affect groundwater.

7. Install back-flow check valves on irrigation wells; authorize construction and operation of pilot saltwater interception systems near Salina to prevent mineral intrusion into groundwater; continue and improve the groundwater-quality studies now being pursued.

8. Develop a stormwater-quality management that provides for stormwater monitoring plans in key cities to address the problem of urban storm runoff (Kansas Department of Health and Environment, 1982).

425

Kentucky

A number of state agencies within the Department of Mines and Minerals and the Kentucky cabinets of Human Resources and Natural Resources and Environmental Protection are responsible for the state's groundwater management. The following state legislation and regulations relate to groundwater protection or management. Water quality standards (401 Kentucky Administrative Regulation No. 5:031) are established for aquatic life, domestic-water supply use, recreational use, and outstanding resource waters. Water Resources laws (Kentucky revised statutes, Chapter 151) require consumptive users of public water to obtain a permit for the withdrawal of 10,000 gallons per day or more. Controls on oil and gas facilities to control water pollution are provided by 401 Kentucky Administrative Regulation 5:090. Under the Kentucky Law 405, (Kentucky Administration Regulation Chapters 8–24), the protection of groundwater quality and recharge capacity associated with surface and underground mining activities is addressed. Kentucky Law 401, Chapter 30, is concerned with solid and hazardous waste management regulations and is administered by the Division of Waste Management. Performance standards for waste disposal sites, including the protection of groundwater contamination, are part of this responsibility. Subsurface sewage disposal regulations (815 Kentucky Administration Regulations 20:141 and 20:160) specify such standards as the minimum size and capacity of private subsurface sewage disposal systems and minimum distance from the systems to drinking water wells. Permits are issued from the Division of Consumer Health Protection, Cabinet for Human Resources. Plugging, casing, and operation of wells are accomplished in accordance with oil and gas regulations (805 Kentucky Administration Regulations 1:020, 1:060, and 1:070) established by the Department of Mines and Minerals. Unreasonable damage to groundwater supplies from waste, oil, and gas is prohibited. The Kentucky Geological Survey maintains a statewide water-data network and the investigation of the state water resources in cooperation with the U.S. Geological Survey. The research, data collection, and analysis provided by this cooperative program form an information base upon which groundwater management decisions are made by appropriate state agencies (USGS, 1985).

426

Louisiana

Five different state agencies have active roles in administering groundwater activities in Louisiana. The Department of Transportation and Development's Office of Public Works (OPW) licenses and regulates drillers of water wells, monitor wells, geotechnical boreholes, and heat-pump wells, as well as those engaged in plugging abandoned wells and boreholes. The OPW registers all water wells drilled in Louisiana and maintains an active computer file of these wells. The OPW-administered Louisiana Water Resources Information Center has the responsibility of indexing all available water resources information for the state. This department is the major state agency that cooperates with the USGS in a groundwater program of data collecting, aerial studies, and research. The Department of Natural Resources has certain regulatory responsibilities relating to the protection of groundwater. The Department's Office of Conservation has jurisdiction over groundwater injection wells and also has regulatory functions relating to the protection of groundwater in areas of lignite mining and oil and gas development. The Louisiana Geological Survey has some groundwater functions, principally in support of the missions of the Department of Natural Resources and other state agencies. The Louisiana Department of Health and Human Resources has the responsibility of ensuring safe and good quality drinking water and also enforces construction standards for public supply wells. The Department of Environmental Quality has responsibilities for monitoring and protecting groundwater as this relates to regulation of solid and hazardous wastes (USGS, 1985).   **427**

Maine

Several state agencies have statutory responsibility for groundwater protection and management. The Department of Conservation, through the Maine Geological Survey and the Land Use Regulation Commission, is responsible for coordinating groundwater research, mapping groundwater availability, performing research about permit-related groundwater problems, and regulating activities that affect groundwater in areas where populations are sparse. The Department of Environmental Protection, through its various bureaus, is responsible for reviewing and licensing activities that affect groundwater. The department also has the responsibility for research into the effects of gasoline leaks and pesticides on groundwater and for groundwater quality assessment and emergency response and cleanup. The Department of Human Services is involved in groundwater protection and management through its Drinking Water Program, Environmental Health Unit, and Public Health Laboratories. It is responsible for reviewing and approving new public water supply sources, monitoring the quality of existing sources, performing research on groundwater-transmitted diseases, and performing water quality analyses of private water supplies. The Maine Land and Water Resource Council is examining the state's present statutes and regulations, agency programs and human resources, and agency activities that pertain to groundwater in an effort to ensure protection of public health and continued availability of groundwater (USGS, 1985). The legal basis for Maine's Groundwater Protection Program is Title 38 (MRSA #38) Article 1-B (attachment #1).

The management of Maine's groundwater quality is set forth by a classification scheme and 15 ordinances for its protection. Maine classifies groundwater as GW-A, which must be of such quality that it can be used for public water supplies, and GW-B. At the present time, no groundwater has been designated as GW-B; the Oliva aquifer is classified as GW-A. The following paragraphs describe the 15 state ordinances that protect groundwater.

1. 22MRSA S2642 gives the municipalities the authority to adopt regulations governing the surface uses of land overlying groundwater aquifers used as public water supply. 30NRSA S2151 enables municipalities to enact police-power ordinances to provide for the protection and conservation of the quality and quantity of groundwater.

2. The Maine Plumbing Code (22 MRSA S42) provides rules and regulations for subsurface sewage disposal systems. Subsurface waste disposal rules (CMR 241 Sections 15 and 16) provide for variances for replacement and new wastewater-disposal systems, respectively.

3. The State Minimum Lot Size Law (12 MRSA S4807) requires a minimum lot size for residential units using subsurface disposal.

4. The Local Subdivision Law (30 MRSA S4956) prohibits the municipal approvement of subdivisions unless they will not result in water pollution and they have sufficient available water for the reasonable foreseeable needs of the subdivision. Furthermore, they must not cause an unreasonable burden on existing water supplies and should not adversely affect the quality and quantity of groundwater. The law provides for exemption (grandfathering) for preexisting subdivisions.

5. The State Site Location of Development Law (38 MRSA S481-491) requires a state permit for any development, and the proposed development must not pose an unreasonable risk for discharge into a significant groundwater aquifer or have any unreasonable adverse effect on groundwater quality and quantity.

6. The State Land Use Regulation Law (12 MRSA S681-689) provides for special protection subdistricts and land-use standards for the protection of aquifers in unorganized and organized townships.

7. The Protection and Improvement of Waters Act (38 MRSA S361-455) requires a license for any subsurface wastewater-disposal system not designated and installed in conformance with the Maine Plumbing Code.

8. The Hazardous Waste Seepage and Solid Waste Management Act (38 MRSA S1301-1310) requires a state license for any waste facility; such licenses are issued only if the facility will not pollute any water in the state. A waste facility shall not overlie a primary

428

recharge area of groundwater supplies. Quarterly monitoring of groundwater is required throughout the life of the facility.

9. The Hazardous Matter Control Law (38 MRSA S1317-1319) prohibits discharges of hazardous matter into or upon any land or water, and if such discharges occur the state, except as licensed by law, is empowered to conduct immediate cleanup operations and recover the costs.

10. The Oil Discharge Prevention and Pollution Control Law (38 MRSA 3541-560, 345-349) prohibits the discharge of oil into or upon any waters of the state except by license under certain prescribed conditions.

429

11. The Underground Oil Storage Facilities in Groundwater Law (38 MRSA S561-570) requires registration of all tanks, sets construction and installation standards, requires installation by a certified installer, and requires secondary containment or leak-detection systems in sensitive geologic areas (over aquifers) for tanks in the marketing and distribution category. It requires daily inventory recording and annual statistical analyses for tanks in the marketing and distribution category. It requires removal of abandoned tanks within 12 months. It authorizes and funds state cleanup or provision of a replacement drinking-water supply and authorizes recovery of costs, clarifies the liability of tank owners, and requires evidence of financial responsibility for owners in the marketing and distribution category.

12. The Pesticide Control Act (22 MRSA S1471) permits the pesticide control board to designate critical areas where pesticide use would present an unreasonable threat to the quality of a water supply, provides for the proper return and disposal of used pesticide containers, and permits the board to accept and store unwanted pesticides.

13. 23 MRSA S652 authorizes the Maine Department of Transportation to reimburse parties for damages caused to their wells by Department of Transportation road-salting operations.

14. The State Waste Discharge License Law (38 MRSA S413) requires owners of salt-storage areas to register their vocations and to make sure that the storage area will not adversely affect any water supply. Storage areas that were registered prior to January 1, 1986, and are already on the Pollution Abatement Priority List will be exempted. The Department of Environmental Protection will assign priorities to sand/salt storage areas prior to November 1, 1986, for pollution abatement action.

15. The State Nuisance Law (17 MRSA S2802, 2701) authorizes

civil action against any offender who corrupts or renders unwholesome or impure the water of an aquifer (personal communication, P. Dutram, State Groundwater Coordinator, 1986).

Maryland and the District of Columbia

The District of Columbia relies mainly on surface waters and has no specific legislation directed toward groundwater. In Maryland, however, groundwater management and planning legislation are extensive. The Department of Natural Resources, through its agencies, has a major role in groundwater resource planning and management. The Maryland Department of Health and Mental Hygiene, through its office of environmental programs, is primarily responsible for regulatory and operational programs with regard to water-quality aspects of groundwater management. This office issues well construction permits (Code of Maryland regulations 10.17.13, implemented in 1945), requires well-completion reports from licensed well drillers, and regulates the disposal of water to groundwater systems (Health-Environmental Article, 9-3222).

The Water Resources Administration provides direction in the development, management, and conservation of the state's water and regulates groundwater use through an appropriation-permit program (Natural Resources Article 8-802, enacted in 1933). This program requires a permit to appropriate ground or surface waters and requires water use reports for withdrawal of 10,000 gallons per day or more. Domestic and farm users (including irrigation use) are exempt from these requirements. The Maryland Geological Survey is responsible for maintaining a statewide water-data network and for investigating the state's water resources. These responsibilities are accomplished in cooperation with the USGS. The research, data collection, and analyses provided by this cooperative program form an information base upon which groundwater management decisions are made by the Water Resources Administration (USGS, 1985).

Massachusetts

All state agencies with groundwater management and planning responsibilities are managed by the Massachusetts Executive Office of Environmental Affairs. The Massachusetts Water

Resources Commission develops and coordinates the water resources planning and management functions of the departments of the Massachusetts Executive Office of Environmental Affairs. Recognizing the interdependency of surface water and groundwater, the commission has recommended a policy of preventing undesirable streamflow depletion through allocation of groundwater withdrawals for public supply. The commission has established criteria and priorities for all water-related cooperative programs with the federal government and with other state agencies.

431

The Massachusetts Department of Environmental Management (MDEM) has two divisions responsible for water activities: the Division of Water Resources and the Bureau of Solid Waste Disposal. The Division of Water Resources collects and disseminates water-resource information and develops state water-resource plans (Water Resource Planning Regulations 313 CMR..00). This Division administers water resources data collection and groundwater assessment programs in cooperation with the USGS and other agencies. It also licenses well drillers and maintains records of well completion reports. The Bureau of Solid Waste Disposal protects groundwater quality by planning for solid and hazardous waste disposal through regional facilities and by dissemination of technical information.

The Massachusetts Department of Environmental Quality Engineering has primary responsibility for groundwater quality through its Division of Water Supply, Division of Environmental Analysis, and Division of Hazardous Wastes. The Division of Water Supply ensures drinking water quality through its well permitting program. It collects and disseminates groundwater quality information and administers programs providing funds for water treatment and aquifer protection.

The Division of Environmental Analysis is the department's analytical laboratory. It collects and analyzes samples of raw and treated public water supplies used for drinking purposes. It is responsible for analyzing groundwater samples suspected of contamination.

The Division of Water Pollution Control is responsible for improving water quality and preventing groundwater pollution. It regulates discharges of polluted material originating from point or other major non-point sources into groundwater by permit. It also administers water resource inventory and water-quality research programs in cooperation with the universities and the U.S. Geological Survey.

The Division of Hazardous Waste regulates activities with a large potential for groundwater contamination. It responds to oil spills and other hazardous waste accidents on an emergency basis, and it investigates illegal disposal activities and supervises the cleanup of hazardous waste sites. Its activities include the approval of groundwater monitoring programs, hydrogeological studies, and evaluation of proposals for cleaning up contaminated groundwater (USGS, 1985).

432 To preserve the quality of groundwater, Massachusetts has developed two regulations that establish the Massachusetts groundwater discharge permit program and the Massachusetts groundwater quality standards. These regulations are identified as Title 314 of the Code of Massachusetts Regulations, Chapters 5 and 6 (31RCMR5.00 and 6.00). The Groundwater Classification program recognizes three classes of groundwater: (1) fresh groundwater found in the saturated zone of unconsolidated geological materials or in porous or fractured rock and bedrock, and designated as real or potential sources of potable water supply; (2) saline waters found in the saturated zone of unconsolidated deposits or consolidated rock and bedrock, and designated as a source of potable mineral waters for conversion to fresh potable waters or as raw water for the manufacturing of sodium chloride or its derivatives or similar products; and (3) fresh or saline waters found in saturated zones of unconsolidated deposits or consolidated rock or bedrock and designated as a source of non-potable water that may come in contact with, but cannot be ingested by, humans (Massachusetts Department of Environmental Quality Engineering, 1986).

Michigan

Two state agencies, the Department of Public Health and the Department of Natural Resources, regulate and manage Michigan's groundwater resources. The Department of Public Health issues permits for domestic and public supply wells and requires well drillers to submit copies of drilling records to the county health departments. This department also monitors the quality of public water supplies. The Department of Natural Resources assists groundwater users by maintaining files of drilling records and by performing hydrogeologic groundwater-quality studies. The department also maps and describes geological formations and monitors mineral wells and subsurface injection of brine (USGS, 1985). The Water Resources Commission of the Department of Natural Resources, by authority conferred by Sections 2 and 5 of Act No. 245 of the Public Acts of 1929, amended R 323.2201–R 323.221 (1980), established rules for the non-degradation of groundwater quality, a permitting system for a discharge into groundwater, and monitoring requirements for existing and new discharges. A procedure for establishing variances from these rules was established (Michigan Department of Natural Resources, 1980).

433

In 1983 it was proposed that the State of Michigan require that environmental impairment liability insurance be obtained by owners or operators of gasoline stations and other facilities known to pose significant risk of contaminating groundwater from accidental spills of chemical products (Michigan Department of Natural Resources, 1983).

Minnesota

Minnesota has extensive groundwater management and planning legislation. The Minnesota Department of Natural Resources has a major role in groundwater-resource planning and management. It provides technical assistance on water supply, conservation, and well interference issues and manages an appropriation permit program. A permit is required to appropriate groundwater, with the exception of domestic use of 25 persons or less. Annual amounts pumped must be reported. The Division of Water is responsible for the maintenance of a statewide observation-well monitoring network and a water-use program and for investigation of the state's water resources. The research, data collection, and analysis provided by this program are carried out in cooperation with the USGS and form part of the data base used by the Minnesota Department of Natural Resources in making groundwater management decisions. The Minnesota Department of Health (MDH) is concerned with the health-related and domestic-supply issues involving groundwater. The MDH approves plans for public supply wells, establishes and enforces well construction standards, and licenses well drillers (Minnesota Statutes, Chapter 156A). The MDH requires well completion reports for new wells and regulates the reinjection of groundwater and groundwater thermal exchange devices (Minnesota Statutes, Chapter 156A.10). The MDH also administers public water-supply regulations in concurrence with the Safe Drinking Water Act (Minnesota Statutes, Chapter 114.381 and 7 MCAR 1.145-1.150). The Minnesota Pollution Control Agency administers programs dealing with groundwater quality issues and pollution control requirements (Minnesota Statutes Chapters 115 and 116). This agency adminsters its programs through the following system of rules:

- Preservation and protection of underground water in the state by preventing any new pollution and by abating existing pollution [6MCAR~4.8022 (WPC-22)]
- Regulation of sewage-sludge land spreading (6MCAR~4.6101-4.6136)
- Regulation of hazardous-waste facilities (6MCAR~4.9001-4.9010)
- Regulation of sanitary landfills (Minnesota Rule SW-6 and SW-12)
- Regulation of septic tanks and drain fields (6MCAR~4.8040)
- Regulation of storage of liquid products (WPC-4)
- Regulation of intrastate and interstate standards for water quality and purity (6 MCAR~4.8014 and 6MCAR~4.8015)

434

The Environmental Response and Liability Act (Minnesota Statutes, Chapter 115b), passed in 1984, authorizes the Minnesota Pollution Control Agency to provide funds for cleanup of contamination sites and to gain reimbursement later. Permits are required for disposal practices and for operating facilities that could affect the quality of groundwater. The agency maintains a network of 400 wells and springs to monitor groundwater quality throughout Minnesota (USGS, 1985).

Mississippi

Mississippi has a twofold strategy for protecting its groundwater. The first part is the exempting of all aquifers or parts of aquifers that meet the criteria for an underground source of drinking water. An underground source of drinking water is an aquifer or a portion of an aquifer that supplies any of the public systems or that contains a sufficient quantity of groundwater to supply a public system and currently supplies water for human consumption or contains fewer than 10,000 mg/l dissolved solids. An aquifer or a portion of one that meets these criteria may be designated as an exempted aquifer by the permit board if it also meets the criteria set forth in Section 146.04 of 40 CFR Part 146 and all amendments set forth in Section 146.04 of 40 CFR Part 146 and all amendments thereto.

The second part of Mississippi's groundwater protection strategy for groundwater is the classification of injection wells. *Class 1* injection wells are used by generators of hazardous wastes or owners or operators of hazardous waste management facilities to inject hazardous wastes beneath the lowermost formation containing an underground source of drinking water within five miles of the well borehole. Other Class 1 wells include industrial and municipal disposal wells that inject fluids beneath the lowermost formation containing an underground source of drinking water within five miles of the well borehole.

435

Class 2 wells are those that inject fluids. These fluids include those that are brought to the surface in connection with conventional oil or natural gas production. They may be commingled with wastewater from gas plants, unless those waters are classified as hazardous at the time of injection. Other Class 2 wells are those injecting fluids for enhanced recovery of oil or natural gas, or for storage of hydrocarbons that are liquid at standard temperatures and pressures.

Class 3 wells are those that inject for extraction of minerals or energy, including: (1) the mining of sulfur by the Frasche process, (2) in situ production of uranium and other metals, (3) solution mining of salts or potash, (4) in situ combustion of fossil fuels, and (5) the recovery of geothermal energy to produce electric power.

Class 4 wells are used for the disposal of hazardous wastes by generators of hazardous wastes, industrial-process wastewater, or radioactive wastes; by owners or operators of hazardous- or industrial-process-waste-management facilities; or by owners or operators of radioactive-waste-disposal sites. Industrial process wastewater or radioactive wastes into or above the formation which within five miles of wells contain an underground source of drinking water.

Class 5 wells are injection wells not included in the first four classes, as defined under Section 146.05 (E 40 CFR part 146 and the amendments thereto) (Mississippi Department of Natural Resources, 1982).

It was not until 1976 that a groundwater bill was enacted. The concept of capacity-use areas, wherein well spacing, well depths, and withdrawal rates are regulated, is a mechanism provided for dealing with areas having identifiable groundwater-supply problems. The major limitations of the 1976 act are that it addresses only withdrawals in excess of 50,000 gallons per day and excludes agricultural, oil, and gas uses. In 1983 the state legislature created the Mississippi Water Management Council to reexamine all state laws pertaining to surface and subsurface waters and to report recommended amendments to the 1985 session. The Mississippi Department of Natural Resources administers and enforces the 1976 Groundwater Statutes (Groundwater Bill codified as Section 5-4-1 *et seq.*, Miss. Code annotated 1972) and the 1966 Water Well Drillers Licensing Act.

436 Primacy in permitting waste injection in Mississippi (other than in connection with oil and gas production) has been assigned to the Department's Bureau of Pollution Control, which also has responsibility for permitting and monitoring hazardous waste sites. Primacy for permitting oil-field waste injection has not been delegated by the U.S. Environmental Protection Agency, although the state of Mississippi now has final authorization to operate its own hazardous-waste program. The department's Bureau of Geology is basically a research organization and is authorized to investigate and report on water resources. The Mississippi State Board of Health ensures that public water supplies meet chemical, bacteriological, and other standards. The water resource investigations in Mississippi are conducted in cooperation with the U.S. Geological Survey and five federal agencies (USGS, 1985).

Missouri

"The Missouri Division of Geology and Land Survey, the Missouri Division of Environmental Quality, and the Missouri Division of Health are the principal state organizations involved in groundwater activities. One of the major responsibilities of the Missouri Division of Geology and Land Survey is the administration of the new Major Water Users Registration Act (revised Statute 256), which requires that withdrawals of more than 100,000 gallons per day be reported annually to the survey." The survey also provides advice to the Department of Environmental Quality on casing depths for public supply wells. "The state geologist, who also is the director of the Division of Geological Land Survey, administers the rules and regulations of the State Oil and Gas Council (RSMO.259.010, 259.020, 259.030, 259.040). In doing so, the state geologist maintains close watch over oil- and gas-drilling practices to ensure protection of groundwater supplies." This division also cooperates with the USGS in maintaining a statewide data network in investigating the state water resources. The Division of Environmental Quality supervises the design and construction of water-supply systems and cooperates with the Missouri Division of Health in monitoring contaminants in water supplies (USGS, 1985).

Missouri has a code of state regulations known as Title 10, Department of Natural Resources Division 20. The codes 10 CSR 20-1.010 through 10 CSR 20-7.015 have various provisions that affect groundwater contamination. These codes extend from June 1974 through February 1986.

437

Montana

The 1973 Montana Water Use Act established a uniform central system for acquisition, administration, and determination of all water rights. The Act also mandates adjudication of all existing rights. To date 10,500 (or about 5%) of the state's existing water rights applications have been adjudicated, all in the Powder River Basin. Appropriation of groundwater supplies for domestic, agriculture, or livestock purposes does not require a water rights permit if the maximum appropriation from the source well is less than 100 gallons per minute. The only requirement is completion of a form within 60 days after the completion of the well. Appropriation of groundwater supplies requires a water rights permit if the maximum yield of the well is a hundred gallons per minute or more or if the well is in a controlled groundwater area. Controlled groundwater areas can be established to protect water rights, an entire water resource, or public health in areas subject to pollution of water supplies. Requirements to be met before issuance of a water rights permit to discharge into groundwater are:

1. Unappropriated water exists that the applicant can use in quantity at the time proposed for application.
2. The rights of prior appropriators will not be adversely affected.
3. The proposed means of construction are adequate.
4. The proposed use is determined as beneficial.
5. The proposed use will not interfere unreasonably with other permitted, planned use for developments or with water previously reserved for other uses.
6. The applicant who proposes to use in excess of 10,000 acre-feet per year (15 cubic feet per second) must provide clear and convincing evidence that the rights of prior appropriators will not be adversely affected.

The Montana Department of Natural Resources and Conservation has responsibility for administering water resources and water rights and assists in the organization and operation of water conservation districts. The Montana Department of Health and Environmental Sciences has the responsibility of regulating the quality of groundwater resources including public water supplies and wastewater management. The Montana Universities Joint Water Resources Research Center conducts and coordinates water studies, sometimes at the specific request of water resource management agencies. The Montana Bureau of Mines and Geology is a non-regulatory agency responsible for conducting applied research projects on all aspects of the State's groundwater resources, maintaining a state wide groundwater information center and data base, and assisting governmental organizations and private citizens with water related problems and requests. The Bureau has an active groundwater cooperative program with the USGS to conduct local and regional hydrogeological investigations throughout the state. The research, data collection, and analysis form an information base that helps regulating agencies make groundwater management decisions and recommendations (USGS, 1985).

Montana manages its groundwater by a classification system, a water quality standard, and a permit system. The classification of groundwater was established October 29, 1982 according to actual quality or actual use in the place where the groundwater had the highest classification.

Class I groundwater is that generally suitable for public and private water supplies, culinary and food processing purposes, irrigation, livestock and wildlife watering and for commercial and industrial purposes with little or no treatment. Class I groundwaters have a specific conductance of less than 1,000 micromhos/cm at 25°C.

Class II waters are generally marginally suitable for a public and private water supplies, culinary and food processing uses and are suitable for irrigation of some agricultural crops, for drinking water for most wildlife and livestock, and for most commercial and industrial purposes. Class II groundwaters may be used for municipal or domestic water supplies in areas where better water quality is not readily available. Class II groundwaters have specific conductance ranging from 1,000 to 2,500 micromhos/cm at 25°C.

Class III groundwaters are suitable for some industrial and commercial uses, as drinking water for some wildlife and livestock, and for irrigation of some salt tolerant crops using special water

438

management practices. In some cases class III groundwaters are the only economically feasible source for a municipal or domestic water supplies. Class III groundwaters have specific conductance ranging from 2,500 to 15,000 micromhos/cm at 25°C.

Class IV groundwaters may be suitable for some industrial commercial and other uses but are unsuitable or for practical purposes untreatable for higher class beneficial uses. These groundwaters have specific conductance greater than 15,000 micromhos/cm at 25°C.

439

Groundwater quality standards are those that are set forth in EPA publication EPA 600/4-79-020. The State of Montana also recognizes a mixing zone in groundwater; that is, discharges of pollutants to groundwater may be entitled to a mixing zone. The size of the mixing zone will generally not extend beyond the property boundaries of the operator of the source. The burden is on the applicant to show, based on the best professional judgment, that beneficial uses of groundwater will not be adversely affected by the allowance of the mixing zone, in order that the groundwater standards will not be violated on the boundaries of the mixing zone.

If local geology, soils, hydrology, groundwater use, the nature of the discharge, or other factors indicate that beneficial uses cannot otherwise be protected, no mixing zone is allowed. For purposes of this rule, only those beneficial uses need be protected which are not owned by the applicant or owned by the applicant but are unrelated to the operations for which this permit is requested (Montana Department of Health and Environmental Science, 1982 a and b).

Nebraska

The State of Nebraska has several agencies actively engaged in groundwater research, planning, regulation, and management. The Conservation and Survey Division of the University of Nebraska's Institute of Agriculture and Natural Resources has the responsibility for maintaining a natural resource data base, conducting research and investigations about most natural resources, reporting its findings, and assisting citizens in resource development management. The State Water Research Institute in Nebraska is the Water Resource Center at the University of Nebraska at Lincoln. The center administers and conducts water resources research, disseminates information, and provides training. The Nebraska Natural Resources Commission is the state's water planning and water resources development funding agency. The commission manages the water planning and review process, including the analysis of state water policy issues and studies of specific water problems. It also participates in groundwater modeling recharge and water quality studies. The Nebraska Department of Water Resources is responsible for regulatory programs relating to groundwater quantity management, registration of all wells except those used solely for domestic purposes, and management of regulations relating to the spacing of wells. Legislative Statute 75-577 provides that the director of the department preside over hearings initiated by the Natural Resource District for creating groundwater control areas. The Nebraska Department of Environmental Control is responsible for the protection and improvement of water quality in the state and administers the National Pollutant Discharge Elimination System permit program and water quality standards. The director is responsible for issuing exemptions to state underground water protection standards. The Nebraska Department of Health administers the National Safe Drinking Water Act and conducts a Public Water System Program to ensure the safety of drinking water. Twenty-four Natural Resource Districts coordinate land- and water management programs with other government entities. Water conservation activities include monitoring water levels and groundwater quality, cooperating in groundwater investigations, and managing groundwater control areas (USGS, 1985).

440

Groundwater is a major source of water for all major users in Nebraska. Nebraska's groundwater quality is generally very good but degradation has occurred throughout the state. Therefore, the state must suggest a strategy that effectively addresses this problem based upon the following substantive propositions:

1. Potential groundwater contaminants are diverse and are often chemically complex and persistent in the environment.
2. Understanding of the health effects of many potential contaminants is limited, particularly regarding long run, chronic effects.
3. Sources of potential groundwater contamination are multifaceted, involving waste handling and disposal and normal product handling, distribution, and use.
4. Detection of groundwater contamination after it occurs or while it is occurring is often difficult.

5. Restoration of contaminated groundwater is often difficult if not technologically or economically prohibitive.
6. Groundwater contamination can persist indefinitely and may be considered essentially irreversible in some cases.

The development of a strategy first involves a better articulation of existing regulatory measures. Second, the coordinated development of additional control measures and program elements where there are significant deficiencies in existing measures that must be incorporated. This orientation largely forges a major role for Nebraska's Department of Environmental Control. Under the Nebraska Act (NEB. REB. STAT. 81-1501) groundwater is recognized as one of the state's most precious resources, and groundwater pollution as a menace to the health and welfare of the public.

441

The Department of Environmental Control is given wide-ranging regulatory authorities to prevent groundwater pollution and to protect water quality. Specifically, the act authorizes the Department of Environmental Control to do the following:

1. Develop comprehensive programs to prevent and control pollution [(81-1504)]
2. Issue orders prohibiting or abating groundwater pollution and require adoption of remedial measures to prevent, control, or abate pollution [81-1504 (7)]
3. Issue permits consistent with Environmental Control Council regulations to prevent, control, and abate pollution [(81-1504)]
4. Require any person engaging in operations that may pollute groundwater to obtain a permit prior to continuing operation [81-1504 (20)]
5. Sue polluters in court on behalf of the state of Nebraska to prevent, control, or abate groundwater pollution [81-1504 (22)]
6. Delegate administration of groundwater-pollution-control programs to local units of government with ECC approval of the local program [81-1504 (23)]

Although the Department of Environmental Control has the lead role, these programs cross over the responsibilities and authorities of other agencies at the federal, state, and local levels (e.g., the Nebraska Department of Health, State Fire Marshal and State Civil Defense Agency, Nebraska Oil and Gas Conservation Commission, Nebraska Department of Water Resources, Nebraska Department of Agriculture, and the Nebraska Natural Resources Commission).

Besides these state agencies and the Natural Resource Districts, the University of Nebraska—as the state's land-grant research institution—also has an important technical and research role in the strategy.

The strategy attempts to incorporate and build upon these federal, state, and local responsibilities and programs. The details of interagency relationships under various strategy elements must be clarified over time. Nebraska does not have an aquifer classification system; rather, it considers all aquifers potential sources of drinking water and seeks to protect them to the highest reasonable degree. (Nebraska Department of Environmental Control, 1985)

442

Nevada

Groundwater use in Nevada is regulated by the Department of Conservation and Natural Resources through the state engineer's office. The concept of safe yield in individual basins is the basis of administration by the state engineer. Basins that have experienced significant water level declines due to groundwater withdrawal have been designated critical basins, thereby effectively limiting additional withdrawals. Currently, two critical basins are located near Reno and Carson City, one near Eureka, and two near Las Vegas. Protection of groundwater pollution is the responsibility of Nevada's Department of Environmental Protection (USGS, 1985).

At present, Nevada does not have an overall strategy for the management of groundwater. However, it does have various state and local agencies that regulate the use of groundwater in compliance with federal regulations. The Department of Human Resources and the Department of Conservation and Natural Resources are the primary agents responsible for protection of Nevada's groundwater. Other state and local agencies play an auxiliary role; these agencies include the Department of Minerals, the Department of Agriculture, and the Clark, Washoe, and Carson City Health Departments. The Division of Health controls the state's drinking water and is responsible for the enforcement of the Safe Drinking Water Act. It tests all existing water supply wells. It also analyzes samples of private domestic wells and is responsible for the regulation of all individual sewage-disposal systems in those districts where the Board of Health has adopted such regulations. The Department of Conservation and Natural Resources has been designated as the state's Water Pollution Control Agency for all purposes of federal water pollution control legislation and implementation of the Nevada Water Pollution Control Law.

The Division of Environmental Protection has been delegated responsibility for implementing and enforcing the Nevada Water Pollution Control Law (NRS 445.131 to 445.354, inclusive). The Division of Environmental Protection has responsibility for administrating several federal programs, including the National Pollution Discharge Elimination System (NPDES) and the Resource Conservation and Recovery Act (RCRA). The division is also developing a pollution management plan and administers loans and grants for the implementing functions under the Nevada Water Pollution Control Law. The Division of Environmental Protection has implemented a permit and enforcement program that meets the mandate of the Nevada Water Pollution Control Law to protect the waters of the state. The permits are NPDES for sewers and industrial surface discharges; Nevada Zero Discharge for heap leaching and lined evaporating ponds; Nevada's Waters of the State for storm runoff infiltration basins, land application of effluents and subdivision fields; and Nevada's Injection Wells, particularly geothermal injection wells. The law specifically prohibits permits that would authorize the discharge of any nuclear, biological, or chemical warfare agent or any high-level radioactive waste.

The Division of Water Resources has two primary mandates: to adjudicate and appropriate all waters of the state for beneficial uses (NRS 533) and to regulate and permit water wells (NRS 534). The state engineer has control of the permitting of water wells. All well drillers are required to be licensed by the state engineer; they must have a 50-foot seal to prevent groundwater contamination from surface runoff.

The Department of Minerals protects fresh waters from pollution resulting from drilling or plugging of wells or from the escape, release, or injection of gas or brine. Wells used to inject gas, air, water, or other fluids into a producing formation must be cased in such a way as to prevent leaks that could damage fresh waters.

The Department of Agriculture continually evaluates the registration and pending registration of pesticides.

County health departments are in charge of inspection of septic systems and are responsible for hazardous spills and for the reporting of leaking underground storage tanks.

The primary groundwater protection program in the state is the underground injection control program of the Nevada Environmental Protection Agency. Thus far, this program has been applied only to wells concerned with oil and gas injection.

443

USGS undertakes a wide range of special studies, such as groundwater recharge or contamination problems in specific basins and statewide studies of deep aquifers. Nevada's number one priority is to seek primacy of the Federal Underground Injection Control Program from EPA.

The Department of Environmental Resources has been designated as the state agency responsible for enforcing the Federal Resource Conservation and Recovery Act, which includes the problem of leaking underground storage tanks.

444

The USGS, in cooperation with the Department of Environmental Protection, is preparing a plan for assigning priorities to groundwater monitoring and developing a complete groundwater data file. Septic tanks, particularly those used by industries, are also regulated.

Groundwater protection is an important element in the planning program of the Department of Environmental Protection and is required by NRS 445.257.

Nevada has long recognized the need for protecting and preserving its limited groundwater resources for the highest possible beneficial use. The policy has always been one of non-degradation. If a discharge is necessary and if the receiving-water quality is better than the standards for drinking water, then the discharge must not lower the quality of the receiving water below the standards for drinking water. If the quality of the receiving body of water is worse than the drinking water standards, then any discharge must not result in further degradation.

The management plan has many short term goals that for the most part can be taken care of under existing laws and regulations; it also has a series of long term goals. The short term goals are to: (1) develop and seek primacy for the underground-injection control program from EPA, (2) develop a program specific to Nevada's needs for underground storage tanks and seek primacy from the federal government, (3) develop memoranda of agreement with appropriate local, state, and federal agencies, (4) evaluate data base needs and the existing data processing program, (5) develop additional septic tank regulations, (6) develop mining regulations, and (7) continue special studies where appropriate.

The long term goals are to: (1) develop a groundwater monitoring network and data base, (2) develop groundwater protection plans for high priority basins, (3) evaluate a groundwater classification system, and (4) seek legislative authority to implement a more comprehensive regulatory program for mining and hazardous product usage and establish a cleanup fund financed by penalties.

Nevada is now seeking to comply with the requirements of RCRA and CERCLA (Nevada Department of Conservation and Natural Resources, 1986).

New Hampshire

New Hampshire manages its groundwater through a series of rules that set forth the policy of the New Hampshire Water Supply and Pollution Control Commission for the protection of groundwater and establish criteria for issuance of permits to discharge or dispose of wastes that may significantly and adversely affect the state's groundwater. The purpose of these rules is to protect groundwaters as a potential source of drinking water. The permits are required for any industrial, municipal, or privately owned waste and sewage facility or for agricultural operations where groundwater degradation is likely to occur.

The maximum contaminant levels (MCLs) adopted by the Commission as drinking water regulations Ws 200–299 shall be the criteria on which the Commission issues groundwater permits. As stated above, groundwater quality shall not be altered in any way which would make it unsuitable for use as a source of drinking water except as follows. The criteria set forth above concerning quality apply beyond the boundaries of the permittee's site, where the site shall include the real property owned by the permittee and operated for the purpose stipulated in the permit along with appropriate easements which abut the real property unless otherwise stipulated in the permit. Under no condition shall the pollutant violate the provision that groundwater must be suitable for drinking water.

The compliance boundary shall be located at one of the following, whichever is the nearest: a) the property boundary; b) five hundred feet from the pollutant source or approved outer edge of waste material; c) a distance specified in the permit depending on the conditions contained in Ws 410.14 which contains the conditions under which groundwater quality for drinking water must be maintained.

Any waste disposal facility must have groundwater monitoring. The requirements for the number, location, and depths of the groundwater monitoring wells shall be dependent upon the geological location of the facility and the actual distance of the compliance boundary as well as the conditions outlined above. Groundwater monitoring within the limited degradation zone and the intervention zone shall be designed to track the occurrence and movement of contaminants (New Hampshire Code of Administrative Rules, 1982).

446   Three agencies are most involved with groundwater activities in New Hampshire: the Water Supply and Pollution Control Commission, the Division of Public Health Services (Department of Health and Welfare), and the Water Resources Board. Responsibility for the protection of ground and surface waters from contamination and for ensuring that the quality of water delivered for public consumption meets minimum safety standards and is tested regularly is generally divided between the Water Supply and Pollution Control Commission and the Division of Public Health Services.

The Water Resources Board, which works with the U.S. Geological Survey, is mainly responsible for determining the availability of water. Water consumption is determined through registration and reporting of water users.

The Council on Resource and Development, composed of members of state agencies, is chaired by the director of State Planning, who adjudicates disagreements among member agencies. Management of groundwater is accomplished by several sections of the Revised Statutes Annotated (RSA). RSA-IV authorizes the New Hampshire Office of State Planning to undertake statewide water-resource planning.

The New Hampshire Water Supply and Pollution Control Commission administers surface and groundwater quality protection through programs set forth in RSA 131, 148, and 149. Under RSA 149:8, III(a), the commission has established a permit program for discharge or disposal of waste that may significantly and adversely affect the groundwater of the state.

Under RSA 147-A, 147-B, and 149-M, the Office of Waste Management within the New Hampshire Division of Public Health Services protects groundwater and monitors its condition. Management and disposal of radioactive waste is authorized by RSA-125, 56-77K.

The New Hampshire Water Resources Board is authorized and directed to investigate groundwater resources in cooperation with the USGS by Chapter 376 of Laws of 1955. RSA-489-B established the Water Well Board to license water well contractors and pump installers and to obtain data on all new well construction through a reporting procedure.

Regulation of commercial excavation of earth by local governments is under a permit system set forth in RSA-155-E (Chapter 481 of 1979 Session Laws). This law prohibits the granting of permits under conditions that would significantly damage sand and gravel aquifers.

Chapter 402 of the 1983 Session Laws amended RSA-481.1 to declare that the groundwaters of the state are an integral part of the overall water resources and that such groundwater resources must be conserved and protected, allocated, and otherwise managed to ensure uses most favorable to the public. Under Chapter 402 the Water Resources Board is authorized to ascertain water use through registration of, and reporting by, water users and to develop and recommend to the General Court policies and a water-resource-management plan to determine priority water uses and an allocation plan to preserve, distribute, and otherwise manage the water resources of the state, including its groundwater.

447

The use of land is a significant factor that affects groundwater quantity and quality. The New Hampshire Office of State Planning and Regional County Commissions provide planning, assistance to municipalities, and technical aid in developing local groundwater management and protection programs.

The New Hampshire Department of Agriculture regulates the use of fertilizers and, in conjunction with the Pesticide Control Board, the use of pesticides and herbicides. The Department of Resources and Economic Development ensures compliance with laws governing forest practices and mining. The state geologists provide geological assistance, including mapping, to resource investigators. The Public Utilities Commission grants public utility rights to supply water in specific service areas after consideration of the source and adequacy of the supply (USGS, 1985).

New Jersey

New Jersey's groundwater-management strategy is the responsibility of the New Jersey Department of Environmental Protection. The program is divided into resource evaluation, quantity production, and quality protection. Resource evaluation consists of assessing and evaluating the state's groundwater resources. The evaluation ranges from description of surface and subsurface geology to analysis of aquifer water quality and recharge rates. Groundwater-resource management strategy is usually developed by the Division of Water Resources. Both the USGS and the New Jersey Geological Survey provide support for preparing management strategies through the development of the necessary groundwater data base. The New Jersey Department of Environmental Protection (1986) has outlined a strategy for collecting groundwater-quality data that emphasizes greater data coordination between data-collecting agencies. Significant groundwater evaluation projects under way include a statewide aquifer mapping and assessment project to be completed in 1990, and the preparation of revised groundwater-quality and quantity-protection standards. The management of New Jersey's groundwaters originated when the Water Supply Commission was created in 1907 to control all public supplies. Today, under the auspices of the Water Supply Management Act (NJSA: 58.1a-1-17), the department's Bureau of Water Allocation requires water diversion permits for all withdrawals of 100,000 gallons per day or more and well permits for all new public and domestic wells. The state's Water Supply Management Act of 1981 serves as the basis for the General Water Supply Management Regulation (NJAC 7:19), which provides for the establishment of water supply critical areas. In these critical areas, severe water supply problems exist. To alleviate these problems, the state is empowered to exercise regional water management controls, not applicable to other areas. In critical areas, regulations exist for the reduction of existing diversion allocations when alternative supplies become available, the promotion of water conservation, and the development of surface or alternative supplies. A determination of safe yields for aquifers is also required.

The Department of Environmental Protection issues permits for groundwater discharges that include surface impoundments, infiltration/percolation lagoons, landfills, injection wells, spray irrigation, and overland flow and land application of residuals for hazardous and non-hazardous wastes. Steps necessary for obtaining a permit range from preapplication conferences and application reviews to public notices and hearings. In addition, all permitted facilities must perform routine discharge in aquifer (upgradient and downgradient) monitoring. Final permits are issued for five years.

The Bureau of Groundwater Quality Management has been revising the state standards for installation and construction of on-site disposal systems. Leaking underground storage tanks are a major contributor to groundwater pollution and well contamination in the state, and underground storage tanks are in the process of being registered. All storage tanks containing hazardous substances and petroleum (1,100 gallons or more) products were required to be registered by May 1986. The Division of Water Resources enforcement element has consistently initiated private remedial actions pursuant to the New Jersey Water Pollution Control Act. This is the lead agency for 12 of the state's 97 sites on the Superfund National Priority List. If the efforts of the agency in these cases are successful, then public funds will not have to be utilized for cleanup (New Jersey Department of Environmental Protection, 1986).

449

The following statutes form the basis for New Jersey's groundwater programs: Department of Environmental Protection Statute NJSA 13:1d-1 *et seq.:* Water Pollution Control Act NJSA 58:10a-1 *et seq.:* Water Quality Planning Act; NJSA 58: 11a-1 *et seq.:* Spill Compensation and Control Act NJSA 58:10-23.11 *et seq.:* Water Supply Management Act NJSA 58:1a-1-17: Safe Drinking Water Act: NJSA 58:12a-1 *et seq.* (New Jersey Department of Environmental Protection, 1983).

The New Jersey Department of Environmental Protection, Division of Water Resources, is the primary agency responsible for managing and regulating water resources in the state. The New Jersey Water Supply Management Act (1981), the Water Supply Bond Act (1981) and the New Jersey Water Supply Authority Act (1981) are elements of the state program to protect and manage groundwater resources. Every five years the State Water Supply Plan is updated by the New Jersey Department of Environmental Protection, Division of Water Resources. The aquifer classification system is described in Chapter 11.

The Water Supply Bond Act provides a water supply fund of $350 million for planning, designing, acquiring, and constructing water supply facilities. A referendum in 1983 changed the Water Supply Bond Act, allowing the bond funds to be used for groundwater studies that do not involve construction (USGS, 1985).

The Assembly Agriculture and Environmental Committee substitute for Assembly Bill No. 1231 imposes a condition on the closure, sale, or transfer of certain properties associated with the manufacturing, refining, transportation, treatment, storage, handling, or disposal of hazardous substances or wastes. The precondition is the execution of an approved cleanup plan which details the measures necessary to detoxify the property or the approval of the Department of Environmental Protection of a declaration that there has been no discharge of hazardous substances or wastes on the property or that any such discharge has been cleaned up in accordance with procedures approved by the department and there remains no hazardous substance or waste on the property. The Department of Environmental Protection is directed to establish minimum standards for soil and water quality which would constitute detoxification, taking into

450 consideration the location of the property and the surrounding conditions. The bill further requires the owners of hazardous substances or waste operations to obtain a surety bond or other financial security which would guarantee the implementation of the cleanup plan. The bill also permits the purchaser or transferee to assume the responsibility to detoxify the property or to continue the operation in essentially the same manner and defer the implementation of the cleanup plan. Violation of any provision of the bill will constitute grounds for avoiding the sale or transfer of real property or business by the transferee, and failure to submit a negative declaration or cleanup plan would constitute grounds voiding the sale or transfer by the department. Furthermore, any person knowingly falsifying information required by the bill would be subject to a penalty of not more than $25,000 for such an offense and a strict liability for all cleanup and removal costs.

The Senate Energy and Environmental Committee amended the bill to correct certain technical and procedural deficiencies of the bill; to require the department to review, approve, or disapprove negative declarations and cleanup plans on a case by case basis prior to the adoption of minimum standards to amend the New Jersey "Spill Compensation and Control Fund" created pursuant to P.L. 1976, c. 141 (C. 5810-23.11 *et seq.*), to permit the expenditure of up to $4,000 annually to administer the bill and to appropriate the first installment.

New Mexico

The first laws regarding water use in New Mexico were established by the New Mexico Territorial Legislature in 1851. In 1931 the state legislature imposed a permit system for the appropriation of groundwater. Groundwater use in New Mexico is regulated by the New Mexico state engineers.

Areas have been designated in which appropriation of additional groundwater is allowed only by permit. At present 31 such areas are designated as declared underground water basins. This represents about 69% of the total area of the state. The basic authority for groundwater quality protection lies with the Water Quality Control Commission, of which the state engineers are members.

Primary responsibility for enforcing the commission's regulations have been delegated by the commission to the New Mexico Environmental Improvement Division. The New Mexico state engineers and the USGS collect groundwater data and conduct cooperative investigations of groundwater resources throughout the state (USGS, 1985).

451

Regulations are implemented to protect all groundwaters of the state of New Mexico that have existing concentrations of 10,000 mg/l or less of TDS. New Mexico is particularly concerned about injection wells and extraction from wells. Therefore, it has a system whereby a plan must be submitted. As to the type of well, how it is to be constructed and what wastes are to be disposed in it. These wells are then monitored. If they are abandoned, they must comply with specific specifications (New Mexico Water Quality Control Commission, 1983).

New Mexico has a two-class system based on total dissolved solids. Groundwater with a TDS content greater than 10,000 mg/l is considered non-potable, and water with a TDS content of less than 10,000 mg/l is protected for potable uses through regulations (New Mexico Water Quality Control Commission, 1983).

New Mexico's management strategy mainly is concerned with the injection of wastes into wells, extraction from wells, and well materials. Anyone wishing to construct such a well must submit a plan. The construction of such wells is regulated, as are the materials to be put into the wells, the casing of the wells, and the monitoring of the wells. In addition, when wells are abandoned, certain procedures are required. Management of these various types of wells currently represents New Mexico's main effort to control groundwater contamination (New Mexico Water Quality Control Commission, 1983).

New York

The two state agencies most directly responsible for groundwater management are the New York State Department of Health (DOH) and the New York State Department of Environmental Conservation (DEC).

Under the Public Health Law and Part 5 of the State Sanitary Code, DOH ensures that public water supply systems are operated and maintained properly to ensure a safe and adequate supply. The program involves regulation, periodic monitoring of water quality, inspection of systems, emergency response to problems, laboratory services, and establishment of drinking water standards. The major elements of the Department of Environmental Conservation water program that affect groundwater management include water-resource planning, ambient water quality standards and classification of groundwater, water discharge permits, and programs that provide for development, operation, and maintenance of municipal wastewater facilities.

452

The Department of Environmental Conservation established a system of groundwater classification and standards in 1967. The most recent revision was made in 1978.

The New York State Pollution Discharge Elimination System Program, administered by DEC, regulates subsurface discharges from point-source municipal, industrial, and commercial facilities.

The State Public Water Supply Permit Program requires that new groundwater withdrawals for public supply be approved by both the Department of Environmental Conservation and the Department of Health (USGS, 1985).

Upper New York State

Groundwater in upper New York State and Long Island is governed by the same agencies. However, the problems in Long Island are much more severe than in upper New York State and, hence, there is a much more detailed management plan. At present, the upper part of New York State is identifying problems of quality and quantity.

The main quality problems are organic solvents, degreasers, and petroleum products. Some 24 upper New York public water supply wells have been closed because of organic contamination. In addition, there have been numerous instances of well contamination by gasoline and petroleum products. Recent findings indicate that pesticides could also be a threat in upstate New York. High nitrate levels have been found in upstate wells in a few locations. This contamination is related to fertilizer use and on-lot septic tanks. It is not a wide-spread problem in upper New York but may be a problem in some specific areas. At the present time, there is no major groundwater quantity problem in New York State.

The water resources issues in upstate New York, which are sometimes referred to as quantity issues, show a need for rehabilitating the water supply infrastructure and the need to plan adequately for drought. These problems relate primarily to the adequacy of water supply facilities and water supply management, rather than the size of the water resource.

Long Island
Long Island has a detailed groundwater management program. The elements of the program are grouped into seven major categories: groundwater source management, groundwater quality protection programs, groundwater quantity management programs, zoning and land development controls, response and remediation programs, regulatory enforcement, and public education and participation. In order for the program needs to be met, the following must be addressed:

453

1. Five key regulatory agencies (New York State Departments of Environmental Conservation and Health, New York City Department of Health, Nassau County Department of Health, and Suffolk County Department of Health Services) must assume primary responsibility for achieving coordinated oversight of groundwater management.
2. A program assessment mechanism must be developed to assist in the evaluation of the program's performance and the conditions of the groundwater resource.
3. Essential interagency coordination among all agencies at all levels of government must be achieved. The resource management framework must consider hydrogeologic zones, groundwater flow regimes, and special management areas.

Fundamentals of groundwater quality management programs are groundwater quality standards and classifications. The groundwater classification of New York State (established by DEC in 1967, revised in 1978) is similar to that of other states.

Class GA. This class is defined as:

1. Best usage of Class GA waters is as a source of potable water supply.

2. The quality standards for Class GA waters shall be the most stringent of the items and specifications applicable to such waters as noted. The maximum contaminant level for drinking water promogated by the Commission of Health is found in 10 NYCRR SUBPART 5-1, Public Water Supplies or any subsequent revision thereof or replacement thereof. You should note: (a) the items and specifications; (b) the maximum contaminant level; (c) the maximum contaminant levels for drinking water promolgated by the administrator under the Safe Drinking Water Act (P.L. 93-523) and

(d) the standards for raw water quality promolgated by the Commissioner of Health as found in 10NYCRR part 170.

Class GSA. The best usage of Class GSA waters is as a source of potable mineral waters, for conversion to fresh potable waters or as raw material for manufacture of sodium chloride or its derivatives or similar products. Such waters are saline waters found in the saturated zone. The following quality standards shall be applicable to Class GSA waters:

454

1. Sewage industrial wastes or other waste; color, taste, or odor producing substances; toxic pollutants; thermal discharge; radioactive substances; or other deleterious matters may not be placed in waters to impair the use of such saline waters for the best usage outlined above or so as to cause or contribute to a condition in contravention of standards for other classified waters of the state.

Class GSB. The best usage of Class GSB waters is as a receiving water for disposal of wastes. Such waters are those saline waters found in the saturated zone which have a chloride concentration in excess of 1,000 mg/l or a total dissolved solids concentration in excess of 20,000 mg/l. Sewage, industrial wastes, or other wastes; color, taste, or odor producing substances; toxic pollutants; thermal discharges; radioactive substances; or other deleterious matter may not be placed into the groundwater that are deleterious, harmful, detrimental or injurious to public health, safety, or welfare or which may cause or contribute to a condition in contravention of standards for other classified waters of the state.

Class GSB shall not be assigned to any groundwaters of the state unless the commissioners find that adjacent tributary groundwaters and the best usage thereof will not be impaired by such classification (New York State Department of Environmental Conservation, 1978, 1983).

It is important that the Department of Environmental Conservation and the Department of Health proceed with all deliberate speed to establish numerical guidelines and/or standards for industrial chemicals commonly used in Long Island and posing a threat to groundwater quality.

Monitoring is a fundamental element of the groundwater program. It ensures both protection of public health and the confirmation of standards. It must also ensure that groundwater withdrawals, recharge, and related activities do not endanger the value of the groundwater resource. Furthermore, such monitoring is able to track down sources of contamination and is therefore fundamental to this program.

A groundwater information system is very important to a groundwater resource management program. Long Island has a massive groundwater management related data base that is probably as extensive as any. Important categories of information include data on the aquifer system itself, water quality, well pumping, important sources of groundwater contamination (particularly those regulated by permit), and others. It is important that this data be stored in such a way that it is readily accessible. The important issue is that the interagency sharing of pertinent data does not adversely affect the necessary integrity of data files.

455

The environmental review processes are in place. They include the requirements of the National Environmental Policy Act (NEPA), the New York State Environmental Quality Review Act (SEQRA), and sole source aquifer provisions of the federal Safe Drinking Water Act.

A very substantial body of areawide planning is already in place for groundwater related programs on Long Island. The primary future emphasis should be on carrying out operational programs as identified in this report, rather than on extensive additional planning. There is a strong need to fill a program gap that exists in the control of hazardous materials that threaten the environment generally and the groundwater source in particular; this is the problem of storage and handling of hazardous materials. In groundwater sensitive areas, there should be instituted consolidated and comprehensive regulatory surveillance of large, highly complex individual industrial facilities and cluster of mixed industrial/commercial facilities.

Groundwater quality protection programs include the following: Hazardous material storage and handling, industrial/commercial waste and hazardous waste, municipal solid waste management.

There is an ongoing need for sound, verifiable data on the quantities, types, and geographical point of generation of hazardous wastes on Long Island. There is a need to evaluate and where necessary remedy existing sites which were used in the past for the disposal of hazardous wastes. There is a need to regionalize the Part 364 Industrial Waste Transportation Permit Program to increase the Department of Environmental Conservation's regional supervision of these waste haulers. There is a need for program resources to increase the inspection, sampling, and effective supervision of facilities regulated within this program area, and to vigorously enforce cases of significant violation.

Municipal solid waste management is a very large problem in

Long Island, and a difficult one. The overall needs for solid waste management include the following: Terminate land disposal of solid wastes in the deep flow recharge areas as soon as possible, in conformance with the Long Island Landfill law. Make resource recovery a more viable management program to serve as the major longterm solution for municipal solid wastes from urbanized areas and deep flow recharge areas. Develop a solid waste management plan for Long Island as a whole. Provide increased program resources to insure the implementation of existing programs to prevent groundwater contamination.

456

The control or elimination of municipal, industrial, and commercial wastewater discharge is the cornerstone of all program efforts to prevent groundwater contamination from these sources.

There are three areas of the program that need special emphasis. First, there must be a shift from the conventional emphasis on surface waters for controlling toxics. Second, the State Pollution Discharge Elimination System (SPDES) must correct deficiencies in the program's protection of groundwater resources. Third, there is a shortfall in the resource base.

The program has certain deficiencies. It has an enforcement policy that places too much emphasis on voluntary compliance and negotiated settlements; therefore, there is poor control on the quality of program data, lack of specific written standards and criteria for the establishment of permit conditions, and lack of program consistency from region to region across the state. More resources must be devoted to these efforts.

On-site sanitary waste disposal is a major problem in Suffolk County. If these systems are not controlled, they can cause unacceptable groundwater quality degradation. Limiting the numbers or the density of on-site systems is necessary; sewering is important in some cases.

In order to protect groundwater from contamination by pesticides and fertilizers, there must be a vigorous program to reduce the use of agricultural chemicals and fertilizers.

There is a special need to protect the groundwater recharge areas in order to maintain high quality, uncontaminated, deep-flowing aquifers (e.g., the Long Island Pine Barrens in Eastern Suffolk County).

The Long Island groundwater quantity management program is mainly controlled by a well permit program that regulates the withdrawal quantities in various regions.

Groundwater flooding is a problem in certain areas, including

Nassau and Suffolk counties. New York State programs that address flooding are tied closely to the applicable federal programs and legislation under the Environmental Conservation Law, Article 16. The Department of Environmental Conservation cooperates with the U.S. Army Corps of Engineers in the development of federally constructed surface water flood control projects.

Water conservation is being managed through a series of measures which in turn effect ground water withdrawals.

457

Proper management of water recharge, which takes advantage of suitable opportunities to recharge water to the aquifer system to the extent feasible, can help to mitigate withdrawal stresses on the groundwater system.

Other important parts of the management program of Long Island groundwater are zoning and land development controls, and response and remediation programs (e.g., spill cleanup, management of contaminated aquifer segments, wellhead treatment, water main extension, water importation, water-quality-treatment districts, and sewering).

As with any management program, enforcement is one of the most important aspects if the program is to be effective (New York State Department of Environmental Conservation, 1986a).

North Carolina

The North Carolina Department of Natural Resources and Community Development implements most of the regulatory and planning procedures related to groundwater resources in the state. Its Division of Environmental Management has the major responsibility for groundwater management and regulatory programs. The Environmental Management Commission has authority over the permitting process and has made the Groundwater Section of the Division of Environmental Management directly responsible for issuing permits for well construction and groundwater withdrawals.

The Commission may designate an area as a capacity-use area whenever it believes the renewal and replenishment of groundwater supplies to be threatened. To date, one such area has been established in east-central North Carolina; however, other areas are being considered.

A permit must be obtained from the Groundwater Section of the Division of Environmental Management for (1) construction of public supply, industrial, and irrigation wells; (2) wells with a designated capacity of 100,000 gallons per day or greater; (3) wells to be used for injection, recharge, or disposal purposes; and (4) a well other than a domestic well located in a designated capacity-use area (North Carolina Well Construction Act of 1967, Article 7-87-88).

458 Injection wells for waste disposal purposes are currently prohibited by state statute. All well drillers must register annually with the North Carolina Department of Natural Resources and Community Development and are required to report all well completions and abandonments.

In addition to water use permits in capacity-use areas for users withdrawing more than 100,000 gallons per day, the North Carolina Department of Natural Resources and Community Development also may require these users to adhere to established maximum withdrawal rates. The agency also can establish minimum water levels resulting from pumping in certain areas.

The Division of Water Resources of the North Carolina Department of Natural Resources and Community Development collects data on the use of groundwater statewide through its water use data program. This division also provides technical assistance to local government water utilities in considering groundwater as a source of supply for the public water system. Technical information on groundwater is also available through seven regional offices of the Department of Environmental Management.

The Department of Human Resources, through its Division of Health Services, has the responsibility for monitoring solid and hazardous waste disposal sites to prevent the contamination of groundwater supplies. It oversees the human health aspects of the public water system.

The North Carolina Department of Natural Resources and Community Development, in conjunction with the U.S. Geological Survey, supports individual and cooperative groundwater research, data collection, and project investigations.

The management of state groundwaters is by a classification of aquifers according to their most important uses and standards governing the maximum amount of pollutants (see Chapter 11). These were established by Statutory Authority, G.S. 143-214 1 and 143-214, 2. (North Carolina Department of Natural Resources, 1985).

North Dakota

North Dakota is in the process of developing its groundwater strategy. Its current management is largely the responsibility of four agencies. The State Water Commission has control of water allocation, groundwater monitoring, and identification of groundwater resources. The Industrial Oil and Gas Commission is in charge of the protection of groundwater associated with oil and gas development. The North Dakota Geological Survey is **459** responsible for protecting groundwater from mineral and geothermal exploration and development. The Public Health Service Commission has the responsibility for groundwater protection and monitoring associated with coal mine development and reclamation (North Dakota Department of Health, 1986b).

The North Dakota water laws do not recognize the doctrine of water ownership or the right to use the water as inherent with ownership of land. Rather, the right to use the water is based on a first-in-time, first-in-right concept and has added qualifications that the use be beneficial.

Chapter 61-01-01, Item 2, of the North Dakota Century Code states, "Water under the surface of the earth, whether such waters flow in defined subterranean channels or are diffused percolating underground waters, belong to the public and are subject to appropriation for beneficial use and the right to use of these waters for such use shall be acquired pursuant to the provisions of Chapter 61-04." In this chapter, the appropriation of water and procedures for acquiring water use permits is described. At present, water use permits are required only for public supply and for irrigation and industrial purposes. Permits are not required for domestic, livestock, or fish and wildlife purposes unless the annual appropriation exceeds 12.5 acre feet. Such permits are issued by the state engineer.

North Dakota has a continuing program designed to ensure safe and orderly development of the state's groundwater resources. In the last 25 years groundwater resources have been identified and described on a county by county basis as part of a cooperative program involving each county of the state, the North Dakota State Water Commission, the North Dakota Geological Survey, and the USGS.

In those counties in which there are large areas of federally owned land, other federal agencies such as the U.S. Forest Service and the U.S. Bureau of Land Management are involved. Digital models of some of the larger and more intensively developed

Table 13-1.   North Dakota Statutes and Rules Governing Management of Groundwater Contamination Sources

Contamination Source	N.D. Century Code Chapter	Rules Under Statute (N.D. Administrative Code Chapter)
1. Waste impoundments	61-28	33-16-01, 33-16-03, Art.33-24
2. Contamination from surface water	61-28	33-16-03
3. Injection wells	38-08, 38-12, 61-28	43-02-05, 43-02-01.1, 33-25-01
4. Solid waste disposal	23-29, 23-20.3	Art.33-20, Art.33-24
5. Well construction	43-35, 38-08.1	33-18-01
6. Subsurface sewage disposal	61-28, 23-19	33-21-01
7. Land application of wastes	61-28, 23-29	33-20-05
8. Accidental spills	61-28, 23-20.3	33-24-04
9. Mining	38-14.1, 61-02	Art.69-05.2, 89-04-08
10. Oil and gas	38-08	43-02-03
11. Underground storage tanks	18-01, 23-20.3	10-07-01, 33-24-05
12. Agricultural chemical application—dry land	4-35	4-35-01
13. Agricultural irrigation	61-28, 61-04, 61-24	
14. Road salting application	61-28	
15. Other sources	61-28.1	33-17-01

Source: North Dakota State Department of Health, 1986

460

aquifers in the state have been developed by the North Dakota State Water Commission and the U.S. Geological Survey (USGS, 1985).

North Dakota recognizes 15 types of sources of groundwater contamination. The statutes and rules governing the management of these sources are listed in Table 13.1.

Ohio

The main state agencies involved in aquifer management in Ohio are the state EPA, the Department of Natural Resources, and the Ohio Department of Health. The Department of Transportation has some responsibilities regarding the design and maintenance of water wells at roadside rest areas. The most important need for a management program is: The Ohio EPA should develop more formalized criteria for siting municipal well fields to include consideration of local geologic conditions to protect the land area for well field expansion. More geologists are needed in EPA, the Department of Health, and the Department of Natural Resources, to assist in formulating a management plan.

The Ohio Environmental Protection Agency is responsible for **461** regulations to protect public water supplies. It issues permits to control waste discharges from public and industrial sources and to regulate landfills and other hazardous waste disposal operations that could affect groundwater resources. To accomplish this work, the agency performs geologic evaluations related to proposed and existing land disposal sites, investigates water well contamination complaints, provides hydrologic information to the general public and to the technical community, and maintains a semiannual water-quality monitoring program of selected wells in principal aquifers.

The Ohio Department of Health regulates drilling of private water wells used for drinking water through rules set forth in 1981. Permits are required and issued by the county boards of health in each of the state's 88 counties.

Since 1949 the Ohio Department of Natural Resources, Division of Water, has required that a copy of drilling records for any newly constructed or modified water well be filed with the division. The division offers recommendations for optimum development of groundwater resource for public supply. A statewide groundwater level monitoring program is conducted cooperatively with the U.S. Geological Survey and the Ohio Department of Natural Resources, Division of Water (USGS, 1985).

Ohio should consider consolidating all groundwater quality management functions under a distinct program within one agency, logically an EPA Division of Groundwater Quality. Functions relating to groundwater quantity, mapping, and aquifer definition should remain within the Ohio Department of Natural Resources, as should groundwater concerns related to strip mining. Responsibility for private water systems should remain with the Ohio Department of Health. If the existing legislative authority proves to be inadequate, new legislation should be passed to develop a groundwater quality management plan. Such a plan would express the intent and concern of the legislature to protect groundwater an would formalize this recognition and commitment. Included in the program should be requirements to develop technical standards for groundwater quality, monitoring requirements, and basic geologic siting criteria for land treatment, storage, and disposal facilities. Monitoring should apply uniformly to all land treatment, storage, and disposal facilities as well as all other major activities that may discharge into groundwater. Perameters, as well as the frequency and duration of monitoring, should be specified. Siting should be an important part of any such management program, particularly in selecting and evaluating any land treatment, storage, and disposal **462** facility and particularly for hazardous wastes. Criteria should be developed for engineering designs that give the characteristics for minimum siting requirements such as soil permeability, depth of bedrock, quality of groundwater, or distance from existing wells. General revenue funds must be appropriated, and alternate funding sources should be considered, including permit and licensing fees or a per-unit fee on all land treatment, storage, and disposal facilities. The data base should be coordinated and computerized to provide on-line access for all agencies. The agencies need to develop and communicate an understanding of the importance of the state's groundwater resources, and a supportive constituency must be developed in the legislature and among the public. The public must understand the importance of groundwater quality and must be part of the strategies for solving groundwater problems (Batelle, 1984).

Oklahoma

"Oklahoma's statutory system to regulate groundwater use underwent major revisions in 1972, and the current system of regulations consists of the 1972 statutory framework with some minor amendments since that date."

"Major features of the current Groundwater Law, codified as 82 O.S. Supplement 1981, 1020.1-1020.22, combine aspects of individual personal property ownership in groundwater and a regulatory aspect of responsible use and regulation of groundwater."

"The Oklahoma Water Resources Board has primary responsibility for regulatory and operational programs with regard to managing groundwater. As part of its responsibilities, the board manages a groundwater appropriation and permit program. Only domestic use is exempt from permit requirements. The board also administers water well drillers' licenses and an enforcement program. It conducts hydrological surveys of each fresh groundwater basin or sub-basin to determine the maximum annual yield" (Oklahoma Water Resources Board, 1985).

The state organizations involved in groundwater activities in support of the management process include the Environmental and Groundwater Institute of the University of Oklahoma, the Oklahoma Geological Survey, the Oklahoma Water Resources Board, and the Water Research Center at Oklahoma State University. The U.S. Geological Survey has participated in cooperative programs with the Oklahoma Geological Survey and the Oklahoma Water Resources Board for research, investigation, and data collection (USGS, 1985).

463

The policy of the State of Oklahoma concerning groundwater is as follows: It is declared to be the public policy of the state of Oklahoma—in the interest of agricultural stability, domestic, municipal, industrial, and other beneficial uses, general economy, and health and welfare of the State and its citizens—to utilize groundwater resources of the State, and for that purpose to provide reasonable regulations for the allocation for reasonable use based on hydrologic surveys of fresh groundwater basins or sub-basins to determine a restriction on the production based upon the acres overlying the groundwater basin or sub-basin (Oklahoma Water Resources Board, 1985).

Oregon

"Oregon law gives the director of the Oregon Water Resources Department the authority to issue permits to appropriate the state's groundwaters for beneficial uses. The director has the authority to take action to limit adverse impact, such as well interference with existing water rights and groundwater pollution, where joint-voluntary action among users is inadequate. The director also regulates licensing of water well drillers and establishes water-well construction criteria." The Oregon Water Resources Department cooperates with the U.S. Geological Survey in investigating the state's groundwater resources. "These activities include data collection, data analysis, and interpretive studies that together form an information base for groundwater resource planning and management."

"The Department of Environmental Quality is responsible for establishing and enforcing rules designed to prevent contamination of Oregon's groundwater resources" (USGS, 1985).

The following are the general policies as outlined in Oregon

Administrative Rules, Chapter 340, Division 41-029, Department of Environmental Quality, General Groundwater Protection Policy 340-41-029. The following statements of policy are intended to guide State Federal agencies and State agencies, cities, counties, industry, citizens, and the Department of Environmental Quality staff in their effort to protect the quality of groundwater.

General policies: a. It is the responsibility of EQC to regulate and control waste sources so that the impairment of the natural quality of groundwater is minimized and to assure beneficial uses of these resources by future generations. b. In order to assure maximum, reasonable protection of public health the public should be informed about groundwater, most particularly its local flow systems or water table aquifers. They should not assume them to be safe for domestic use unless quality testing demonstrates a safe supply. Domestic water drawn from an aquifer should be tested frequently to assure its continual safety. c. The Department will concentrate its control strategy development and implementation efforts in areas where disposal practices and activities regulated by the Department have the greatest potential for degrading groundwater qualities. d. The Department will seek the assistance and cooperation of the Water Resources Department to design an ambient monitoring program adequate to determine long term quality trends for significant groundwater flow systems. e. The EOC recognizes and supports the authority and responsibilities of the Water Resources Department and Water Policy Review Board in the management of groundwater and the protection of groundwater quality.

Source control policies: a. Consistent with general policies for protection of surface water, highest and best practical treatment and control of sewage, industrial wastes, and landfill leachates shall be required so as to minimize potential pollutant loading to groundwater.

b. Establishment of controls more stringent than those identified in Sub-section A of this section may be required by the EQC in situations where DEQ demonstrates that such controls are needed to assure protection for beneficial use. 1. The Water Resources Director declares a critical area for reasons of quality. 2. EPA designates a sole source aquifer pursuant to the Federal Drinking Water Act.

c. Less stringent controls than those identified in Sub-section A of this section may be approved by the EQC for a specific area if a request, including technical studies, showing that lesser controls will adequately protect beneficial uses, that this study is made by

464

representatives of the area and if the request is consistent with other state laws and regulations.

d. Disposal of wastes on or into the ground in a manner which allows potential movement to groundwater shall be authorized and regulated by existing rules of the Department of Water Pollution Control Facility permit, Solid Waste Disposal Facility permit, or On-site Sewage Disposal System Construction permit, whichever is appropriate. 1. WPCF shall specify appropriate groundwater quality protection requirements and monitoring and reporting requirements. 2. Solid Waste Disposal Facility permit shall be used for landfills and sludge disposal not covered by NNPDES or WPCF permits. 3. On-site sewage disposal system construction permit shall be issued in accordance with adopted rules.

e. In order to minimize groundwater quality degradation potentially resulting from non-point sources it is the policy of the EQC that activities associated with land and animal management, chemical application and handling, and spill prevention be conducted using the appropriate state-of-the-art management practices.

Problem abatement policies: a. It is the intent of EQC to see that groundwater problem abatement plans are developed in a timely fashion. To implement this they will utilize the following but will not be limited to these methods. They are: permits, special permit conditions, penalties, fines, commission orders, compliance schedules, moratoriums, department orders, and geographic rules. b. In areas where groundwater quality is being degraded as a result of existing individual source activities or waste disposal practices the Department may establish the necessary control and abatement schedule requirements. These will be implemented through the existing permit authorities, department orders, or commission orders issued pursuant to ORS Chapter 183. c. In urban areas where groundwater is being degraded as a result of on-site sewage disposal practices and an area-wide solution is necessary, the Department may propose a rule for adoption by the Commission and incorporation into the basin section of the State Water Quality Management (OAR Division 41). d. The Department shall notify all known impacted or potentially affected units of government of the opportunity to comment on the proposed rule at a scheduled public hearing and of their right to request a contested case hearing pursuant to ORS Chapter 183 prior to the Commission's final order adopting the rule. (Oregon Department of Environmental Quality, 1985; Oregon Administrative Rules, 1984).

The classification system recommended has 3 classes: Class I,

Unique Aquifer; Class II, Drinking Water Aquifer; Class III, Managed Aquifer.

Pennsylvania

"The Pennsylvania Department of Environmental Resources is the state agency responsible for developing water management policies and practices. A comprehensive "State Water Plan" forms the basis for water management in the state. Several offices and bureaus within the department conduct hydrologic studies of groundwater resources independently and in cooperation with the USGS. Existing state statutes and regulatory programs do not comprehensively address the allocation of groundwater or provide for long-term management of groundwater resources. At present, Pennsylvania's statutes provide that each adjoining landowner has an equal and correlative right to make reasonable use of groundwater below his or her land."

"Two statutes focus on groundwater aspects of water resource management: The Water Well Drillers License Act and the Clean Streams Law. The Water Well Drillers License Act is essentially a driller registration program. The Clean Streams Law is designed to control and prevent pollution of all state waters." It includes springs and underground waters and prohibits the discharge of sewage or industrial wastes unless authorized by permit and done in accordance with regulations adopted by the Pennsylvania Department of Environmental Resources. The Delaware River Basin Commission and the Susquehanna River Basin Commission, which are interstate compacts, play an increasingly important role in managing the groundwaters in the eastern two-thirds of Pennsylvania. These two agencies require approval of proposed groundwater activities that may have "substantial effect" on basin waters to ensure consistency with commission-adopted comprehensive plans and to ensure "proper conservation, development, management or control of water resources of the basins."

Both of these agencies usually limit their review to projects involving withdrawals exceeding 100,000 gallons per day. The Delaware River Basin Commission (DRBC) has exercised authority through the designation of groundwater protected areas in southeastern Pennsylvania, particularly during the 1960s and 1980 and 1981.

The Groundwater Protective Program is intended to improve

management of groundwater in a 1,500-square-mile section of predominantly Triassic lowland formations in southeastern Pennsylvania. The protected area comprises all or portions of Montgomery, Bucks, Chester, Berks, and Lehigh counties. Within the designated areas the groundwater withdrawals are carefully regulated to accomplish the most effective, long-term utilization of the resource. Under the DRBC regulations, any new withdrawal or increase in withdrawal from an existing well of 10,000 gallons per day or more requires a DRBC permit.

467

Owners of existing or proposed wells from which withdrawals of more than 10,000 gallons per day are expected must consult with the DRBC at least one month before exploratory drilling and must submit hydrological reports as part of the DRBC permit-application process (USGS, 1985).

The management of Pennsylvania waters is described above. Pennsylvania has an extensive permitting system for the discharge of wastes, clearly defining concentrations and amounts that can be discharged. It also has a well developed water quality criteria program that sets forth the concentrations of organisms, metals, organic substances, etc., that may be in discharged wastes. In addition, it has ambient water quality standards. Most of these apply more directly to surface waters, but it is implied that the same standards apply to groundwater.

Rhode Island

The General Assembly set forth the following policies regarding groundwater: 1. The groundwaters of the state are a critical, renewable resource which must be protected to insure the availability of safe and potable drinking water for present and future needs. 2. It is a paramount policy of the state to protect the purity of present and future drinking water supplies by protecting aquifers, recharge areas, and watersheds. 3. It is the policy of the State to restore and maintain the quality of groundwater to a quality consistent with its use for drinking supplies and other designated beneficial use without treatment facilities. 4. All groundwater of the state shall be restored to the extent practical to a quality consistent with this policy. 5. It is a policy of the state not to permit the introduction of pollutants into the groundwater of the state in concentrations which are known to be toxic, carcinogenic, mutagenic, or teratogenic. To the maximum extent practical efforts shall be made to require the removal of such pollutants from discharges where such discharges are shown to have already occurred. 6. The existing and potential sources of groundwater shall be maintained and protected where existing qualities and adequate to support certain uses, such quality shall be upgraded if feasible, to protect the present potential uses of the resources. 7. The groundwaters of the state are to be protected for the use as agriculture, industrial, and potable water supplies and other reasonable uses; and as a supplement to surface water for recreation, wildlife, fish and other aquatic life, agriculture, industry and potable water supplies. 8. Discharges to groundwater which substantiall discharge into surface waters, and which will cause a contravention of surface water quality or standards shall not be permitted. 9. No degradation of the state's groundwater shall be permitted unless the state chooses to allow lower water quality as a result of essential, desirable and justifiable economic, commercial, industrial or social development" (Rhode Island Groundwater Protection Act, 46-13.1-1 to 46-13.1-4).

468

The groundwater classification program in Rhode Island is as follows. *GAA:* Groundwater sources which without treatment are suitable for public drinking water use. *GA:* Groundwater sources that may be suitable for public or private drinking water without treatment. *GB:* Groundwater sources that may not be suitable for public or private drinking water without treatment due to known or presumed degradation. *GC:* Groundwater which may be suitable for waste disposal practices because of past or present land use or hydrological conditions which make the groundwater more suitable for receiving permitted discharges than for development as public or private water supplies. It is the duty of a Director of a State Department of Environmental Management to develop water quality standards for each classification with specific maximum levels of contaminants for each classification.

In Rhode Island, groundwater management and planning is the responsibility of several state agencies. The Rhode Island Statewide Planning Program prepares and updates policies in regards to development, management, and protection of groundwater resources, but does not have legal authority to enforce them.

The Rhode Island Department of Health is required by statute (Rhode Island General Laws 46-13-1. *et seq.*) to ensure the quality of water in public supplies, including groundwater. There are more than 400 public water supplies that serve 25 people or more for 60 days or more during the year. The Department of Health is also authorized to order the abatement of pollution that poses a threat to public supply.

The Rhode Island Water Resources Board and the Rhode Island Department of Environmental Management have the principal legal authority for development, planning, management, and protection of the quality of groundwater in Rhode Island. The Rhode Island Water Resources Board is charged under Rhode Island General Laws (46-15-1, *et seq.*) to formulate and implement long range plans for the development of water resources, including groundwater, needed for public water supply. Under this statute, plans by public-supply systems for acquiring additional groundwater supplies from new sources must be approved by the Rhode Island Water Resources Board.

469

Groundwater withdrawals, other than those for public supply, are not regulated. This statute also authorizes registration of well drillers by the Water Resources Board, which requires drillers to submit well-completion reports. Provisions of Chapter 46-15 of this law empower the Water Resources Board to function as a steward of all state water resources and to develop policies controlling allocation, interbasin transfers, and conservation of water resources.

The regulation of waste discharges to groundwater is the responsibility of the Department of Environmental Management under Rhode Island General Law (46-12-1, *et seq.*). The Department of Environmental Management is authorized to classify groundwater and surface waters and to establish rules and regulations for the protection of both. A classification system and rules and regulations for protection of surface water are in place (see Chapter 11). A comprehensive strategy for protecting groundwater quality was being developed in 1984 by the Department of Environmental Management.

Hydrogeological data is obtained by the U.S. Geological Survey in cooperation with the Department of Environmental Management and the Rhode Island Water Resources Board (USGS, 1985).

South Carolina

The South Carolina Department of Health and Environmental Control and the South Carolina Water Resources Commission are responsible for protecting the quality of the state's groundwater resources. Programs include review and permitting of all public supply wells for proper design and construction; regulation of the water well drilling industry to ensure compliance with minimum well construction standards; and regulation of all sites of potential groundwater contamination, such as pits, ponds, lagoons, feedlots, and injection wells, to ensure compliance with proper monitoring and cleanup activities.

The Water Resources Commission water-management program is authorized by the Groundwater Use Act of 1969. This program is designed to protect aquifers in designated areas by regulating the design, construction, spacing, and abandonment of wells to protect the aquifers from saltwater intrusion and overpumping.

470

All groundwater users that withdraw more than 100,000 gallons per day must obtain a permit from the commission and must report monthly water use on a quarterly basis. Under the act, the commission is authorized to regulate groundwater withdrawals within the capacity-use areas. The program is designed primarily to minimize the effect of intense localized pumping. Groundwater data and technical assistance are provided to groundwater users by the U.S. Geological Survey in cooperation with the South Carolina Department of Health and Environmental Control and the South Carolina Water Resources Commission (USGS, 1985).

All groundwater of the state is classified GB (S. C. Pollution Control Act 48-1-10 *et seq.*, S. C. Code of Laws, 1976) for use as drinking water without treatment. Class GA is established for future classification as exceptional water, and Class GC for future classification as having little potential use as drinking water (South Carolina Department of Health and Environmental Control, 1985).

South Dakota

The management of the state's groundwater resources is accomplished through a water record and permit system and the State Water Plan, administered by the South Dakota Department of Water and Natural Resources.

The Office of Water Policy, within the Department of Water and Natural Resources, provides the technical policy analyses needed to implement the State Water Plan. The Division of Geological Survey is charged with studying and mapping the groundwater resources of the state. The Division of Water Development has the responsibility to coordinate development and management of South Dakota's water resources for maximum public benefit. The Division of Water Quality reviews groundwater quality data to determine if contamination is occurring or if additional legal authority is required to protect the quality of groundwater. The Division of Water Rights is charged with licensing and other functions concerned with regulation and management of the waters of the state, including groundwater. Although state law does not allow withdrawals from an aquifer to exceed the average annual recharge, it does not regulate the effects of pumping on flowing wells (USGS, 1985).

471

Tennessee

The Tennessee Department of Health and Environment, Office of Water Management, is responsible for groundwater management. The Groundwater Protection Division issues licenses to qualified well drilling contractors, requires conformance with well construction regulations, and receives reports of well completions, as mandated by the Water Wells Drillers Act (Tennessee Code Annotated 69-11-101, *et seq.*). The Water Supply Division requires community suppliers to submit designs for new facilities to the state for review and approval. The state requires compliance with design and construction guidelines. It requires water treatment and treatment plant operators to be trained and licensed (Tennessee Code Annotated 68-13-701, *et seq.*). The Office of Water Management requires that the state be notified if a user withdraws more than 50,000 gallons per day (Tennessee Code Annotated 69-8-105, *et seq.*). The Office of Water Management supports investigations related to groundwater protection and human-health effects.

The Tennessee Division of Geology is responsible for enforcement of regulations of the Tennessee Oil and Gas Board. These regulations are intended to protect the quality of fresh groundwater in areas of oil and gas development. Tennessee intends to seek primacy in the implementation of the Underground Injection Control Program within the state (Tennessee Water Control Board, 1983). If this program does become the responsibility of the state, the Groundwater Protectionv Division will be responsible for enforcement. The state cooperates with the U.S. Geological Survey in collecting hydrological data and doing research as to groundwater occurrence, movement, and quality.

Other federal agencies involved are the Department of Energy, the Army Corps of Engineers, and the Tennessee Valley Authority (USGS, 1985).

Texas

"Groundwater in Texas is the property of the landowner, and its use is subject to very few limitations in accordance with the 'English,' or common law doctrine of riparian rights. Owners of land overlying defined groundwater reservoirs or aquifers may adopt voluntary well regulations through mutual association in "underground water conservation districts." Section 52.001 of the Texas Water Code provides the framework for these districts, and to date, 12 districts have been created, but only 9 are currently active. The act creating the groundwater districts was amended by the 63rd Legislature in 1973 (House Bill 935) primarily to allow for the control of land subsidence caused by the withdrawal of groundwater."

472

The three state agencies that are actively engaged in various phases of the state's water resources program are the Texas Department of Water Resources, the Railroad Commission of Texas, and the Texas Department of Health. The Texas Department of Water Resources has been given primary responsibility for implementing the provision of the state's constitution and laws relating to the conservation, protection, and development of both surface water and groundwater resources. "Groundwater data are collected and analyzed independently and in cooperation with the USGS or other federal and state agencies. The Railroad Commission of Texas has the responsibility for protecting groundwater from possible pollution that may result from the exploration for and development and production of petroleum, natural gas, and geothermal resources, as well as from surface mining of lignite, coal, and uranium."

The Texas Department of Water Resources assists the Railroad Commission by making recommendations for the protection of usable quality groundwater in connection with the exploration for and production of oil, gas, and other minerals, as well as the disposal of oil field brines by injection into subsurface formations. "The Texas Department of Health regulates the disposal of municipal and mixed municipal-industrial solid wastes, establishes drinking-water standards for public supplies, and has primacy in administering the provisions of the Federal Safe Drinking Water Act" (USGS, 1985).

Utah

"Groundwater use in Utah is regulated by the Utah Department of Natural Resources, Division of Water Rights." Several areas have been designated where additional appropriation of groundwater is not allowed. "Withdrawals for irrigation from four southwestern basins have been limited by court decree, and discharge from irrigation wells in these basins is metered to verify compliance."

"In other areas the appropriation of groundwater is not allowed because of its potential effect on surface waters, and in some areas appropriation of groundwater only for domestic use is allowed. This appropriation of groundwater is restricted in more than half the state."

473

The Utah Department of Health, Division of Environmental Health, is responsible for protecting groundwater quality and prevention, control, and abatement of groundwater pollution (USGS, 1985).

Vermont

Vermont's declared policy of maintaining high quality groundwater is to be realized by limiting "human activities that present an unreasonable risk"; this is balanced, however, with "the need to maintain and promote a healthy and prosperous agricultural community." The Secretary of the Agency of Groundwater Preservation is directed to develop a comprehensive management plan. The primary components of the state's management program are classificaition of aquifers based on their character and exposure risk, and technical criteria and standards relating to classification and activities that pose risk to groundwater and are therefore to be regulated or precluded. Vermont's classification system is described in Chapter 11 (Commons, 1985).

Certification is required for the discharge of effluents. Vermont's protection program also includes certification of the discharge of hazardous wastes, monitoring, and rules and regulations for enforcement. Vermont recognizes or has designated aquifer protection areas. The analyses for establishing these protection areas involve groundwater use, land-use surveys, and categories of aquifer protection areas. There are also criteria and regulations concerning wells (Vermont Agency of Environmental Conservation, 1983, 1984).

The management of groundwater in Vermont is divided primarily among three state agencies: the Department of Agriculture, the Department of Health, and the Department of Water Resources and Environmental Engineering, which is a unit of the Agency of Environmental Conservation.

The Department of Agriculture regulates the use and storage of pesticides. The Department of Health protects drinking water supplies, and the Department of Water Resources and Environmental Engineering protects, regulates, and, where necessary, controls groundwater resources.

The three departments are represented on the Groundwater Coordinating Committee, which serves as a clearinghouse for the exchange of information relating to groundwater. The committee recommends policies to member agencies that have statutory authority. The Groundwater Management Unit, which is one of the major units within the Department of Water Resources and Environmental Engineering, addresses the broadest range of groundwater issues. Program areas include: water well drilling licenses and well reporting, groundwater level monitoring, aquifer protection area mapping, underground injection control, data management, special studies, technical assistance, application review for land use and development, injection-well permits, and public information, education, and administration.

The state geologist, the University of Vermont, and the Agency of Transportation also have roles in the management of groundwater (USGS, 1985).

474

Virginia

The policy enunciated by the General Assembly of Virginia recognizes that all groundwaters of the state belong to the public, and in order to protect and beneficially utilize and ensure the preservation of public welfare, safety, and health, provisions must be made to control groundwater resources (Sect. 62.1-44.84— Groundwater Act 1973, Chapt. 3.4, Title 62.1).

Virginia's groundwater management plan states that groundwater quality standards will apply statewide and will apply to all groundwaters occurring at or below the uppermost season limits of the water table. In order to prevent the entry of pollutants into groundwater occurring in any aquifer, a soil zone or alternate protective measurement sufficient to preserve and protect present and anticipated uses of groundwater shall be maintained at all times. Zones for mixing wastes with groundwater may be allowed upon request, but shall be determined on a case by case basis and shall be kept as small as possible. It is recognized that natural groundwater quality varies statewide. Four physiographic provinces have been determined for application of standards, namely, the Coastal Plain, Piedmont and Blue Ridge, Valley and Ridge, and Cumberland Plateau.

Virginia has an anti-degradation policy. If the concentration of any constituent in groundwater is less than the limit set forth by groundwater standards, the natural quality for the constituent shall be maintained; natural quality shall also be maintained for all constituents, including temperature, not set forth in groundwater standards. If the concentration of any constituent in groundwater exceeds the standard for that constituent, no addition of that constituent to the naturally occurring concentration shall be made. Variance to this policy will not be made unless it has been affirmatively demonstrated that a change is justifiable to provide necessary economic or social development, that the necessary degree of waste treatment cannot be economically or socially justified, and that the present and anticipated uses of such water will be preserved and protected. Virginia also has a system of standards for groundwater quality. Registration and reporting are required for water withdrawals (Virginia State Water Control Board, 1982).

475

Groundwater in Virginia is managed by the State Water Control Board, which is authorized by the Groundwater Act of 1973. The act places with the board the responsibility for designating groundwater management areas to control the rate of groundwater withdrawal when excessive declines are observed in groundwater levels or artesian pressures, when there is substantial interference between wells, when the available groundwater supply is being or is about to be withdrawn, or when actual or potential pollution of groundwater supplies occurs.

Two areas have designated groundwater-management areas: the Eastern Shore Peninsula and southeastern Virginia. Within these areas withdrawals of more than 50,000 gallons per day must be permitted and reported. Public, domestic, and agricultural users are exempt—only industrial and commercial users must comply. Elsewhere in the state, users who withdraw more than 10,000 gallons per day are required to report annual withdrawals to the board. This is authorized by Regulation II, enacted in 1982.

The Virginia State Department of Health, in cooperation with the Virginia State Water Control Board, is authorized to regulate the use and quality of groundwater to protect public health. The Department of Health regulates public supply systems, domestic supply systems with on-site septic systems, and solid waste disposal facilities. Virginia has an interstate cooperative agreement with Maryland and North Carolina to exchange information about wells and pumping near their mutual boundaries (USGS, 1985).

Washington

Groundwater in Washington is regulated chiefly by the Washington Department of Ecology and the Washington Department of Social and Health Services. The Washington Department of Ecology is responsible for administering all of the state's groundwaters and for issuing water rights based on Chapter 90.44 of the Revised Code of Washington (RCW). Potential users of groundwater who wish to withdraw more than 5,000 gallons per day must make application to the Washington Department of Ecology, which then determines if the proposed use is of public interest. The prime considerations include the effects of the proposed withdrawal on groundwater levels and surface water bodies. If the proposed withdrawal threatens to lower groundwater levels more than 10 feet annually, the application is usually denied. Recently the Washington Department of Ecology denied many applications in the Odessa–Lind area, where groundwater levels have declined significantly because of intensive irrigation pumping. This department also regulates well drillers and well drilling activities. It also conducts technical investigations unilaterally and in cooperation with the U.S. Geological Survey. The protection of groundwater quality is the concern of both the Department of Ecology and the Department of Social and Health Services. Under Chapter 90.48 of the RCW, the Department of Ecology has been designated as the state's water pollution control agency and is responsible for administration of the Underground Injection Control provisions of the Safe Drinking Water Act of 1974 (Public Law 93-523) and any other groundwater provisions of the federal Clean Water Act.

The Department of Social and Health Services is charged with administering the drinking water protection aspects of the Federal Safe Drinking Water Act, and under Chapter 43.20 of the RCW, it regulates public water systems (USGS, 1985).

West Virginia

Water law in West Virginia is based on a modification of the riparian doctrine. State organizations such as the Water Resources

Board, the Department of Natural Resources, the Division of Water Resources, the State Department of Health, the Department of Mines, the Division of Oil and Gas, and the State Geological and Economic Survey implement most of the regulations and planning and research programs for the protection and management of groundwater in the state.

The State Natural Resources Law of 1933, as revised by Chapter 133 of the Act of 1961, created the Water Resources Board and the Division of Water Resources. The Water Resources Division administers and enforces all laws relating to conservation, development, protection, and use of the state's groundwater resources. Further revision by Chapter 20 of the Acts of 1964 places the responsibility for enforcement of water-pollution legislation with the Division of Water Resources.

477

The State Department of Health, under authority of the public health laws of West Virginia, Chapter 16, Article 1, Section 9, regulates public supply systems operated by individuals, companies, corporations, institutions, and county and municipal governments. Through its Division of Sanitary Engineering and through the State Board of Health, the Department of Health regulates installation of public supply systems and adherence to public-supply-quality standards.

Permit applications for the drilling of oil and gas wells, and protection of freshwater aquifers from contamination are the responsibility of the Division of Oil and Gas, Department of Mines, as established in Article 4, Chapter 22, of the Code of West Virginia, 1931.

The State Geological Survey and Economic Survey, in cooperation with the USGS, maintains a statewide water data network and is responsible for investigating the state's water resources. Research, data collection, and analysis provided by this cooperative program form an information base upon which groundwater management decisions are made by the West Virginia Department of Natural Resources and by other state agencies charged with the protection and management of the state's groundwater resources (USGS, 1985).

Wisconsin

"In 1984 Wisconsin passed groundwater legislation designed to (1) set groundwater quality standards, (2) provide funds for replacement of contaminated water supplies, (3) provide an environmental repair fund, (4) develop a water quality monitoring network, and (5) certify laboratories to be used to analyze groundwater quality.

The monitoring network includes the following four classifications: (1) problem assessment monitoring, (2) regulatory monitoring, (3) at-risk well monitoring, and (4) management practice monitoring. The details for these components are still being developed with the assistance of the newly created Ground Water Coordination Council of State Agencies. Prior to the passage of this legislation, Wisconsin had many regulations designed to protect groundwater. Injection of wastes into wells was prohibited, septic tank system installers and well drillers required licenses, and large-capacity wells required permits. These and other regulations are still in effect.

Several Wisconsin agencies are concerned with developing and enforcing rules related to groundwater protection. The principal agency is the Department of Natural Resources" (USGS, 1985).

The plan for groundwater protection is specifically addressed in Chapter 160 of the Wisconsin Statutes, 1983 Wisconsin Act #10. This law requires the Department of Natural Resources to adopt state groundwater standards, and Chapter NR 140 of the Wisconsin Administrative Code was enacted to meet this requirement. All state agencies that regulate sources of groundwater contamination are required to comply with the groundwater standards. NR 140 establishes two levels of groundwater standards. (The enforcement standard is the maximum concentration of a substance allowable in groundwater and a preventive action standard is a limit set at a percentage of the enforcement standard.) Standards have been established for 39 substances of health concern and 10 substances of welfare concern. In addition, methodology has been established for preventive-action limits for 15 indicator parameters based on background quality.

Under Chapter 160 of the Wisconsin Statutes, the Department of Natural Resources is required to develop and operate a system for monitoring and sampling groundwater. The department is required to notify the appropriate regulatory agencies when monitoring data indicates that a substance from a contaminating source controlled by that agency has been found in groundwater.

Wisconsin is developing a three-tiered groundwater planning program consisting of (1) a statewide plan, (2) a groundwater-reports-and-water-quality-management plan, and (3) county groundwater-management plans. This management of groundwater is coordinated under Chapter 160, which establishes the duties of the Wisconsin Groundwater Coordinating Council.

Various state agencies have various responsibilities concerning groundwater. The Wisconsin Department of Natural Resources has the lead role in groundwater management in Wisconsin. It manages water resources, technical services, water supply, wastewater, and solid and hazardous waste. The Wisconsin Department of Transportation has responsibility for groundwater protection under 1983 Wisconsin Act 410. The Department of Transportation now regulates bulk storage of road salts throughout the state. The University of Wisconsin has responsibility for education, basic and applied research, and technical assistance in groundwater management. The Wisconsin Geological and Natural History Survey is responsible for inventorying and mapping the geologic and hydrologic resources of the state. The Wisconsin Department of Health and Social Services has the primary environmental, epidemiologic, and toxicologic expertise for the state of Wisconsin. The Wisconsin Department of Industry, Labor and Human Relations has several programs relating to groundwater management, including regulation of private sewage-disposal systems and underground petroleum storage tanks. The Wisconsin Department of Agriculture, Trade, and Consumer Protection has responsibilities for pesticides, fertilizer storage, and animal-waste management that relate to groundwater.

479

Water resource planning is an important activity in the state. The areawide water quality management plans are carried out under the guidelines of NR 121 (Wisconsin Administrative Code) and Sections 208 and 303 of the Clean Water Act and include such elements and information as a groundwater quality data summary; critical land uses and contamination sources; actual and potential groundwater problems; groundwater-management options; best management practices; recommendations for county plans; and management recommendations. Wisconsin has a groundwater monitoring plan (Wisconsin Act 410) (Wisconsin Department of Natural Resources, 1986).

Wyoming

The Wyoming state engineer administers laws and regulations pertaining to the state's groundwater and is charged with providing for orderly development of groundwater and its protection from waste and contamination. The state engineer issues permits for groundwater diversion and may recommend designation of an area as a groundwater control area. New wells may be prohibited in control areas and withdrawals regulated after due process. Three control areas have been designated, all of which are in southeastern Wyoming and have wells that withdraw water from the High Plains Aquifer.

The State Department of Economic Planning and Development and the Farm Loan Board provide technical assistance and loans **480** for groundwater development. This financial and technical assistance has provided considerable impetus to the use of groundwater for irrigation. The Water Quality Division of the Wyoming Department of Environmental Quality is a primary agency for groundwater quality protection.

The Oil and Gas Conservation Commission regulates injection of groundwater for secondary recovery of petroleum. This commission also regulates the re-injection of water produced with the oil (USGS, 1985).

Chapter 11 of the rules of the Wyoming Department of Environmental Quality, Water Quality Division, includes the rules dealing with the Groundwater Pollution Control Permit System. These rules were adopted in accordance with WS 9-4-101 through 9-4-115, WS 35-11-112(A)(I), and WS 35-11-302. This extensive permitting system controls subsurface discharges by permit, environmental monitoring programs for the groundwaters of the state, and special process discharges/in situ mining (Wyoming Department of Environmental Quality, 1984).

Wyoming's standards and groundwater classification system are interrelated. The standards are prescribed to protect the natural quality of underground water; the groundwaters are classified by use and ambient quality, so that standards can be applied to protect the water quality.

Waters that are known sources of supply and appropriated for uses identified in WS 35-11-102 and 103 are classified as domestic waters, waters for fish and aquatic life, and waters for industry (see Chapter 11). Discharges or activities that impact the underground sources of water for existing uses identified in WS 35-11-102 and 103(c)(i) shall be such that the intended use or uses of the water are not unsuitably affected at any place or places of withdrawal or natural flow to the surface (Wyoming Department of Environmental Quality, 1984).

Unappropriated waters are classified by ambient water quality. In general, classifications of groundwater shall be based on the water quality standards for each class of water, except for class I which is based upon ambient water quality. *Class I* groundwaters are suitable for domestic use. Their ambient quality is suitable for domestic purposes and does not have concentrations in excess of any of the state standards for class I groundwaters. *Class II* groundwaters are suitable for agricultural use where soil conditions and other factors are adequate. This groundwater does not have concentrations in excess of any of the state standards for class II groundwaters. *Class III* groundwaters are suitable for livestock, and **481** do not have concentrations in excess of the state standards for class III groundwaters.

Special class groundwaters are suitable for fish and aquatic life. The ambient quality of these underground waters does not have concentrations in excess of state standards for class special A groundwaters. Waters with these special classifications shall not contain biologically hazardous or toxic or potentially toxic material, or substances in amounts or concentrations that exceed maximum allowable concentrations. (These are based upon EPA guidelines stated in the Federal Register for December 24, 1975, part 4, these in turn are based upon the best available scientific information as determined by the administrator.) The water shall not be impaired for its intended use or suitability, or contribute to a condition in contravention of groundwater quality standards or to any toxic or hazardous effect on the natural biota. No discharge into waters of these classes shall cause a variation in the range of any of the parameters inconsistent with the standards for these classes.

Class IV(A) are suitable for industry. These quality requirements range widely and almost every industrial application has its own standards. This groundwater has a total dissolved solids concentration not in excess of 10,000 mg/l. *Class IV(B)* groundwaters may have a total dissolved solids concentration in excess of 10,000 mg/l. A discharge into an aquifer containing Class IV (A)/(B) groundwater shall not result in the water being unfit for its intended use. Discharges into class IV waters are closely regulated.

Class V groundwaters are closely associated with commercial deposits of hydrocarbons or other minerals, or are considered a geothermal resource. Discharges into class V groundwaters shall be for mineral production and not result in degradation or pollution of associated or other groundwater (unless the affected groundwater quality can be returned to background or better quality after mining by reduction or elimination of pollution), or in the waste of other water resources. A discharge into class V groundwater may also be for the purpose of production of geothermal resources.

Class VI groundwater may be unusable or unsuitable for use due to excessive concentrations of dissolved solids or specific conductuance (Wyoming Department of Environmental Quality, 1980a).

Bibliography

Agrelot, J. C., J. J. Malot, and M. J. Visser. 1985. *Vacuum: Defense System for Ground Water UOC Contamination.* Presented at the Fifth National Symposium on Aquifer Restoration and Ground Water Monitoring, May 21–24. Columbus, Oh. 10 pp.

Alabama Department of Consumer and Regulatory Affairs. Environmental Control Division. 1986. *Water Quality Report to Congress for Calender Years 1984 and 1985.* April.

Aller, L., T. Bennett, J. H. Lehr, and R. J. Petty. 1985. *Drastic: A Standardized System for Evaluating Groundwater Pollution Potential Using Hydrogeologic Settings.* Robert S. Kerr Environmental Research Laboratory. Office of Research and Development. Ada, Okla. EPA/600/2-85/018. May. 163pp.

American Petroleum Institute (API). 1975. *Annual Statistical Review.* Washington, D.C.: American Petroleum Institute. 79pp.

———. 1980. *Underground Spill Cleanup Manual.* API Pub. 1628. Washington, D.C.: American Petroleum Institute. 34pp.

Anderson, M. P. 1984. Movement of Contaminants in Groundwater: Groundwater Transport—Advection and Dispersion. In *Groundwater Contamination: Studies in Geophysics.* Washington, D.C.: National Academy Press. 179pp.

Arizona Division of Health Services. 1979. *Arizona Surface Impoundment Assessment—Groundwater Contamination Cases.* Chap. 6. Phoenix, Ariz. 47pp.

Arkansas Department of Pollution Control and Ecology. 1986. *Arkansas Water Quality Inventory Report—1986.*

Bibliography

——. 1986. *Draft-Elements of Arkansas Groundwater Quality Protection Strategy.* 96pp.

Association of State and Interstate Water Pollution Control Administrators (ASIWPCA). 1985a. *America's Clean Water. The States' Nonpoint Source Assessment 1985.* Washington, D.C.: the Association.

——. 1985b. *America's Clean Water. The States' Nonpoint Source Management Experience 1985.* Washington, D.C.: the Association.

Ballentine, R. K., S. R. Rezner, and C. W. Hall. 1972. *Subsurface Pollution Problems in the United States.* Washington, D.C.: U.S. EPA Office of Water Programs. Technical studies report: TS-00-72-02. 24pp.

Banks, H. O. 1981. Management of Interstate Aquifer Systems. *Journal of Water Resources Planning and Management Division. Proceedings of the American Society of Civil Engineers* 107(2):563–77.

Battelle Institute. 1984. *Executive Summary. Final Report to the Ohio Environmental Protection Agency. Needs and Recommendations for an Ohio Groundwater Quality Strategy.* Columbus, Oh. March. 9 pp.

Bear, J. 1972. *Dynamics of Fluids in Porous Media.* New York: American Elsevier. 764pp.

Bekure, S. 1971. An Economic Analysis of the Intertemporal Allocation of Groundwater in the Central Ogallala Formation. Ph.D. Diss., Oklahoma State University at Stillwater.

Berg, J. W., and F. Burbank. 1972. Correlations Between Carcinogenic Trace Metals in Water Supplies and Cancer Mortality. *Annals of the New York Academy of Science* 199:249–64.

Bittinger, M. W. 1981. The Ogallala Story—What Have We Learned? *Ground Water* 19(6):586–87.

Bouwer, H. 1981. Protecting the Quality of Our Groundwater: What Can We Do? *Groundwater Monitoring Review* 1(2):22–26.

Boyle, W. 1981. Control Measures—Subsurface Disposal. In *National Center for Ground Water Research (NCGWR). Proceedings of a Conference on Microbial Health Consideration of Soil Disposal of Domestic Wastewaters,* pp. 209–26. Norman, Okla. May.

Braids, O. C. 1981a. Behavior of Contaminants in the Subsurface. In *Seminar on the Fundamentals of Ground Water Quality Protection.* Presented by Geraghty and Miller, Inc., and American Ecology Services, Inc. Cherry Hill, N.J. 5–6 October.

Bredehoeft, J. D., and T. Maini. 1981. Strategy for Radioactive Waste Disposal in Crystalline Rocks. *Science* 213 (17 July):293–96.

Bunch, S. E., and P. Jacobs. 1979. Health Costs Due to Environmental Hazards. *Journal of Environmental Health* 41(5):267–69.

Burge, W. D., and P. B. Marsh. 1978. Infectious Disease Hazards of Landspreading Sewage Wastes. *Journal of Environmental Quality* 7(1):1–9.

Cabelli, V. 1981. Epidemiological Approach Toward Determining Endemic Transmission of Waterborne Disease. In *National Center for Ground*

Water Research (NCGWR). Proceedings of a Conference on Microbial Health Considerations of Soil Disposal of Domestic Wastewaters, pp. 177–87. Norman, Okla. May.

California Department of Health Services. 1986. *Organic Chemical Contamination of Large Public Water Systems in California.* Sacramento. April. 152pp.

California Department of Water Resources. 1975. *California's Ground Water.* Bulletin 118. September. 135pp.

———. 1980. *Ground Water Basins in California: A Report to the Legislature in Response to Water Code Section 12924.* Bulletin 118–80. January. 73pp.

California State Water Resources Control Board, Surveillance and Monitoring Section. 1981. *Assessment of Ground Water Problems and Potential Issues for Use at Program Conference.* Internal memo, 30 January.

———. 1984. *Water Quality Inventory for Water Years 1982 and 1983.* Water Quality Monitoring Report No. 84-3TS. June.

Canter, L. W., R. C. Knox, R. P. Kamat. 1982. *Evaluation of Septic Tank System Effects on Ground Water Quality.* Norman, Okla.: National Center for Ground Water Research. May. 158pp.

Canter, L., C. H. Ward and R. C. Knox. 1985. *Ground Water Pollution Control.* Chelsea, Mich.: Lewis Publishers, Inc. 526pp.

Carter, L. J. 1983. The Radwaste Paradox. *Science.* 219(4580):33–36.

Centers for Disease Control (CDC). 1982. *Water Related Disease Outbreaks: Annual Summary 1981.* Department of Health and Human Services. Publication No. (CDC) 82-8385. 13pp.

———. 1983. *Water Related Disease Outbreaks: Annual Summary 1982.* Department of Health and Human Services. Publication No. (CDC) 83-8385. 15pp.

———. 1984. *Water Related Disease Outbreaks: Annual Summary 1983.* Department of Health and Human Services. Publication No. (CDC) 84-8385. 15pp.

———. 1985. *Water Related Disease Outbreaks: Annual Summary 1984.* Department of Health and Human Services. Publication No. (CDC) 85-2510. 15pp.

Chemical Manufacturers' Association (CMA). 1987. *Program for State Groundwater Management.* Washington, D.C.: the Association. January.

Cherry, J. A. 1984. Contaminant Migration in Groundwater: Processes and Problems. In *Proceedings of the Second National Water Conference. The Fate of Toxics in Surface and Ground Waters.* The Academy of Natural Sciences, Philadelphia, Pa., pp.69–94. January 24 and 25.

Cherry, J. A., R. W. Gillham, and J. F. Barker. 1984. Contaminants in Groundwater: Chemical Processes. In *Groundwater Contamination: Studies in Geophysics.* Washington, D.C.: National Academy Press. 179pp.

485

Bibliography

Colorado Department of Health. 1973. *Criteria Used in the Review of Waste Water Treatment Facilities.* Denver.

————. Office of Health Protection. 1984. *Ground Water Protection for Colorado.*

Commons, Geoffrey. 1985. Vermont's New Groundwater Law. *Vermont Law Review.* 10(2):481–485.

Commonwealth of Massachusetts Ground Water Quality Standards. *Code of Massachusetts Regulations* Title 314, Chapter 6.

Comstock, G. W. 1979. Water Hardness and Cardiovascular Diseases. *American Journal of Epidemiology* 110(4):375–400.

Connecticut Department of Environmental Protection. 1980. *Connecticut Water Quality Standards and Criteria.* 9 September. 28pp.

————. (Water Compliance Unit). 1986. *State of Connecticut 1986 Water Quality Report to Congress.*

Conolly, R. 1980. *Toxic Chemical Contamination of Ground Water.* Testimony given in the U.S. EPA oversight hearings, 24–25 July. 18 September.

Conservation Foundation. 1987a. A Guide to Groundwater Pollution: Problems, Causes, and Government Responses. In *Groundwater Protection.* Washington, D.C.: Conservation Foundation. Pp.47–232.

————. 1987b. Groundwater: Saving the Unseen Resource. The Final Report of the National Groundwater Policy Forum. In *Groundwater Protection.* Pp. 1–46. Washington, D.C.: Conservation Foundation.

Cooper, R., and A. Olivieri. 1981. Public Health Risk Evaluation of Wastewater Disposal Alternatives. In *National Center for Groundwater Research (NCGWR). Proceedings of a Conference on Microbial Health Considerations of Soil Disposal of Domestic Wastewaters,* pp. 189–208. Norman, Okla. May.

Council on Environmental Quality (CEQ). 1981a. *Contamination of Ground Water by Toxic Organic Chemicals.* Prepared by David E. Burmaster. 85pp.

————. 1981b. *12th Annual Report.* Washington, D.C.: U.S. GPO. 291pp.

————. 1984. 15th Annual Report of the Council on Environmental Quality. Washington, D.C.: U.S. GPO. 719pp.

Craun, G. F. 1984. Health Aspects of Groundwater Pollution. In *Groundwater Pollution Microbiology.* ed. G. Bitton and C. P. Gerba. Somerset, New York: John Wiley and Sons, Chapter 7.

————. 1985. A Summary of Waterborne Illness Transmitted through Contaminated Groundwater. *Journal of Environmental Health* 48(3):122–127.

————. 1986. Chemical Drinking Water Contaminants and Disease. In *Waterborne Diseases in the United States.* ed. G. F. Craun. Boca Raton, Fl.: CRC Press. Chapter 4.

Davis, S. N., and R. J. M. DeWiest. 1966. *Hydrogeology.* New York: John Wiley and Sons. 463pp.

Delaware Comprehensive Water Resources Management Committee. 1983.

The Management of Water Resources in Delaware. Groundwater Quality Management. Document No. 40-08/83/10/03. July. 37pp.

Delaware Department of Natural Resources and Environmental Control. Division of Water Resources. 1986. *1986 Delaware Water Quality Inventory.* Vols. I, II, and III. April.

Demopoulos, H. B., and E. G. Gutman. 1980. Cancer in New Jersey and Other Complex Urban/Industrial Areas. *Journal of Environmental Pathology and Toxicology* 3:219–35.

Deutsch M. 1963. *Ground Water Contamination and Legal Controls in Michigan.* U.S. Geological Survey Water Supply Paper 1691. Washington, D.C. 79pp.

District of Columbia Department of Consumer and Regulatory Affairs. Environmental Control Division. 1986. *Water Quality Assessment.* April.

Dolmatch, T. B., ed. 1982. *Information Please Almanac.* New York: Simon & Schuster. 958pp.

Domenico, P. A. 1972. *Concepts and Models in Ground Water Hydrology.* New York: McGraw-Hill. 405pp.

Engberg, R., and R. F. Spalding. 1978. *Groundwater Quality Atlas of Nebraska. Resource Atlas no. 3.* Lincoln: University of Nebraska. 39pp.

Environ Corp. 1983. *Approaches to the Assessment of Health Impacts of Groundwater Contaminants.* Prepared for the Office of Technology Assessment. Washington, D.C.: August. 207pp.

Environmental Reporter. 1986. "Nuclear Waste—Three Sites Selected for Further Review: Plans Postponed for Second Waste Repository." 17(5):110.

Everett, L. G. 1980. *Ground Water Monitoring.* Schenectady, N.Y.: General Electric Co. Technology Marketing Operation. 440pp.

Exner, M. E., and R. F. Spalding. 1979. Evolution of Contaminated Ground Water in Holt County, Nebraska. *Water Resources Research* 15(1):139–47.

Florida Department of Environmental Regulation. 1980. *Florida Surface Impoundment Assessment: Final Report.* Tallahassee. January. 298pp.

———. 1981a. *Hazardous Waste Inventory, June 1981.* Tallahassee. 59pp.

———. (Ground Water Section). 1981b. *Summary of Known Cases of Ground Water Contamination in Florida.*Tallahassee. September. 12pp.

———. 1982. Public Drinking Water Systems. In *Rules of the Florida Department of Environmental Regulation.* Chapter 22. Tallahassee. 8 July.

———. 1984. *Sites List—Petroleum Contamination Incidents.* Tallahassee. September.

———. 1985a. *Fact Sheet—Florida's Groundwater Protection Program.* Tallahassee. March. 21pp.

487

Bibliography

————. 1985b. Permits. In *Rules of the Florida Department of Environmental Regulation.* Chapter 17-4. Tallahassee, 19 December.

————. 1985c. Public Drinking Water Standards. In *Rules of the Florida Department of Environmental Regulation.* Chapter 17-22. Tallahassee. 8 July.

————. Bureau of Operations. 1985d. *The Sites List—Summary Status Report—July 1, 1985–December 31, 1985. Tallahassee. September.*

————. 1985e. *Underground Injection Control. In Rules of the Florida Department of Environmental Regulation.* Chapter 17–28. Tallahassee. 8 May.

————. 1985f. Water Quality Standards. In *Rules of the Florida Department of Environmental Regulations.* Chapter 17–3. Tallahassee.

————. Bureau of Water Quality Management. 1986. *1986 Water Quality Inventory for the State of Florida.* Tallahassee. June.

Freeze, R. A., and J. A. Cherry. 1979. *Groundwater.* Englewood Cliffs, N.J.: Prentice-Hall. 604pp.

Frick, D., and L. Shaffer. n.d. *Assessment of the Availability, Utilization and Contamination of Water Resources in New Castle County, Delaware.* Prepared by Office of Water and Sewer Management, Department of Public Works, New Castle County, Del., for U.S. EPA Office of Solid Waste Management Programs. Newark. Contract no. WA-6-99-2061-J. 215pp.

Fuhriman, D. K., and J. R. Barton. 1971. *Ground Water Pollution in Arizona, California, Nevada and Utah.* U.S. EPA. Office of Research and Monitoring. Report no. 16060ERU. Washington, D.C. 249pp.

Georgia Department of Natural Resources. Environmental Protection Division. 1984. *A Ground Water Management Plan for Georgia.* 26pp.

Geraghty, J. J. 1981. Containment of a Plume of Contaminated Ground Water. In *Seminar on The Fundamentals of Ground Water Quality Protection.* Presented by Geraghty and Miller, Inc., and American Ecology Services, Inc. Cherry Hill, N.J. October.

Geraghty, J. J., D. W. Miller, F. van der Leeden, and F. L. Troise. 1973. *Water Atlas of the United States.* 3d ed. Syosset, N.Y.: Water Information Center, Inc. 122 plates.

Geraghty and Miller, Inc. 1979. *Investigations of Ground Water Contamination in South Brunswick Township, N.J.* Syosset, N.Y.: Geraghty and Miller, Inc. 49pp + 10 appendices.

Gerba, C. 1981. Virus Occurrence in Ground Water. In *National Center for Ground Water Research (NCGWR). Proceedings of a Conference on the Microbial Health Considerations of Soil Disposal of Domestic Wastewaters,* pp. 144–57. Norman, Okla. May.

Gillham, R. W., M. J. L. Robin, J. F. Barker and J. A. Cherry. 1983. *Groundwater Monitoring and Sampling Bias.* Prepared for the American Petroleum Institute. Washington, D.C. June. 206pp.

Glatt, D. C. 1986. *Pesticide and Herbicide Survey of Selected Municipal*

Drinking Water Systems in North Dakota. North Dakota State Department of Health. Bismarck. 7 February.

Goad, M. S. 1982. *New Mexico's Experience in Setting and Using Groundwater Quality Standards.* New Mexico Health and Environment Department. Environmental Improvement Division. September. 23pp.

Gormly, J. R., and R. F. Spalding. 1979. Sources and Concentrations of NO$_3$-Nitrogen in Ground Water of the Central Platte Region, Nebraska. *Ground Water* 17(3):291–300.

Greenberg, M., T. Burke, J. Caruana, G. W. Page, and K. Ohlson. 1981. Approaches and Initial Findings of a State-Sponsored Research Program on Population Exposure to Toxic Substances. *The Environmentalist* 1:53–63.

Gunnell, R. D. 1986. *State of Utah Water Quality Report.* August.

Gutentag, E. D., F. J. Heimes, N. C. Krothe, R. R. Luckey, and J. B. Weeks. 1984. *Geohydrology of the High Plains Aquifer in Parts of Colorado, Kansas, Nebraska, New Mexico, Oklahoma, South Dakota, Texas, and Wyoming.* High Plains RASA Project. U.S. Geological Survey Professional Paper 1400-b. Washington, D.C.: U.S. GPO, 63pp.

Hadeed, S. J. 1979. *DBCP Well Sampling Program for Yuma County, Arizona* (7 June–26 July 1979). Phoenix: Arizona Department of Health Services. 33pp.

Hagedorn, C. 1981. Transport and Fate: Bacterial Pathogens in Ground Water. In *National Center for Ground Water Research (NCGWR). Proceedings of a Conference on the Microbial Health Considerations of Soil Disposal of Domestic Wastewaters,* pp. 84–102. Norman, Okla. May.

Hajali, P. A., and L. W. Canter. 1980. *Rehabilitation of Polluted Aquifers.* National Center for Ground Water Research Report no. NCGWR 80-12. Norman, Okla. 46pp.

Hall, C. W. 1984. Quality Protection: The Issue in Perspective. *Environmental Professional* 6:46–51.

Hallberg, G. R. 1985. *Nonpoint Source Contamination of Groundwater by Agricultural Chemicals—Technical Comments.* Prepared for the United States Senate Committee on Environment and Public Works, Subcommittee of Toxic Substances and Environmental Oversight. Iowa Geological Survey. Iowa City. December. 33pp.

Harris, J. R. 1986. Clinical and Epidemiological Characteristics of Common Infectious Diseases and Chemical Poisonings Caused by Ingestion of Contaminated Drinking Water. In *Waterborne Diseases in the United States,* ed. G. F. Craun. Boca Raton, Fl.: CRC Press. Chapter 2.

Harris, R. H. 1982. Health Effects Associated with Organic Chemical Contaminants in Ground Water. In *Abstracts of Papers Presented at American Association for the Advancement of Science Meeting,* p.335. Washington, D.C. January.

489

Bibliography

Harrison, E. Z. and M. A. Dickinson. 1984. *Protecting Connecticut's Groundwater—A Guide to Groundwater Protection for Local Officials.* Connecticut Department of Environmental Protection. September. 37pp.

Hawaii Department of Health. 1986. *305(b) Report on Water Quality: Report for the Year Ended December 31, 1986.* April.

Heath, R. C. 1985. Introduction to State Summaries of Ground-Water Resources. In *National Water Summary 1984.* Water Supply Paper 2275, pp.118–121. U.S. Geological Survey. Washington, D.C.

490 Henderson, T. R., J. Trauberman, and T. Gallagher. 1984. *Groundwater: Strategies for State Action.* Washington, D.C.: Environmental Law Institute. 353pp.

High Plains Associates: Camp, Dresser & McKee, Inc., Black & Veatch, and Arthur D. Little, Inc. 1982. *Six State High Plains–Ogallala Aquifer Regional Resources Study.* March.

Hileman, B. 1982. Nuclear Waste Disposal: A Case of Benign Neglect? *Environmental Science and Technology* 16(5):271A–75A.

Hubert, J. S., and L. W. Canter. 1980a. *Health Effects from Ground Water Usage.* Norman, Okla.: National Center for Ground Water Research (NCGWR). Report no. NCGWR 80-17. 300pp.

———. 1980b. *Acid Rain and Ground Water Quality.* Norman, Okla.: National Center for Ground Water Research. Report no. NCGWR 80-25 (TP). 15pp.

Idaho Department of Health and Welfare. Division of Environment. 1985. *Groundwater Quality Management Plan for Idaho.* Boise. 23pp.

———. Division of Environment. 1986. *Idaho Water Quality Status Report 1986.* September.

Idaho Department of Health and Welfare and Idaho Department of Water Resources. 1985. *Snake Plain Aquifer Technical Report.* September. 117pp.

Illinois Environmental Protection Agency. 1971. *Design Criteria Used in the Review of Waste Water Treatment Facilities.* Springfield.

———. Division of Water Pollution Control. 1986. *Illinois Water Quality Report, 1984–1985.* July.

Indiana Department of Environmental Management. 1984–85. *1984–85 305(b) Report.*

Inside EPA. 1980. Washington, D.C.: Inside Washington Publishers, pp. 7–8. 9 May.

Iowa Department of Natural Resources. Environmental Protection Commission. 1987. *Iowa Groundwater Protection Strategy—1987.* 106pp.

Jercinovic, D. E. 1982. *Assessment of Refined Petroleum Product Contamination Problems in Surface and Ground Waters of New Mexico.* New Mexico Health and Environment Department and Natural Resources Department. November.

Junk, G. A., R. F. Spalding, and J. J. Richard. 1980. Areal, Vertical and Temporal Differences in Ground Water Chemistry II—Organic Constituents. *Journal of Environmental Quality* 9(3):479–83.

Kansas Department of Health and Environment. 1982. *Groundwater Quality Management Plan for the State of Kansas.* Bulletin 3-4. 1 January. 77pp.

Kasabuck, H. F. and W. F. Athoff. 1983. An Overview of New Jersey's Groundwater Quality Program. *Ground Water* 21(5): Sept–Oct.

Keeley, J. W. 1976. Ground Water Pollution Problems in the United States. In *Proceedings of a Water Research Conference, "Ground Water Quality—Measurement, Prediction and Protection,"* pp. 17–32. University of Reading, Berkshire, England. 6–8 Sept.

——. 1977. Magnitude of the Ground Water Contamination Problem. In *Public Policy on Ground Water Protection,* ed. W. R. Kerns, pp. 2–10. Proceedings of a national conference at Virginia Polytechnic Institute and State University, Blacksburg, Va. 13–16 April.

Kelley, R. D. 1985. *Synthetic Organic Compound Sampling Survey of Public Water Supplies.* Iowa Department of Water, Air, and Waste Management. April.

Kentucky Natural Resources and Environmental Protection Cabinet. Division of Water. 1986. *1986 Kentucky Report to Congress on Water Quality.*

Kerns, W. R., ed. 1977. *Public Policy on Ground Water Protection.* Proceedings of a national conference at Virginia Polytechnic Institute and State University, Blacksburg, Va. 13–16 April. 163pp.

Knowlton, H. E., and E. J. Rucker. 1979. Land Farming Shows Promise for Refinery Waste Disposal. *The Oil and Gas Journal* 77 (May 14):108–16.

Knox, R. C., L. W. Canter, D. F. Kincangon, E. L. Stover, and C. H. Ward. 1984. *State-of-the-Art Aquifer Restoration. Volume I, Sections 1–8.* Robert S. Kerr Environmental Research Laboratory. U.S. EPA. Office of Research and Development. Ada, Okla. November. 399pp.

Krone, R. B., P. H. McGauhey, and H. B. Gotars. 1957. Direct Discharge of Ground Water with Sewage Effluents *ASCE Journal of the Sanitary Engineering Division* 83(SA4):1–25.

Krone, R. B., G. T. Orlab, and C. Hodgkinson. 1958. Movement of Coliform Bacteria Through Porous Media. *Sewage Industrial Wastes* 30:1–13.

Krueger, J. 1984. *Nebraska Groundwater Quality Protection Strategy Problem Assessment.* Nebraska Department of Environmental Control. Water and Waste Management Division. June.

Lance, J. C. 1981. Microbial Health Considerations of Soil Disposal of Domestic Wastewaters: Soil Considerations. In *National Center for Ground Water Research (NCGWR). Proceedings of a National Conference on the Microbial Health Considerations of Soil Disposal of Domestic Wastewaters,* pp. 11–23. Norman, Okla. May.

491

Bibliography

Last, J. M. 1980. *Public Health and Preventive Medicine,* ed. Maxcy-Rosenau. 11th ed. Norwalk, Conn.: Appleton-Century-Crofts.

League of Women Voters Educational Fund. 1985. *The Nuclear Waste Primer. A Handbook for Citizens.* New York: Nick Lyons Books. 90pp.

Lehr, J. H. 1975. Ground Water Pollution—Problems and Solutions. In *Water Pollution Control in Low Density Areas,* ed. W. J. Jewell and R. Swan, pp. 111–20. Hanover, N.H.: University Press of New England.

———. 1982. How Much Ground Water Have We Really Polluted? *Ground Water Monitoring Review,* Winter, pp. 4–5.

———. 1985. Groundwater: Its Protection is Within Reach. *Environmental Geology and Water Sciences* 7(3):125–127.

Lehr, J. H., W. A. Pettyjohn, M. S. Bennett, J. R. Hanson, and L. E. Sturtz. 1976. *A Manual of Laws, Regulations and Institutions for the Control of Ground Water Pollution.* Prepared for U.S. EPA, EPA-440/9-76-006. 381pp.

Lemmon, J. 1980. Drums Along the Salt. In *Proceedings of the Arizona Section of the American Water Resources Association Symposium on Water Quality Monitoring and Management,* pp. 7–12. Tucson, Ariz. 24 October. 138pp.

Leopold, L. B. 1974, *Water: A Primer.* San Francisco: W. H. Freeman and Co. 172pp.

Lindorff, D. E., and K. Cartwright. 1977. *Ground Water Contamination: Problems and Remedial Actions.* Illinois State Geological Survey, Environmental Geology Notes no. 81. May. 58pp.

Louisiana Department of Environmental Quality. Office of Water Resources. 1986. *1986 Louisiana Water Quality Inventory.*

MacKichan, K. A., and J. C. Kammerer. 1961. *Estimated Use of Water in the United States, 1960.* USGS Circular 456. 26pp.

Madison, R. J. and J. O. Brunett. 1985. Overview of the Occurrence of Nitrate in Ground Water in the United States. In *National Water Summary 1984.* Water Supply Paper 2275, pp. 93–105. Washington, D.C.: U.S. Geological Survey.

Magnuson, P. 1981. *Ground Water Classification.* Syosset, New York: Geraghty and Miller, Inc. 56pp.

Malot, J. J. 1985. Unsaturated Zone Monitoring and Recovery of Underground Contamination. Presented at the Fifth National Symposium on Aquifer Restoration and Ground Water Monitoring. Columbus, Ohio, 21-24 May.

Maine Department of Environmental Protection. 1986. *State of Maine 1986 Water Quality Assessment.*

Maine Revised Statutes Annotated. 1979. *Classification of Ground Water.*

Mann, L. J. 1985. Ground-Water-Level Changes in Five Areas of the United States. In *National Water Summary 1984.* Water Supply Paper 2275, pp. 106–113. Washington, D.C.: U.S. Geological Survey.

Mapp, H. P., and V. R. Eidman. 1976. A Bioeconomic Simulation Analysis of Regulating Ground Water Irrigation. *American Journal of Agricultural Economics* 58(3):391–402.

Marshfield Engineering Services. 1982. *Vermont Underground Injection Practices and Groundwater Contamination Incidents Survey.* Prepared for the Vermont Department of Water Resources and Environmental Engineering. Groundwater Management Section. May.

Maryland Department of Health and Mental Hygiene. Office of Environmental Programs. 1986. *Maryland Water Quality Inventory: A Report on the Progress Toward Meeting the Goals of the Clean Water Act.* April.

493

Mason, T. J., and F. W. McKay. 1974. *U.S. Cancer Mortality by County: 1950–1969.* Washington, D.C.: U.S. Dept. of Health, Education and Welfare Publication no. (WIH) 74-615.

Massachusetts Department of Environmental Quality Engineering. Division of Water Supply. 1984. *Draft—Public Water Supplies Which Have Been Closed Due to Contamination.* 20 March.

———. Division of Water Pollution Control. 1986. *Commonwealth of Massachusetts Summary of Water Quality 1986.* March.

McQuillan, D. M. 1984. *Water Quality Concerns in the Albuquerque South Valley.* Prepared for the Governor's South Valley Public Health Response Team. New Mexico Health and Environmental Department. February. 13pp.

Meyer, C. F., ed. 1973. *Polluting Ground Water: Some Causes, Effects, Controls and Monitoring.* U.S. EPA Environmental Monitoring Series. EPA-600/4-73-001b.

Michigan Cabinet Council on Environmental Protection. 1984. *Ground Water Protection Initiatives.* September.

Michigan Department of Natural Resources. Water Resources Commission. 1980. *General Rules.* Parts 1, 4, 5, 6, 9, 10, 11, 21, and 22.

———. 1983. *Groundwater Management Strategy for Michigan.* p. 13.

———. 1985. *Progress Report on Implementation of Michigan's Groundwater Protection Initiatives.* September.

Miller, D. W. 1981a. Basic Elements of Ground Water Contamination. In *Seminar on the Fundamentals of Ground Water Quality Protection.* Presented by Geraghty and Miller, Inc., and American Ecology Services, Inc. Cherry Hill, N.J. 5–6 October.

———. 1981b. Tools for Investigating Ground Water Contamination. In *Seminar on the Fundamentals of Ground Water Quality Protection.* Presented by Geraghty and Miller, Inc., and American Ecology Services, Inc. Cherry Hill, N.J. 5–6 October.

———. 1981c. Planning On-Site Investigations. In *Seminar on the Fundamentals of Ground Water Quality Protection.* Presented by Geraghty and Miller, Inc., and American Ecology Services, Inc. Cherry Hill, N.J. 5–6 October.

Miller, D. W., F. A. DeLuca, and T. L. Tessier. 1974. *Ground Water Contamination in the Northeast States.* Prepared for U.S. EPA Office of Research and Development. EPA 660/2-74-056. 328pp.

Bibliography

Miller, D. W., and J. P. Sgambat. 1985. Abatement and Restoration—
  Overview of Remedial Actions at Waste Disposal Sites. In *The
  Fundamentals of Ground-Water Contamination.* Presented by
  Geraghty and Miller, Inc., and American Ecology Services, Inc.
  Syosset, N.Y. October. 25pp.

Miller, J. C., P. S. Hackenberry, and F. A. DeLuca. 1977. *Ground Water
  Pollution Problems in the Southeastern United States.* Prepared for
  U.S. EPA Office of Research and Development. EPA 600/3-77-012.
  361pp.

Minnesota Pollution Control Agency. 1984. *Permanent List of Priorities.*
  August.

———. 1986. *Minnesota Water Quality: Water Years 1984, 1985.*

Mississippi Department of Natural Resources. Bureau of Pollution Control.
  1982. *Underground Injection Control Program Regulations.* February.
  19pp.

———. 1986. *Mississippi Water Quality Report 1986.*

Missouri Department of Natural Resources. 1986. *Water Pollution Control
  Program—Missouri Water Quality Report 1986.*

Montana Department of Health and Environmental Sciences. Environmental
  Sciences Division. 1982a. *Administrative Rules of Montana.* Title 16,
  Chapter 20, Water Quality. Subchapter 2. Public Water Supplies.

———. Environmental Sciences Division. 1982b. *Administrative Rules of
  Montana.* Title 16, Chapter 20. Water Quality. Montana Groundwater
  Pollution Control System.

———. Environmental Sciences Division. *Montana Water Quality 1984.*
  October.

———. Water Quality Bureau. 1986. *Montana Water Quality 1986.*

Murray, C. R., and E. B. Reeves. 1972. *Estimated Use of Water in the
  United States in 1970.* U.S. Geological Survey Circular no. 676.

———. 1977. *Estimated Use of Water in the United States in 1975.* U.S.
  Geological Survey Circular no. 765. 39pp.

Mutch, R. D. Jr., J. H. Clark and J. V. Rouse. 1987. Recent Advances in the
  In-Situ Management of Uncontrolled Waste Disposal Sites. In
  Proceedings of The Third National Water Conference. Philadelphia.
  13–15 January.

National Academy of Sciences (NAS). 1977. *Drinking Water and Health.*
  Vol. 1. Washington, D.C.: National Academy Press. 939pp.

———. 1979. *The Geochemistry of Water in Relation to Cardiovascular
  Disease.* Washington, D.C.: National Academy Press. 98pp.

———. 1980a. *Drinking Water and Health.* Vol. 2. Washington, D.C.:
  National Academy Press. 393pp.

———. 1980b. *Drinking Water and Health.* Vol. 3. Washington, D.C.:
  National Academy Press. 415pp.

———. 1981a. *Drinking Water and Health.* Vol. 4. Washington, D.C.:
  National Academy Press. 226pp.

———. 1981b. *Coal Mining and Ground Water Resources in the United States.* Washington, D.C.: National Academy Press. 197 pp.

———. 1981c. *The Health Effects of Nitrate, Nitrate and N-Nitroso Compounds.* Part 1. Washington, D.C.: Committee on Nitrate and Alternative Curing Agents in Food. 529pp.

———. 1983. *Drinking Water and Health.* Vol. 5. Washington, D.C.: National Academy Press. 157pp.

———. 1986. *Drinking Water and Health.* Vol. 6. Washington, D.C.: National Academy Press. 457pp.

National Center for Ground Water Research (NCGWR). 1981. *Microbial Health Considerations of Soil Disposal of Domestic Wastewaters. Proceedings of the Conference.* Norman, Okla. 11–12 May. 242pp.

National Research Council. Committee on Ground-Water Quality Protection. 1986. *Ground Water Quality Protection: State and Local Strategies.* Washington, D.C.: National Academy Press. 309pp.

Nebraska Department of Environmental Control. 1978. *Ground Water Protection Standards.* 30 October. Lincoln. 11pp.

———. 1980a. *Water Quality Report. Pursuant to Section 305(b) of the Clean Water Act.* Lincoln. 303pp.

———. 1980b. *Final Report: Nebraska Surface Impoundment Assessment.* January. Lincoln. 72pp.

———. 1981. *An Investigation of the Causes of Nitrate Contamination in the Ground Water of the Lower Big Nemaha Drainage Basin.* Ground Water Quality Protection Program. Lincoln. 8pp.

———. 1985. Draft—*Nebraska Ground Water Quality Protection Strategy— A Technical Summary.* 43pp.

———. Water Quality Division. 1986. *Nebraska Water Quality Report.* April.

Neri, L. C., D. Hewitt, and G. B. Schreiber. 1974. Can Epidemiology Elucidate the Water Story? *American Journal of Epidemiology* 99(2):75–88.

Nevada Department of Conservation and Natural Resources. Division of Environmental Protection. 1986. *Ground-Water Quality Management Plan for Nevada.* Carson City. 125pp.

New England Interstate Water Pollution Control Commission. Newsletter. 1985. *Water Connection* 2(3) November.

New Hampshire Code of Administrative Rules. 1982. *Part Ws 410, Protection of Groundwaters of the State.*

New Hampshire Water Supply and Pollution Control Commission. 1986. *New Hampshire Water Quality Report To Congress (305b).* April.

New Jersey Department of Environmental Protection. Division of Water Resources. 1981a. *New Jersey Ground Water Pollution Index, 1975– June 1981.* Trenton. 55pp.

———. 1981b. Ground Water Quality Standards. In *New, Revised, and Amended Rules Concerning Water Quality Standards.* Chapter 6.

———. 1983. Draft—*Ground Water Programs in New Jersey. A guide to Rules, Programs and Officials.* April. 15 pp.

Bibliography

———. 1984. *New Jersey Ground Water Pollution Index, September, 1974–April, 1984.* New Jersey Geological Survey Open File Report No. 84-1.

———. 1986. *New Jersey 1986 State Water Quality Inventory Report.* July.

New Mexico Environmental Improvement Division. 1980. *New Mexico Surface Impoundment Assessment.* February. Sante Fe. 157pp.

New Mexico Water Quality Control Commission. 1983. *New Mexico Water Quality Control Commission Regulations as Amended through November 17, 1983.* 70pp.

———. 1986. *Water Quality and Water Pollution Control in New Mexico 1986.* October.

New York Department of Environmental Conservation. 1978. *Ground Water Classifications, Quality Standards and/or Limitations.* Effective 1 September. 16pp.

———. 1983. Groundwater Problems. In *Draft—Up State Groundwater Management Program.* Chapter 2.

———. Division of Water. 1986a. *Final—Long Island Groundwater Management Program.* June. 237pp.

———. 1986b. *New York State Water Quality Report 305b.*

North Carolina Department of Natural Resources and Community Development. Division of Environmental Management. 1986. *Water Quality Progress in North Carolina, 1984–85.*

North Carolina Environmental Management Commission. 1979. *Classifications and Water Quality Standards Applicable to the Ground Water of North Carolina.* Effective 10 June. 7pp.

———. Division of Environmental Management. 1985. North Carolina Administrative Code, Title 15. *Classifications and Water Quality Standards Applicable to the Groundwaters of North Carolina.* 1 March. 18pp.

North Dakota State Department of Health. 1986a. *Computer Print-out of Groundwater Contamination Sites. 1982–86.* 13 March.

———. Division of Water Supply and Pollution Control. 1986b. *The Status of Water Quality in the State of North Dakota, 1984–85.* June.

Nyer, E. K. 1985. *Groundwater Treatment Technology.* New York: Van Nostrand Reinhold Co. 188pp.

Office of Technology Assessment (OTA). 1984. *Protecting the Nation's Ground Water from Contamination*—Vols. I and II. Washington, D.C. OTA-0-233. October. 503pp.

Oklahoma Water Resources Board. 1985. *Rules, Regulations and Modes of Procedure.* Publication No. 126. Oklahoma City. 118pp.

Okun, D. A. 1980. Water Quality Management. In *Public Health and Preventive Medicine,* ed. Maxcy-Rosenau. 11th ed., pp. 975–1018. Norwalk, Conn.: Appleton-Century-Crofts.

Oregon Administrative Rules. 1984. Chapt. 340, Division 41-Department of Environmental Quality. General Groundwater Quality Protection Policy.

Oregon Department of Environmental Quality. Hazardous and Solid Waste
     Division. 1985. *Draft—State Groundwater Quality Protection Program.*
     December. 58pp.
——. 1986a. *Oregon 1986 Water Quality Program Assessment and
     Program Plan for 1987.* June.
——. 1986b. *Draft—Inventory of Groundwater Quality Problems and
     Areas of Concern.* 7pp.

Page, G. W. 1981. Comparison of Ground Water and Surface Water for
     Patterns and Levels of Contamination by Toxic Substances.
     Environmental Science and Technology 15(12):1475–81.
——. 1982. Maximum Contaminant Levels for Toxic Substances in Water:
     A Statistical Approach. *Water Resources Bulletin,* 18(6):955–964.
Paige, S., C. Morgan, H. Bryson, G. Hunt, P. Roqoschewski, P. Spooner, D.
     Twendell, and R. Wetzel. 1980. Preliminary Design and Cost
     Estimates for Remedial Actions of Hazardous Waste Disposal Sites.
     In *Management of Uncontrolled Hazardous Waste Sites.* Proceedings
     of the U.S. EPA National Conference, pp. 202–7. Washington, D.C.
     15–17 October.
Pennsylvania Department of Environmental Resources. Bureau of Water
     Quality Management. 1972. *Spray Irrigation Manual.*
——. 1984. *1984 Water Quality Inventory Section 305b and Appendices.*
——. Bureau of Water Quality Management. 1986. *1986 Water Quality
     Assessment.* April.
——. Title 25. *Rules and Regulations.* Subpart C. Protection of Natural
     Resources Article I: Land Resources, Article II: Water Resources, and
     Article VI: General Health and Safety.
Peters, H. J. 1982. *Groundwater Management in California.* American
     Society of Civil Engineers. Las Vegas, Nev. April 26-30. Preprint 82-
     035. 13 pp.
Pettyjohn, W. A., ed. 1972. *Water Quality in a Stressed Environment:
     Readings in Environmental Hydrology.* Minneapolis, Minn.: Burgess
     Publishing Co. 309pp.
Piskin, R., L. Kissinger, M. Ford, S. Colantino, and J. Lesnak. 1980.
     Inventory and Assessment of Surface Impoundments in Illinois.
     Prepared by State of Illinois EPA Div. of Land/Noise Pollution.
     January. 160pp.
Pories, W. J., E. G. Mansour, and W. H. Strain. 1972. Trace Elements That
     Act to Inhibit Neoplastic Growth. *Annals of the New York Academy of
     Science* 199:265–71.
Press, F., and R. Siever. 1978. *Earth.* San Francisco: W. H. Freeman and
     Co. 613pp.
Pye, V. I., R. Patrick and J. Quarles, 1983. *Groundwater Contamination in
     the United States.* Philadelphia: University of Pennsylvania Press.

Rhode Island Department of Environmental Management. Division of Water

497

Resources. 1984. *The State of the State's Water Resources—Rhode Island: A Report to Congress.* April.

———. Division of Water Resources. 1986. *The State of the State's Water Resources—Rhode Island: A Report to Congress.* April.

Rhode Island Groundwater Protection Act. 1985. Title 46. Chapter 13.1.

Robertson, F. N., 1975. Hexavalent Chromium in the Ground Water in Paradise Valley, Arizona. *Ground Water* 13(6):516–27.

Romero, J. C. 1972. The Movement of Bacteria and Viruses through Porous Media. In *Water Quality in a Stressed Environment, Readings in Environmental Hydrology,* ed. W. A. Pettyjohn, pp. 200–223. Minneapolis, Minn.: Burgess Publishing Co.

Saar, R. A. 1985. Behavior and Movement of Contaminants. In *The Fundamentals of Ground-Water Contamination.* Presented by Geraghty and Miller, Inc., and American Ecology Services, Inc. Syosset, N.Y. October. 45pp.

Salt Institute. 1980. *Survey of Salt, Calcium Chloride and Abrasive Use in the United States and Canada for 1978–1979.* Salt Institute RP-2-80-2M. Alexandria, Va. 57pp.

Scalf, M. R., J. W. Keeley, and C. J. LaFevers. 1973. *Ground Water Pollution in the South Central States.* Prepared for U.S. EPA. EPA-R2-73-268, 181pp.

———, and W. J. Dunlap, 1977. *Environmental Effects of Septic Tank Systems.* Prepared for U.S. EPA. EPA-600/3-77-096.

Schmidt, K. D. 1972. Ground Water Contamination in the Cortaro Area, Pima County, Arizona. In *Hydrology and Water Resources in Arizona and the Southwest,* Vol. 2. Proceedings of the 1972 meetings of the Arizona Section of the American Water Resources Association and the Hydrology Section of the Arizona Academy of Science. Prescott, Ariz. 5–6 May.

———. 1973. Ground Water Quality in the Cortaro Area Northwest of Tucson, Arizona. *Water Resources Bulletin* 9(3):598–606.

Schneider, W. J. 1972. Hydrologic Implications of Solid Waste Disposal. In *Water Quality in a Stressed Environment: Readings in Environmental Hydrology,* ed. W. A. Pettyjohn, pp. 130–45. Minneapolis, Minn.: Burgess Publishing Co.

Schroeder, H. A. 1969. The Water Factor. *New England Journal of Medicine* 280:836–83.

Schroeder, L., ed. 1985. *The Journal of Freshwater. Vol. 9.* Nevarre, Minn.: Freshwater Foundation. 40pp.

Sharefkin, M., M. Shechter, and A. Kneese. 1984. Impacts, Costs, and Techniques for Mitigation of Contaminated Ground Water: A Review. *Water Resources Research* 20(12):1771–1783.

Sharrett, A. R. 1979. The Role of Chemical Constituents of Drinking Water in Cardiovascular Diseases. *American Journal of Epidemiology* 110(4):401–19.

Sills, M., J. Struzziery, and P. Silbermann. 1980. Evaluation of Remedial

Treatment, Detoxification and Stabilization Alternatives. In *Management of Uncontrolled Hazardous Waste Sites,* pp. 192–201. U.S. EPA National Conference. Washington, D.C.

Sobsey, M. 1981. Transport and Fate of Viruses in Soil. In *National Center for Ground Water Research (NCGWR). Proceedings of the Conference on Microbial Health Considerations of Soil Disposal of Domestic Wastewaters,* pp. 103–25. Norman, Okla. May.

Solley, W. B., E. B. Chase, and W. B. Mann IV. 1983. *Estimated Use of Water in the United States in 1980.* Geological Survey Circular 1001. Alexandria, Va. 56pp.

Souder, K. 1983. *Gasoline Contamination of Ground Water in the Village of Pecos, New Mexico: Technical Findings and Recommendations.* New Mexico Health and Environment Department. October. 10pp.

South Carolina Department of Health and Environmental Control. 1980. *Inventory of Ground Water Contamination Cases in South Carolina.* March. 58pp.

————. 1981a. *Inventory of Known Ground Water Contamination Cases and Generalized Delineation of Five Ground Water Recharge Areas in South Carolina.* Draft. November. 122pp.

————. Division of Water Supply. 1981b. *State Primary Drinking Water Regulations.* May. 135pp.

————. 1983. *A Report to the General Assembly on Groundwater Contamination in South Carolina.* November.

————. Office of Environmental Quality Control. 1985. *Water Classifications and Standards (Regulation 61-68) Classified Waters (Regulation 61-69).* Columbia. June. 58pp.

————. Office of Environmental Quality Control. 1986. *Water Quality Assessment FY 84–85.* Columbia. June. 58 pp.

Spalding, R. F., J. R. Gormly, B. H. Curtiss, and M. E. Exner. 1978a. Nonpoint Nitrate Contamination of Ground Water in Merrick County, Nebraska. *Ground Water* 16(2):86–95.

Spalding, R. F., G. A. Junk, and J. J. Richard. 1978b. Pesticides in Ground Water Beneath Irrigated Farmland in Nebraska. *Pesticides Monitoring Journal* 14(2):70–73.

Spalding, R. F., M. E. Exner, J. J. Sullivan, and P. A. Lyon. 1979. Chemical Seepage from a Tail Water Recovery Pit to Adjacent Ground Water. *Journal of Environmental Quality* 8(3):374–83.

Spalding, R. F., and M. E. Exner. 1980. Areal, Vertical and Temporal Differences in Ground Water Chemistry. Part I: Inorganic Constituents. *Journal of Environmental Quality* 9(3):466–79.

Spofford, W. O., Jr. 1986. Giardiasis: A Return of Waterborne Disease? *Resources* no. 83. Spring, pp. 5–9. Washington, D.C.: Resources for the Future, Inc.

Stengel, R. 1982. Ebbing of the Ogallala. *Time,* 10 May.

Suflita, J. M. 1987. The Biodegradation of Ground Water Pollutants When Oxygen is Unavailable. In *Proceedings of The Third National Water Conference.* Philadelphia. 13–15 January.

499

Bibliography

Tennessee Department of Health and Environment. Office of Water
     Management. 1986. *1986 305(b) Status of Water Quality in
     Tennessee.* August.
Texas Department of Water Resources. 1984a. *Computer Print-out of
     Groundwater Investigations.* November. 27pp.
———. 1984b. *Water for Texas: A Comprehensive Plan for the Future.* Vol.
     1. Austin. November. 72pp.
———. 1984c. *Water for Texas: Vol. 2—Technical Appendix.* Austin.
Texas Water Commission. 1986. *The State of Texas Water Quality
     Inventory.* 8th ed.
Thomas, H. E., and D. A. Phoenix. 1976. *Summary Appraisals of the
     Nation's Ground Water Resources—California Region.* U.S.
     Geological Survey Professional Paper 813-E. 51pp.
Thompson, D. R. 1977. Surface and Subsurface Mining: Policy
     Implications. In *Public Policy on Ground Water Quality Protection,*
     ed. W. R. Kerns, pp. 22–29. Proceedings of a national conference at
     Virginia Polytechnic Institute and State University, Blacksburg, Va.
     13–16 April.
Tinlin, R., ed. 1976. *Monitoring Ground Water Quality: Illustrative
     Examples.* Santa Barbara, Calif.: General Electric Co./TEMPO. EPA-
     600/4-76-036. 81pp.
Todd, D. K. 1959. *Ground Water Hydrology.* New York: John Wiley and
     Sons. 336pp.
Toups, J. M. 1974. *Water Quality and Other Aspects of Ground Water
     Recharge in Southern California.* Paper presented at the annual
     conference of the American Water Well Association, 15 May.
Tucker, R. K. 1981. *Ground Water Quality in New Jersey: An Investigation
     of Toxic Contaminants.* New Jersey Dept. of Environmental
     Protection, Office of Cancer and Toxic Substances Research. 60pp.

UNESCO. 1980. *Aquifer Contamination and Protection.* International
     Hydrological Programme. Project 8.3 of the Studies and Reports in
     Hydrology 30. Paris. 440pp.
United States Congressional Research Service. 1980. *Resource Losses
     from Surface Water, Ground Water and Atmospheric Contamination:
     A Catalog.* Report prepared for the U.S. Senate Committee on
     Environment and Public Works. 246pp.
United States Department of Commerce. 1974. *Statistical Abstract of the
     United States, 1974.* Social and Economic Statistics Administration,
     Bureau of the Census. 1028pp.
———. 1977. *Transport Statistics in the United States for the year ended
     31 December 1976.* Part 6: Pipelines. Prepared by the Bureau of
     Accounts, Interstate Commerce Commission. U.S. GPO Stock No.
     026-000-01101-9. 29pp.
———. 1980. *Census of Agriculture—State Summary Data.* Bureau of the
     Census. Washington, D.C.
———. 1982. *Liquid Pipeline Incident Summary Data, 1968–1981.*

Prepared by the Information Systems Division, Research and Special Programs Administration. January. 56pp.

———. 1986. *National Data Book and Guide to Sources—1986.* Statistical Abstract of the United States. 106th ed. Bureau of the Census. Washington, D.C. 985pp.

U.S. Environmental Protection Agency. (U.S. EPA). 1973. *An Environmental Assessment of Potential Gas and Leachate Problems at Land Disposal Sites.* Hazardous Waste Management Division. Open File Report (SW 110 of).

———. 1976. *National Interim Primary Drinking Water Regulations.*

———. 1977. *The Report to Congress. Waste Disposal Practices and Their Effects on Ground Water.* Prepared by the Office of Water Supply and Office of Solid Waste Management Programs. 512pp.

———. 1978a. *Executive Summary: Surface Impoundments and Their Effects on Ground Water Quality in the United States—A Preliminary Survey.* EPA-570/9-78-005. June. 30pp.

———. 1978b. *Surface Impoundments and Their Effects on Ground Water Quality in the U.S.—A Preliminary Survey.* EPA/9-78-004. 275pp.

———. 1980a. *Ground Water Protection. A Water Quality Management Report.* November. 36pp.

———. 1980b. *Planning Workshops to Develop Recommendations for a Ground Water Protection Strategy. Appendices.* Office of Drinking Water. June. 171pp.

———. 1981a. *Computer Print-out of Disease Outbreaks Attributed to Ground Water between 1948 and 1980.* Cincinnati, Oh.

———. 1981b. *Ground Water Research Plan.* Office of Research and Development. EPA-600/9-81-031. 34pp.

———. 1983. *Surface Impoundment Assessment National Report.* Office of Drinking Water. Washington, D.C. EPA 570/9-84-002. December. 116pp.

———. 1984a. *Ground-Water Protection Strategy.* Office of Ground-Water Protection. Washington, D.C. August. 56pp.

———. 1984b. *National Statistical Assessment of Rural Water Conditions.* Office of Drinking Water. Washington, D.C. EPA 570/9-84-004. June. 111pp.

———. 1984c. *National Statistical Assessment of Rural Water Conditions Executive Summary.* Office of Drinking Water. Washington, D.C. EPA 570/9-84-003. June. 21pp.

———. 1984d. *A List of Some Groundwater Contamination Problems that have Occurred in Region 10.* Seattle, Washington.

———. 1985a. *Ground-Water Monitoring Strategy.* Office of Ground-Water Protection. Washington, D.C. December. 54pp.

———. 1985b. *National Water Quality Inventory 1984 National Report to Congress.* Office of Water Regulations and Standards. Washington, D.C. EPA-440/4-85-029. 173pp.

———. 1985c. *Overview of State Ground-Water Program Summaries* Vol. I. Office of Ground-Water Protection. Washington, D.C. March. 28pp.

———. 1985d. *Selected State and Territory Ground-Water Classification Systems.* Office of Ground-Water Protection. Washington, D.C. May. 105pp.

———. 1985e. *State Ground-Water Program Summaries* Vol. II. Office of Ground-Water Protection. Washington, D.C. March. 660pp.

———. 1986a. *Guidelines for Ground-Water Classification under the EPA Ground-Water Protection Strategy Final Draft.* Office of Ground-Water Protection. Washington, D.C. December, 137pp.

502

———. 1986b. *Pesticides in Ground Water: Background Document.* Office of Ground-Water Protection. Washington, D.C. May. 72pp.

———. 1986c. *Septic Systems and Ground-Water Protection: A Program Manager's Guide and Reference Book.* Office of Ground-Water Protection. Washington, D.C. July. 72pp.

———. 1986d. *Septic Systems and Ground-Water Protection: An Executive's Guide.* Office of Ground-Water Protection. Washington, D.C. July. 13pp.

———. 1986e. *State Program Briefs—Pesticides in Ground-Water.* Office of Ground-Water Protection. Washington, D.C. May. 93pp.

———. 1986f. *Summary of State Reports on Releases from Underground Storage Tanks.* Office of Solid Waste. Washington, D.C. EPA 600/M-86/020. July. 95pp.

———. 1986g. *Underground Motor Fuel Storage Tanks: A National Survey.* Vol. I: *Technical Report.* Office of Pesticides and Toxic Substances. Washington, D.C. EPA 560/5-86-013. May. 200pp.

United States General Accounting Office (U.S. GAO). 1978. *Waste Disposal Practices—A Threat to Health and the Nation's Water Supply.* CED-78-120. June. 34pp.

———. 1980. *Ground Water Overdrafting Must be Controlled. A Report to the Congress of the U.S. by the Comptroller General.* CED-80-96. 12 September. 52pp.

———. 1984. *Federal and State Efforts to Protect Groundwater.* Washington, D.C. February. 80pp.

U.S. Geological Survey (USGS). 1985. *National Water Summary 1984.* Water Supply Paper 2275. Washington, D.C. 467pp.

U.S. Library of Congress. Congressional Research Service. 1983. *Underground Injection of Wastes.* Report No. 83-196 ENR. by D. V. Feliciano. Washington, D.C. October. 21pp.

———. 1984. *Groundwater: What It Is, and How It Is Being Protected.* Report No. 84-16 EPR. by D. V. Feliciano. Washington, D.C. January. 52pp.

U.S. Water Resources Council. 1978a. *The Nation's Water Resources, 1975–2000.* Vol. 1: *Summary. Second National Water Assessment.* 86pp.

———. 1978b. *The Nation's Water Resources, 1975–2000.* Vol. 2: *Water Quantity, Quality and Related Land Considerations.* 618pp.

———. 1978c. *The Nation's Water Resources, 1975–2000.* Vol. 3: *Analytical Data Survey.* 89pp. Appendix I: Social, Economic and

Environmental Data; Appendix II: Annual Water Supply and Use Analysis; Appendix III: Monthly Water Supply and Use Analysis; Appendix IV: Dry Conditions Water Supply and Use Analysis; Appendix V: Streamflow Conditions.

———. 1980. *Essentials of Ground Water Hydrology Pertinent to Water Resources Planning.* Bulletin 16 (revised).

University of Nebraska, Lincoln. 1980. *Maps of Ground Water Nitrate-Nitrogen Concentrations. Reconnaissance Sampling of the National Uranium Resource Evaluation Program.* Prepared by Nebraska Conservation and Survey Div. Inst. of Agriculture and Natural Resources.

Utah Department of Health. Bureau of Water Pollution Control. 1986. *State of Utah 305(b) Biennial Water Quality Report.* August.

van der Leeden, F., L. A. Cerrillo, and D. A. Miller. 1975. *Ground Water Problems in the Northwestern United States.* Prepared for U.S. EPA, Office of Research and Development. EPA-660/3-75-018. 361pp.

Van Everdingen, R. O., and R. A. Freeze. 1971. *Subsurface Disposal of Waste in Canada.* Inland Waters Branch Tech. Report. Dept. of Environment, Canada.

Vermont Agency of Environmental Conservation. Department of Water Resources and Environmental Engineering. 1983. *Vermont Aquifer Protection Area Reference Document.* March. 21pp.

———. 1984. *Vermont Statutes Annotated Environmental Protection Regulations.* Chapter 6: Solid Waste Management; Subchapter 6: Hazardous Waste Management Regulations as amended on September 13, 1984. Title 10.

———. Department of Water Resources and Environmental Engineering. 1985a. 1985—Banner Year for the Environment. *Conservation Quarterly.* Autumn.

———. 1985b. *Vermont Statutes Annotated.* Chapter 48. Groundwater Protection. Title 10.

———. Department of Water Resources and Environmental Engineering. 1986. *1986 Water Quality Assessment.*

Virginia State Water Control Board. 1974. *Rules of the Board and Standards for Water Wells. The Groundwater Act of 1973.* Richmond. May. 33 pp.

———. 1982. *Water Quality Standards.* Publication no. RB-1-80. Richmond. April. 78pp.

———. 1986. *Virginia Water Quality Assessment.* Volume 1 of 2. Information Bulletin 565. April.

Walker, W. H., 1969. Illinois Ground Water Pollution. *Journal of the American Water Well Association* 61(1):31–40.

Ward, C. H. and J. M. Thomas. 1987. How Clean is Clean Ground Water Remediated by In-Situ Biorestoration? In *Proceedings of The Third National Water Conference.* Philadelphia. 13–15 January.

503

Bibliography

Warren, J., H. Mapp, D. Ray, D. Kletke, and C. Wang. 1982. Economics of Declining Water Supply in the Ogallala Aquifer. *Ground Water* 20(1):73–79.

Washington Department of Ecology. 1986. *Water Quality Assessment and Program Strategy (305-b) Report.* Appendix 2.

Weeks, J. B. 1986. High Plains Regional Aquifer-System Study. In *Regional Aquifer-System Analysis Program of the U.S. Geological Survey. Summary of Projects, 1978–84.* U.S. Geological Survey Circular 1002. Alexandria, Va., pp. 30–49.

Weimar, R. A. 1980. Prevent Ground Water Contamination Before It's Too Late. *Water and Wastes Engineering.* February: 30–33, 63.

Wendt, C. W., A. B. Onken, O. C. Wilke, and R. D. Lacewell. 1976. *Effects of Irrigation Methods on Ground Water Pollution by Nitrates and Other Solutes.* EPA-600/2-76-291. December. 331pp.

Werth, E. P. 1985. Setting Up Monitoring Programs. In *The Fundamentals of Ground-Water Contamination.* Geraghty and Miller, Inc., and American Ecology Services, Inc. Syosset, N.Y. October. 33pp.

West Virginia Department of Natural Resources. Division of Water Resources. 1986. *State of West Virginia 1983–85 305(b) Report.*

Wilson, J. 1982. *Groundwater: A Non-Technical Guide.* Philadelphia, Pa.: The Academy of Natural Sciences. 84pp.

Wisconsin Department of Natural Resources. Bureau of Water Resources Management. 1986. *Wisconsin Water Quality: 1986 Report to Congress.* June. 144pp.

World Health Organization. 1979. *Human Viruses in Water, Wastewater and Soil.* World Health Organization Technical Report Series 639. Geneva. 50pp.

Wyoming Department of Environmental Quality. Water Quality Division. 1980a. Quality Standards for Wyoming Ground Waters. *Water Quality Rules and Regulations.* Chapter 8. 13pp.

———. 1980b. Wyoming Groundwater Pollution Control Permits. *Water Quality Rules and Regulations.* Chapter 9. 24pp.

———. 1984. Design and Construction Standards for Sewerage Systems or Other Facilities Capable of Causing or Contributing to Pollution and Mobile Home Park and Campground Sewerage and Public Water Supply Distributions. *Water Quality Rules and Regulations.* Chapter 11. May. 152 pp.

Index

509

513